Maritime Pollution Liability and Policy

Energy and Environmental Law & Policy Series
Supranational and Comparative Aspects

VOLUME 13

The aim of the Editor and the Editorial Board of this series is to publish works of excellent quality that focus on the study of energy and environmental law and policy.

Through this series the Editor and Editorial Board hope:

- to contribute to the improvement of the quality of energy/environmental law and policy in general and environmental quality and energy efficiency in particular;
- to increase the access to environmental and energy information for students, academics, non-governmental organizations, government institutions, and business;
- to facilitate cooperation between academic and non-academic communities in the field of energy and environmental law and policy throughout the world.

Kluwer Law International

Maritime Pollution Liability and Policy

China, Europe and the US

Edited By

**Michael G. Faure, Han Lixin &
Shan Hongjun**

Wolters Kluwer
Law & Business

AUSTIN BOSTON CHICAGO NEW YORK THE NETHERLANDS

Published by:
Kluwer Law International
PO Box 316
2400 AH Alphen aan den Rijn
The Netherlands
Website: www.kluwerlaw.com

Sold and distributed in North, Central and South America by:
Aspen Publishers, Inc.
7201 McKinney Circle
Frederick, MD 21704
United States of America
Email: customer.service@aspenpublishers.com

Sold and distributed in all other countries by:
Turpin Distribution Services Ltd.
Stratton Business Park
Pegasus Drive, Biggleswade
Bedfordshire SG18 8TQ
United Kingdom
Email: kluwerlaw@turpin-distribution.com

Printed on acid-free paper.

ISBN 978-90-411-2869-0

Printed in Great Britain.

List of Contributors

Boyd, James. Resources for the Future, the United States of America
Chen, Qi. Dalian Maritime University, China
Chu, Beiping. Dalian Maritime University, China
Faure, Michael. Maastricht University and Erasmus University Rotterdam, the Netherlands
Force, Robert. Tulane University Law School, United States of America
Guo, Ping. Dalian Maritime University, China
Han, Lixin. Dalian Maritime University, China
Hu, James Zhengliang. Shanghai Maritime University and Shanghai Wintell & Co Law Firm, China
Huybrechts, Marc. Catholic University of Leuven and University of Antwerp, Belgium
Jiang, Yuechuan. Dalian Maritime University, China
Johnston, Jason Scott. University of Pennsylvania Law School, the United States of America
Li, Lianjun. Richards Butler in Association with Reed Smith LLP., Chartered Institute of Arbitrators, China Maritime Arbitration Commission, China International Economic and Trade Arbitration Commission, China Maritime Law Association
Li, Tiansheng. Dalian Maritime University, China
Li, Zhonghua. Zhuhai Maritime Safety Administration, China
Liu, Nengye. Ghent University, Belgium
Maes, Frank. Ghent University, Belgium
Mukherjee, Proshanto. World Maritime University, Sweden and Colombo International Nautical and Engineering College, Sri Lanka
Qin, Mu. Richards Butler in Association with Reed Smith LLP.
Shan, Hongjun. Dalian Maritime University, China

Somers, Eddy. Ghent University, Belgium
Song, Ying. Peking University, China
Soyer, Baris. Swansea University, United Kingdom
Sun, Guang. Dalian Maritime University and Dalian Maritime Court, China
Wang, Dapeng. Dalian Maritime University, China
Wang, Hui. Catholic University of Leuven, Belgium
Wang, Yuh-Ling. Pacific Star Taipei Office, Taiwan
Zhang, Jinlei. Dalian Maritime University, China
Zhang, Liying. China University of Political Science and Law, China
Zhou, Zhujun. Zhuhai Maritime Safety Administration, China

Summary of Contents

Table of Contents

List of Abbreviations

AIS	Automatic Identification System
APPS	Act to Prevent Pollution from Ships
ARDW	Administrative Regulation on Dumping Wastes at Sea
AREPOE	Administrative Regulation on the Environmental Protection on Ocean Oil Exploration and Exploitation
ARPSDR	Administrative Regulation of Preventing Pollution from Ship Dismantling and Repair
ARPSP	Administrative Regulation on Preventing Ships from Polluting Sea Areas
BIMCO	Baltic and International Maritime Council
C.D. Cal.	Central District of California
CA	Court of Appeal
CCS	China Classification Society
CERCLA	Comprehensive Environmental Response, Compensation, and Liability Act
CFR	Code of Federal Regulations
CISG	Contracts for the International Sale of Goods
CLB	International Convention on Civil Liability for Bunker Oil Pollution Damage
CLC	The Convention on Civil Liability for Oil Pollution Damage
CMAC	China Maritime Arbitration Commission
CMC	China Maritime Code
CMI	Comité Maritime International
CMLA	Canadian Maritime Law Association
CMLR	Common Market Law Reports
CMST	Central Maritime Safety Tribunal
CNY	Chinese Yuan

COLREG	Convention on the International Regulation for Prevention Collisions at Sea
COPC Fund	China Oil Pollution Compensation Fund
CRTD	Convention on Civil Liability for Damage Caused during Carriage of Dangerous Goods by Road, Rail and Inland Navigation Vessels
CV	Contingent Valuation
CWA	Clean Water Act
DEFRA	Department for the Environment, Food and Rural Affairs
DEPI	Division of Environmental Policy Implementation
DK	Danish Kroner
DMU	Dalian Maritime University
DOI	Department of Interior
DOJ	Department of Justice
DPA	Deepwater Port Act
Duke L.J.	*Duke Law Journal*
DWT	Deadweight Tonnage
E.L. Rev.	*European Law Review*
EC	European Community
ECJ	European Court of Justice
EDI	Economic Damage Indicator
EEZ	Exclusive Economic Zone
EIA	Environmental Impact Assessment
EPA	Environmental Protection Agency
EPL	Environment Protection Law
EQB	Environmental Quality Board
EU	European Union
EWCA	East-West Center Association
EWHC	High Court of England and Wales
FAO	Food and Agriculture Organization
Fed. Reg.	Federal Register
FSA	False Statement Act
FWPCA	Federal Water Pollution Control Act
GAO	General Accounting Office
GATT	General Agreement on Trade and Tariffs
GDP	Gross Domestic Product
GMSA	Guangdong Maritime Safety Administration
GPA	Global Programme of Action for the Protection of the Marine Environment from Land-Based Activities
GPOFA	Guangdong Provincial Oceanic and Fishery Administration
GT	Gross Tonnage
HBL	Hydrostatic Balance Loading
HCE	Heavy Construction Equipment
HKD	Hong Kong Dollar
HKEC	Hong Kong East Cluster

HKLRD	*Hong Kong Law Reports & Digest*
HKSAR	Hong Kong Special Administrative Region
HKSOA	Hong Kong Shipowners' Association
HNS	Hazardous and noxious substances
IAPH	International Association of Ports and Harbors
ICS	International Chamber of Shipping
IG	International Group of Protection and Indemnity Clubs
IIML	Institute of International Maritime Law
ILO	International Labour Office
IMDG	International Maritime Dangerous Goods
IMEC	International Maritime Employers' Committee
IMO	International Maritime Organization
IMST	Incheon Maritime Safety Tribunal
INTERCARGO	International Association of Dry Cargo Ship Owners
InterManagers	International Ship Managers' Association
INTERTANKO	International Association of Independent Tanker Owners
IOPC Fund	International Oil Pollution Compensation Fund
ISF	International Shipping Federation
ISM	International Safety Management
ISO	International Organization for Standardization
ISU	International Salvage Union
ITF	International Transport Workers' Federation
ITLOS	International Tribunal for the Law of the Sea
ITOPF	International Tanker Owners Pollution Federation Limited
JHA	*Journal of Humanitarian Assistance*
JHA	Justice and Home Affairs Council
KMPRC	Korean Marine Pollution Response Corporation
KMST	Korean Maritime Safety Tribunal
KRW	South Korean Won
LEP	Law on Environmental Protection
LGDJ	Librairie générale de droit et de jurisprudence
LLMC	Limitation of Liability for Maritime Claims
LMAA	London Maritime Arbitrators Association
LNG	Liquefied Natural Gas
LOF	Lloyd's Open form
MARDEP	Marine Incident Investigation Section of the Marine Department of Hong Kong Special Administrative Region
MARPOL	International Convention for the Prevention of Pollution from Ships
MBTA	Migratory Bird Treaty Act
MCL	Michigan Compiled Laws
MEP	Ministry of Environmental Protection
MEPC	Marine Environment Protection Committee
MEPL	Marine Environmental Protection Law of the People's Republic of China

METRO	Maastricht European Institute for Transnational Legal Research
MOA	Ministry of Agriculture
MOMAF	Ministry of Maritime Affairs and Fisheries
MPLP	Marine Pollution Liability and Policy
MSA	Maritime Safety Administration
MSA	Maritime Safety Agency
MSA	Merchant Shipping Act
MT	Metric Ton
NEPA	National Environmental Policy Act
NJDEP	New Jersey Department of Environmental Protection
NMMBA	National Museum of Marine Biology and Aquarium
NMSA	National Marine Sanctuaries Act
NOAA	National Oceanic and Atmospheric Administration
NOK	Norwegian Krone
NOWPAP	Action Plan for the Protection, Management and Development of the Marine and Coastal Environment of the Northwest Pacific Region
NRD	Natural Resource Damage
NRDA	Natural Resource Damage Assessments
OCIMF	Oil Companies International Marine Forum
OCSLA	Outer Continental Shelf Lands Act
OECD	Organisation for Economic Co-operation and Development
OILPOL	Oil Pollution
OPA	Oil Pollution Act
OPLR	Oil Pollution Liability Regulations
OPRC	Oil Pollution Preparedness
ORB	Oil Record Book
OSCAR	Oil Spill Contingency and Response
OSG	Overseas Shipholding Group, Inc.
OSIR	Oil Spill Intelligence Report
OSLTF	Oil Spill Liability Trust Fund
P&I Club	Protection and Indemnity Club
PC	Privy Council
PPP	Polluter Pays Principle
PRC	People's Republic of China
Pub.L.	Public Law
QB	Queens Bench
RCCL	Royal Caribbean Cruises, Ltd
RGDIP	Revue Générale de Droit International Public
RINA	Royal Institution of Naval Architects
RMB	Renminbi
RPPSAS	Regulation on Preventing Pollution of Sea Areas from Ships
RSP	Regional Seas Programme
SBT	Segregated Ballast Tanks

SCLR	Scottish Council of Law Reporting
SDR	Special Drawing Right
SEPA	State Environmental Protection Administration
SHI	Samsung Heavy Industries
SJD	Doctor of the Science of Law
SLT	Scots Law Times
SOA	State Oceanic Administration
SOLAS	the International Convention for the Safety of Life at Sea
SOSREP	Special Representative of the Secretary of State
STCW	International Convention on Standards of Training, Certification and Watchkeeping for Seafarers
TAPAA	Trans-Alaska Pipeline Authorization Act
TVE	Chinese Township and Village Enterprises
TWD	Taiwan Dollar
UK	United Kingdom
UNCLOS	United Nations Convention on the Law of the Sea
UNCTAD	United Nations Conference on Trade and Development
UNEP	United Nations Environment Programme
UNESCO	United Nations Educational, Scientific and Cultural Organization
UNTS	United Nations Treaty Series
US	United States
USC	United States Code
USCA	United States Code Annotated
USCG	United States Coast Guard
USD	United States Dollar
VLCC	Very-Large Crude Carriers
WLR	*Weekly Law Reports*
WTO	World Trade Organization
ZMSA	Zhuhai Maritime Safety Administration

Chapter 1

Introduction

Michael Faure, Han Lixin** & Shan Hongjun****

1 ORIGIN OF THIS BOOK

Dalian Maritime University (DMU) (Dalian, China) celebrated its 100th anniversary in June 2009. At the occasion of this anniversary the Institute of International Maritime Law and Law School of Dalian Maritime University organized jointly with the Maastricht European Institute for Transnational Legal Research (METRO, the faculty research institute of the law faculty of Maastricht University) a conference on Marine Pollution Liability and Policy on 4 and 5 June 2009.

The focus on Marine Pollution Liability and Policy was related to a previous conference organized jointly by METRO and (at that time) DMU professor Hu Zhenliang. That earlier conference took place from 2–5 July 2004 in Nanjing and resulted in a book.[1]

The Nanjing conference and the subsequent book focused on prevention and compensation of marine pollution damage. That focused mainly on issues of civil

* Maastricht University and Erasmus University, Rotterdam, the Netherlands.
** Dalian Maritime University, China.
*** Dalian Maritime University, China.

1. M.G. Faure & J. Hu (eds), *Prevention and Compensation of Marine Pollution Damage. Recent Developments in Europe, China and the US*, Comparative Environmental Law & Policy Series (Alphen aan den Rijn: Kluwer Law International, 2006).

Michael G. Faure, Han Lixin & Shan Hongjun, *Maritime Pollution Liability and Policy: China, Europe and the US*, 1–10.

liability and compensation. The conference that was held on the occasion of the 100th anniversary of DMU and gave rise to the current book, which is based on expertise and experience but also in many aspects goes beyond the topics discussed at the Nanjing conference. This time, central issues in prevention and compensation of marine pollution are discussed, and considerable attention was paid to the Bunker Convention, which recently entered into force, as well as to the topic of criminal law. Specific attention is paid to developments in the United States and China, and case studies are presented.

2 DEFINING THE PROBLEM

2.1 Social Relevance

It hardly needs any explanation that marine pollution damage causes significant turmoil in Europe as well as in China. Europe has already been awakened by the dangers of marine pollution damage, first with the *Torrey Canyon* incident, then with the *Amoco Cadiz* and more recently with the *Erica* incident, all of these causing huge ecological and economic damage in Western Europe. Also, China has recently become aware of the risks of marine pollution damage and is all the more interested in taking adequate measures to prevent and compensate for marine pollution damage, given its huge coastline and therefore increased exposure to the risk for marine pollution damage.

The many incidents with oil tankers leaking enormous quantities of crude oil have shown the huge and devastating consequences of marine pollution damage. The world has been exposed probably in the most explicit way to these horrible consequences of the *Exxon Valdez* incident in Alaska in 1989, when the pictures of birds and mammals such as seals covered with oil went worldwide. In addition to the ecological disaster such an incident caused, many also point to the huge economic consequences of these (but also of smaller) incidents. Many working in the coastal zones are dependant upon fish, clean water and clean beaches for their income, not only because of fishery activities, but sometimes also because of tourism. It often takes many years before the damaged coastlines are fully recovered, if at all.

It therefore hardly needs to be demonstrated that all pollution incidents are of high social relevance. The same is true, moreover, for the role the law could play in fighting these incidents. Indeed, since the 1960s, conventions have been drafted with precisely the aim to on the one hand prevent marine pollution damage and on the other hand provide adequate compensation if such an incident would occur. However, many of the earlier mentioned greater incidents have shown that these conventions were not able to provide adequate compensation. The more recent incidents involving the *Erika* and the *Prestige* provided dramatic evidence of this. Therefore, serious amendments were made to certain international conventions, but still the question arises as to whether the current legal framework is adequate to deal with marine pollution damage.

2.2 Legal Relevance

First, the question must be asked whether the international framework that exists today to on the one hand prevent and on the other provide compensation for marine pollution damage is indeed adequate. Other serious questions are posed, as well, such as in relation to the financial caps (limitations) that traditionally exist in maritime law. Traditionally financial limitations on liability have been defended in maritime law because the usual victim in maritime law is the charterer, who can seek protection via first-party insurance in the case of loss of cargo. However, this justification for financial caps is not available if a third party, rather than a contracting party, suffers the loss. Thus it must be questioned whether the current financial caps, even after recent revisions, are adequate to provide compensation for marine pollution damage. Recent incidents, such as those described earlier, have shown that even after implementation of the amendments to the conventions there is a question whether the amounts available are sufficient to provide adequate compensation.

In addition, the question arises as to whether compensation for marine pollution damage should merely be provided via liability rules and traditional insurance. Fund solutions were implemented previously to (partially) deal with marine pollution damage, as well. Furthermore, how can an optimal combination of liability rules, insurance and fund solutions be put into place to provide adequate compensation?

Finally, marine pollution damage is a domain that is highly important from a legal perspective because it clearly shows that sometimes a wide gap may exist between the 'law in the books' and the 'law in practice'. Indeed, many countries have (usually as a result of international conventions) implemented legislation with the aim to prevent (intentional or accidental) marine pollution damage. The major problem, however, is that many of these incidents occur on the high seas where various aspects regarding enforcement are concerned, including in the areas of the competence of authorities, enforcement strategies to be followed and potential sanctions to be applied. Many have argued that effective enforcement of marine pollution prevention legislation is often lacking.

2.3 Comparative Research

Research into legal aspects of marine pollution damage is of great importance for both Europe and China; this book therefore is both legal multi-disciplinary and a comparative study of the contemporary legal issues for China, America and Europe. This comparative approach is developed further at paragraph 3.3. Many European countries have (usually as a result of international conventions) implemented oil pollution legislation in the areas of prevention of marine pollution damage and compensation via liability rules and (usually mandatory) insurance. The same is the case in China, which has, with its 18,000-km coastline, an important incentive to prevent marine oil pollution. China is at the point of redrafting its maritime code (a process already partially concluded). Within this redrafting process, China is considering introduction of a separate chapter in the maritime code, dealing more specifically with regulation of marine pollution damage. It should be of interest to China and Europe to learn from each other's experiences with legal remedies against marine pollution damage, particularly to China in its redrafting process. However, as indicated earlier, recent incidents in Europe have cast doubt on the adequacy of the current legal framework. Therefore, reforms to improve protection against marine pollution damage are being considered in Europe. The United States is an interesting source for comparison. It is a particularly interesting jurisdiction as far as oil pollution is concerned because it basically chose not to join the existing international conventions but rather to develop its own Oil Pollution Act 1990 (OPA). This may provide an interesting example for China and Europe.

This book addresses legal remedies against marine pollution damage in a comparative perspective, from the angles of China, America and Europe. The goal of this comparative approach is to promote mutual learning and new theoretical and empirical insights that may lead to improvement of current regulatory structures in Europe and China.

2.4 Recent Evolutions

It is important to stress that there are many reasons to pay attention to the legal remedies to marine pollution damage that have emerged in 2009, because there are evolutions in both China and Europe and at the international level, making it worthwhile to address these issues. On the one hand, there is considerable debate between China and the International Maritime Organization (IMO) as to whether China should join the International Oil Pollution Compensation Fund (IOPC Fund). China is considering setting up a national fund, but at the same time IMO pressuring China to also join the IOPC Fund. Changes are also taking place at the international level as far as maritime safety and pollution prevention are concerned, for example, with respect to the increasing shift to port state control. Also, increasingly, international organizations (e.g., Comité Maritime International (CMI)) pay attention to marine pollution caused by sources other than oil. The carriage of dangerous substances by sea is now as subject of intense discussion.

More attention is being directed to specific remedies that should be applied to marine pollution damage. A classic topic, but of increasing importance, is the role of civil liability for marine pollution and the role of insurance (through the so-called protection and indemnity (P&I) clubs) in covering damage caused by marine pollution. However, safety regulation tends to be enforced through criminal sanctions. Notably, in the European Union (EU) there is a clear tendency to force Member States to use criminal sanctions in non-compliance with regulations concerning prevention of marine pollution damage. This merits more generally the question as to the role of criminal law, monitoring and inspections in preventing marine pollution damage.

Yet another important recent evolution is that the International Convention on Civil Liability for Bunker Oil Pollution Damage of 2001, also known as the 'Bunker Convention', entered into force on 21 November 2008. The Bunker Convention came into effect for China on 9 March 2009. This new legal instrument raises a variety of interesting legal questions, specifically with respect to the delineation between on the one hand the well-known International Convention on Civil Liability for Oil Pollution Damage (the CLC) and on the other hand the Bunker Convention. The most recent oil spill from the offshore drilling unit Deepwater Horizon in the Gulf of Mexico, US in April 2010 made people alert that there are still loopholes in the existing system as far as pollution from offshore oil exploration and production is concerned. The current international regime on civil liability and compensation for pollution would not cover pollution damage from offshore facilities even when the international conventions would apply (which is not the case since the US is not a party to any of these international conventions). Under the existing US law, the available compensation under OPA (the limited liability of the operator British Petroleum and the finances from the Oil Spill Liability Trust Fund) is likely to be insufficient to cover the expected damage. Litigations from various infected states including New Orleans and Louisiana have been filed in courts. However, the crucial issue of how much damage to be awarded to the victims remains to be seen. Moreover, the US Congress has been in hot debate concerning the increase of limitation of liability or easement of such a mechanism as a whole. This incident not only sheds doubts on the civil liability for offshore facilities discharging oil, but also the safety regulations, the criminal liability and all other aspects of pollution caused by such offshore facilities.

3 METHODOLOGY

3.1 Focus

Even though the topic of marine pollution damage is potentially very broad and obviously has connections to traditional topics in maritime law (such as the carriage of goods), the contributions in this book focus on specific legal remedies related to prevention and compensation for marine pollution damage. The major advantage of focusing on a few topics (which will be further discussed below in 5) is that the various contributions in this book can be ordered into specific parts and deal with clearly related issues. Notwithstanding the focus on marine pollution prevention and compensation, a broad variety of topics will be discussed. As well, a variety of legal remedies (liability, administrative law, contracts and criminal law) will be considered in regard to both prevention and compensation.

3.2 Legal Multidisciplinary

A legal multi-disciplinary approach will be followed in this study because it is thought that various legal disciplines should be employed to provide a complete picture of the optimal ways to remedy oil pollution damage. The following disciplines will therefore be used and combined:

Maritime law: It is obvious that one should begin with maritime law, given the fact that marine oil pollution has its source in ships, particularly tankers.

Environmental law: In environmental law, oil pollution damage has been analysed in detail. Many scholars writing on environmental liability and transboundary pollution have analysed legal schemes regarding oil pollution damage.

Liability law: Thus far the schemes chosen to compensate oil pollution damage originated in liability law. The conventions usually rely upon a strict liability rule, albeit that the amount of compensation due to the victim is limited (financial caps). Liability law should be used to ascertain the functions and goals of tort law and determine whether these goals can be achieved within the framework of oil pollution damage.

Insurance law: Most conventions have introduced not only strict liability but also mandatory insurance. Mandatory insurance always raises important questions, from insurance, legal and economic perspectives. The type of insurance market to provide the insurance market must be determined. The questions also arise as to whether the insurance contract itself should be regulated and whether exceptions between the insurance company and the insured can be opposed to the third party (the victim).

Criminal law: Effective prevention of oil pollution damage will be possible only if an adequate system of enforcement can also be organized. In this case, also, important questions arise, for example, as to the territorial scope of jurisdiction for States' powers to control tankers and to what extent States have the power to prosecute for oil pollution damage committed within (or outside) its territories. In addition, the question arises as to whether enforcement will take place merely on the basis of administrative law or whether criminal law enforcement is possible as well. Finally, criminal law is needed to address the optimal sanctions in oil pollution damage. This is all the more important because it has been indicated in the literature that the current insurers (the P&I clubs) in practice provide full insurance for the fines imposed in the case of oil pollution damage. The insurance of fines may well jeopardize an effective deterrence of oil pollution damage. In that case it must be determined if sanctions other than fines could be imposed and how they are to be executed.

3.3 The Comparative Approach

As indicated earlier, this book not only engages a legal multi-disciplinary approach but will also rely on a comparative approach. Thus, the relevant international conventions will be examined, considering their implementation in China and Europe, as well as the particular importance of the US regime. The importance of comparative research in this domain has already been discussed. The comparison in this book will relate to conventions and experiences in Europe as well as (specifically) the United States and China. Moreover, an integrated comparative approach will be followed, meaning that attention is also paid to the question, for example, of whether solutions or evolutions in the United States could also provide interesting information and insights for the revision of legislation in China.

3.4 Empirical Research

It may be clear that any legal research, especially when it is devoted to an important domain such as marine pollution, should not merely remain 'law in the books' but should also focus on the 'law in action' by examining how in legal practice enforcement of oil pollution compensation takes place. Considerable important empirical material has been gathered, particularly through a variety of case studies. These deal with important cases in China; some of the cases illustrate the dangers and limitations of particular approaches (e.g., of the criminal law) in legal practice. This empirical research, reflected in a variety of case studies, gives important insights into the working of international and national prevention and compensation mechanisms. It is by analysing the working of these mechanisms via case studies in legal practice that additional insights can be obtained concerning the effectiveness of those instruments.

4 TOPICS

The broad question of how to optimally prevent and compensate oil pollution damage in China, Europe and the United States is too broad for an even focused book like this one. Therefore, various topics will be defined in which several aspects can be examined. The methodological approaches we have developed (3 above) will be applied in discussion of these topics. Some issues that could therefore be analysed at the conference discussed in the following sections:

4.1 Implementation of Conventions in Europe and China

An initial issue to consider is the state of legislation in international conventions in regard to s prevention and compensation of oil pollution damage. The conventions should be described, including both the legal history and recent reforms. In addition, the implementation of the conventions (if this is the case) in Europe and China should be discussed, with particular attention given to the relative differences between the two regimes.

4.2 Critical Analysis of the Current Structures

A critical review of the current structure of prevention and compensation for oil pollution damage is necessary. In this second project, particularly, approaches that allow for a critical analysis, such as 'law and economics', will be used, for example, in considering whether the current structure of liability rules, combined with insurance and regulation, provides for optimal incentives to prevent oil pollution damage. Financial caps on liability will be critically analysed, as well.

4.3 Alternative Routes of Compensation

Focus particularly will be on the question of how an optimal compensation of oil pollution damage could be given, in light of the shortcomings that can be seen today. Indeed, as mentioned previously, all agree that the financial limits on liability, even after the recent revisions, do not guarantee full compensation for oil pollution damage in the case of major accidents. Therefore, critical analysis is necessary in the area of why current insurance markets (particularly P&I clubs) are not able to generate higher amounts of compensation. In addition, whether further restructuring of conventions (or national legislation) is possible and who should bear the costs of increased compensation must also be considered. Moreover, alternative routes of compensation (other than liability and liability insurance) should be examined as well. The option of first party insurance by victims (coastal zones) could be examined, as well as the possibility of (supplementary) compensation funds and maybe even State intervention.

4.4 Prevention and Enforcement

Finally, also a focus on the question how on optimal enforcement of oil pollution legislation can be achieved is indicated. Aspects of criminal law will play an important role in this respect. Thus, the issue of the appropriate jurisdiction in the case of oil pollution crimes has to be examined, as well optimal control techniques and sanctions. Empirical material with respect to the best legal techniques to prevent oil spills will play an important role, as will the economic analysis of criminal law, which may indicate how public resources can be optimally allocated to prevent oil pollution crimes.

5 FRAMEWORK OF THE PROJECT

This book resulted from close cooperation between the organizer of the conference, DMU, and METRO of Maastricht University. The partners worked closely in preparation of the conference as well as in the subsequent editorial process. The conference enjoyed the support of the Ius Commune Research School and the Institute International Maritime Law of Dalian Maritime University. Both the Maritime Law Research Society of the Liaoning Law Academy and the Admiralty Law Committee of the China Maritime Law Association supported the organization of the conference that constituted the basis of this book.

6 STRUCTURE OF THE PRESENTATION

This book is divided in six parts. Part I presents the general issues in prevention and compensation of marine pollution. The contributors focus on the general structures available for prevention and compensation of marine pollution, as well as, for example, the right of entry in marine ports for ships in distress and specific questions concerning the measurement of marine damages and the actionability of claims. Also, the question of whether pure economic loss can be compensated plays a large role in that respect.

Part II deals with the important Bunker Convention, which recently entered into force. A general overview of the convention is provided, with special attention paid to the interface between the Bunker Convention and other conventions. Also considered is the limitation of liability and compulsory insurance for bunker oil pollution damage. In addition, the importance of the Bunker Convention for China is critically examined.

Part III focuses on criminal liability. An economic analysis of the rationale for criminal liability is provided, and recent evolutions in the EU and in China are discussed, as well as the situation in Hong Kong. Critical questions are asked with respect to the expanding criminal liability of masters and crew in oil pollution cases.

Part IV focuses on the compensation system for oil pollution damage as regulated in the US Oil Pollution Act 1990. The extent to which this US example and the experience in the United States pose a challenge for the reform of Chinese legislation is considered.

Part V is devoted solely to evolutions in China. The question is addressed *inter alia* whether China complies with the international conventions it ratifies and incorporates, and discussions concerning the application of the CLC in China and related judicial authentication of marine oil pollution are also reviewed.

Part VI finally presents interesting case studies dealing *inter alia* with the well-known *Hebei Spirit* oil pollution incident, the *Amorgos* oil pollution incident and the *Ilona* oil spill.

The final part contains a set of comparative conclusions made by the editors.

7 CONTRIBUTORS

The contributors to this book come from Europe (Michael Faure, Marc Huybrechts, Frank Maes, Proshanto Mukherjee, Eddy Somers and Baris Soyer), from the United States (James Boyd, Robert Force and Jason Scott Johnston), from Hong Kong (Li Lianjun and Qin Mu), from Taiwan (Wang Yuh-Ling) and of course from China (too many to name here). Many of the Chinese contributors are connected to the Dalian Maritime University or to a provincial Maritime Safety Administration. In addition, various judges (e.g., Sun Guang) have contributed to the book as well. A complete list of contributors and their affiliation is provided after the Table of Contents.

8 ACKNOWLEDGEMENTS

As editors of this book, we are very grateful to many people who made the book and the preceding conference possible. We would like to thank the many sponsors of the international conference on marine pollution liability and policy held in Dalian on the 4 and 5 June 2009: Penavico Dalian, Cosco Logistics Dalian, Shandong Far East Marine Shipping Group, Dalian Chun An Ship Management Company, Heng Xin Law Office, Tytop Law Firm and ZM & Co., as well as Wang Jing & Co. Metro, the Ius Commune Research School, the Science Committee of the Law Faculty of Maastricht University and especially Dalian Maritime University Law School have provided generous financial support that made this project and hence this book possible. In this respect, we would especially like to mention Huang Fen, Lin Yanyan, Sha Xiancen and Zhang Jinlei for kind help and support in the organization of the preceding conference and the collection of the papers for this volume. In addition, we are grateful to everyone else who made both the conference and the publication of this book possible. In this respect, we owe special thanks to Tumaini Gurumo and Michael Wells-Greco, who revised the English of the non-native English speakers.

We owe special thanks to the administrative centre of METRO and especially to Joke Declercq and Marina Jodogne for their editorial assistance in reviewing the texts, footnotes and referencing and for editorial assistance in the preparation of the publication of this book. Wang Hui performed an intense check to control the uniformity of the Chinese Laws and Regulations cited.

The texts were finalized in September 2009, and for that reason developments after that date have not been treated. Especially in view of the ongoing debates surrounding offshore facility discharging oil into sea, it shows the marine pollution legislations and regulations are still in dynamic development. Even when the international or regional, national systems have been modified and adapted to such an extent that it would be considered adequate to cope with the majority of the claims, the technical boundaries illustrated by catastrophes show again that further changes and improvements are necessary. Therefore, it remains important to pay attention to the question how legal regimes can adapt to changing technical realities of oil pollution catastrophes, also in the future.

Reference

Faure, M.G. & J. Hu (eds). *Prevention and Compensation of Marine Pollution Damage. Recent Developments in Europe, China and the US*. Comparative Environmental Law & Policy Series. Alphen aan den Rijn: Kluwer Law International, 2006.

Part I

General Issues in Prevention and Compensation of Marine Pollution Damage

Chapter 2

Prevention and Compensation for Marine Pollution

*Wang Hui**

1 INTRODUCTION

The ocean has used by human beings for centuries to transport passengers and various cargoes, including oil and all sorts of chemical substances. With developments in technology and trade, shipping has become increasingly important for global commerce. Sometimes cargo transported on board is released into the sea; this occurs for various reasons, especially when an accident occurs (e.g., the collision of two ships) that may result in loss of the cargo or fuel oil being discharged. It is very likely that the discharge of certain substances into the sea causes pollution to the marine environment. When the risk for marine pollution from shipping was recognized, many conventions were adopted to deal with the problem.

A large number of international conventions tackle the issue of marine pollution from various perspectives. Indeed, pollution in the marine environment may come from different sources and in various situations; it may result from accident or simply through daily operation. Moreover, the polluting agents may vary. Hence, preventive measures may also differ according to the characteristics of the polluting agents. The legal regimes addressing pollution prevention measures for

* Catholic University of Leuven, Belgium.

Michael G. Faure, Han Lixin & Shan Hongjun, *Maritime Pollution Liability and Policy: China, Europe and the US*, 13–40.

different polluting substances may therefore seem very technical in nature. Despite the existence of legal regimes related to pollution prevention, pollution incidents nevertheless happen. This raises the question of whether these technical standards are sufficient to protect the marine environment. If it appears from analysis that the legal standards established in the international conventions are not adequate, a related question is whether the contents and the design of the regulations are adequate or whether the lack of effective enforcement leads to the failure of the conventions whose purpose is to protect the marine environment.

Another aspect of the marine pollution issue is that when a pollution incident (especially of a large scale) occurs, many victims may suffer damage to different extents. They will of course require compensation. International legal regimes have been developed to provide a mechanism for victims to obtain compensation. A critical issue is whether these victims can be compensated sufficiently under the current regime or whether another legal regime can be used to provide compensation to the pollution victims. Again, it is the case at international level that there are various compensation regimes for pollution victims, which set up different liability limits for different polluting substances. This means pollution victims who suffer the same (high) amount of damage may, in the end, receive different amounts of compensation simply because the polluting substances are different. In that case, the question arises as to whether such a differentiation is justified.

The goal of this chapter is to critically examine two important aspects of the international legal regimes related to marine environmental protection – prevention of pollution incidents and civil liability and compensation for pollution damages. Moreover, as just indicated, the legal regimes applicable may differ according to the substances that have caused the damage. For this comparative study, only certain polluting substances are discussed, mainly oil, certain hazardous and noxious substances (HNS) and bunker oil. In contrast to the existing literature, which addresses the pollution caused by oil, HNS and bunker oil separately, this chapter adopts an integrated approach. This means that not only are both prevention and compensation regimes analysed but also that the legal regimes concerning different polluting substances will be submitted to comparative analysis. Discussion concerning the interaction among regulations regarding pollution prevention and civil liability regimes designed to provide compensation may teach us how to make the best use of each of the two sides of the coin and may provide useful insight for improvement of the current regimes. Comparative study of the legal regimes dealing with polluting substances may disclose advantages and disadvantages of each regime. This may again be useful for the design of an effective or efficient legal regime tackling the marine pollution problem.

This chapter is limited to a discussion of the legal regimes at the international level. However, especially in recent years, regional initiatives play an increasingly important role in contributing to the development of the international legal scenario. Therefore, these interesting regional efforts will be discussed in view of how they lead to legal changes in the international regimes.

This chapter is structured as follows, after a short introduction in section 1, the historical development of the international legal regime on marine pollution will be briefly discussed in section 2. Prevention regimes is discussed in section 3. In section 4, compensation regimes dealing with selected polluting substances are submitted to comparative analysis. Section 5 focuses on the relationship between regulations aiming at *ex ante* prevention and civil liability regimes aiming at compensation and presents critical analysis of the current regime. Finally, in section 6 contains concluding remarks.

2 LEGAL HISTORY

Shipping activity probably began when a log was used to cross a stream,[1] but the risk for pollution caused by shipping did not attract the attention of human beings until the twentieth century, when shipping became frequently employed for global trade and commerce. Also as a result of technological developments, the variety of cargoes transported on ships has increased, among which some substances pose risks to the marine environment once released into sea.

Among the early attempts to tackle the marine pollution problem at international level, one important step was the International Convention for the Prevention of Pollution of the Sea by Oil 1954 (OILPOL 1954).[2] OILPOL 1954 specifically recognized the potential risk of oil as a pollutant to the marine environment. It laid down certain discharge standards and established some 'special areas' where dumping of oily wastes was prohibited.

However, the growth in oil trade and developments in industrial practices showed that further action was needed. The international legislation on marine pollution remained a slow process. It seems at that time that the major and prevailing concern remained the maritime tradition of freedom of the sea.[3] Moreover, when a maritime incident occurred, the protection of human life enjoyed priority, as illustrated by the International Convention for the Safety of Life at Sea (SOLAS), which was adopted in 1914 as a reaction to the *Titanic* disaster.[4]

It was probably not until the *Torrey Canyon* incident in 1967 that the whole world began to realize the seriousness and urgency of the problem of marine pollution, and hence more initiatives were undertaken. As a result, more influential international conventions were concluded in the years following the *Torrey Canyon* incident. At the diplomatic conference in 1969, two conventions were adopted

1. Background document for World Maritime Day 1998, IMO.
2. Entered into force on 26 Jul. 1958, subsequently amended in 1962, 1969 and 1971.
3. T. Alcock, 'Ecology Tankers and the Oil Pollution Act of 1990: A History of Efforts to Require Double Hulls on Oil Tankers', *Ecology Law Quarterly* 19 (1992): 127.
4. The late nineteenth and early twentieth centuries is the golden age of passenger travel by sea. Given the technical restrictions at the time (there were no aircraft yet), movement of people including massive emigration from Europe to America was largely carried out by ships. Hence, casualties occurred frequently. The liner *Titanic* on her maiden voyage in April 1912 sank, and 1,500 passengers and crew members died in the disaster. Against this background, SOLAS 1914 was adopted, and its major goal was indeed as suggested by the title of the convention, for the safety of human life.

to deal with the deficiencies brought to light by the *Torrey Canyon* incident, namely, the International Convention relating to the Intervention on the High Seas in Cases of Oil Pollution Casualties (the Intervention Convention) and the International Convention on Civil Liability for Oil Pollution Damage (CLC). Later, in 1971, the International Convention on the Establishment of an International Fund for Compensation for Oil Pollution Damage (the Fund Convention) was adopted to complete the civil liability and compensation system for marine oil pollution. Therefore, a two-tiered compensation system was established, the first tier to be paid by the ship owner and the second tier to be paid by the Fund through contributions by the oil interests.

When the civil liability regime for marine oil pollution was considered resolved at the time, the attention of the legislation turned to other issues such as the prevention of pollution. Hence, after the conclusion of the international civil liability regime for marine oil pollution, other deficiencies indicated by the *Torrey Canyon* incident. Prevention of marine pollution was targeted at an international conference in 1973, with the conclusion of the International Convention for the Prevention of Pollution from Ships, 1973 (MARPOL 73).[5] In 1974, the International Convention for the Safety of Life at Sea (SOLAS 74) was adopted to address the safety of navigation.[6]

During the period between 1976 and 1977, a series of tanker accidents occurred, mostly in or near US waters, including the stranding of the tanker *Argo Merchant*. *Argo Merchant* ran aground off Massachusetts in December 1976. Although it was a relatively small tanker, carrying only 27,000 tonnes of oil, it caused huge public concern because the oil spill threatened New England resorts and Georges Bank fishing ground. The United States hence took the lead in requesting the International Maritime Organization (IMO) to consider further regulations concerning tanker safety. Therefore, a conference was convened in February 1978 to address tanker safety and pollution prevention. This conference adopted the 1978 MARPOL Protocol. MARPOL 73 had not entered into force then, so the 1978 MARPOL Protocol absorbed the parent convention, with the combined instrument often referred to as MARPOL 73/78.

One month after the 1978 conference, the grounding of *Amoco Cadiz* off Brittany made more countries alert to the risk for oil pollution. In a relatively short period, sufficient states ratified MARPOL by October 1982 and the MARPOL 73/78 entered into force on 2 October 1983.

Another problem indicated by the *Amoco Cadiz* incident was the inadequacy of the international regime on marine pollution compensation. International conferences were organized by the IMO in 1984 to discuss the possibility of improving the current oil pollution compensation regime. This led to the adoption of the 1984 Protocols to the CLC 1969 and the Fund Convention 1971 (1984 Protocols).

---

5. MARPOL 1973 was adopted on 2 Nov. 1973. It covered pollution by oil, chemicals, harmful substances in packaged form, sewage and garbage.
6. SOLAS has been amended several times since its adoption in 1914, mainly to incorporate the new technology to improve safety at sea.

The 1984 Protocols rely very much on the accession of the United States in order to become effective. When the *Exxon Valdez* incident happened in US waters, and in reaction to the public out cry, the US Government quickly decided to adopt its own domestic legislation, the Oil Pollution Act of 1990 (OPA 90), which addresses both prevention and compensation of marine pollution caused by oil. The civil liability regime adopted in OPA is largely incompatible with the CLC/Fund regime. Moreover, the political pressure from states to maintain the superiority of state law made it impossible for the US Government to ratify the international conventions.

The withdrawal of the United States from the international regime made clear that the 1984 Protocols would never come into force.[7] Nevertheless, the changes proposed in the 1984 Protocols were needed, especially the increase of the compensation amounts available. Therefore, the IMO adopted in 1992 two other Protocols (the 1992 Protocols) to eliminate the need for US ratification of the international regime but incorporate the changes proposed in the 1984 Protocols. The 1992 Protocols came into effect in 1996 with an expanded scope of compensation and an increased amount of compensation.

Meanwhile, MARPOL has been amended to take into account the new technology and techniques that contribute to reduce pollution. One important provision in OPA 90 is the mandatory requirement of double hulls for all tankers calling in US ports. As a result of US lobbying, the IMO adopted in 1992 requirements related to double hull designs.

The attempt to establish an international compensation regime for HNS was again picked up by the IMO in 1996. This led to the adoption of the International Convention on Liability and Compensation for Damage in connection with the Carriage of Hazardous and Noxious Substances by Sea (HNS Convention) on 3 May 1996. However, because of some technical difficulties, this convention has so far never entered into force.

Again in 1999 and 2001, more oil spill disasters were caused by the grounding of tankers – the *Erika* (1999) and the *Prestige* (2001). These triggered further changes in the MARPOL Convention, mainly the speeded phasing out of single hull tankers and the changes in the compensation regime, mainly, further increase of the compensation amount under the CLC and the Fund Convention (through the 2000 Protocols) and establishment of a Supplementary Fund (through the 2003 Protocols).

---

7. The entry into force conditions of the 1984 Protocols were laid down in such a way that the Protocols could come into force only when the United States ratified it. The 1984, the CLC required to be ratified by at least ten states and at least six of which had a tanker fleet of over 1 million tonnes. By June 1990, only six states (Australia, France, Germany, Peru, Saint Vincent and the Grenadines and South Africa) had ratified the 1984 CLC and only one of these six states, France, had tanker tonnage of over 1 million tonnes. The 1984 Fund Protocol required ratification by at least eight states with a total contributing oil import of at least 600 million tonnes. By June 1990, it was only ratified by France and Germany, which had combined imports of only 164 million tonnes. The US participation would have made it a major oil contributor to the International Oil Pollution Compensation Fund.

The issue of marine pollution caused by fuel (bunker) has received world attention only since 2000. A new compensation regime for marine pollution caused by bunker was quickly adopted in 2001, effective since 2008. Structural standards were included in MARPOL in 2004 to require protected location of fuel oil tanks.

This overview of the historical development of the international regime concerning marine pollution prevention and compensation shows that the major changes of the international regime often came as a reaction to certain pollution incidents.[8] Moreover, the occurrence of pollution incidents never seems to stop. The ensuing question is whether the international regime formulated in such a way is effective and adequate. If indeed the international regime is not adequate in addressing the pollution issue, the question arises as to whether it is the design of the conventions (e.g., pollution standards or compensation amount) or the enforcement and the implementation of the existing measures that constitutes the root of the problem. This requires further study and a more in-depth analysis.

3 PREVENTION

3.1 General Overview of MARPOL

The main international convention dealing with the prevention of marine pollution by ships is MARPOL 73/78 and its subsequent amendments and SOLAS 74. The Articles of the convention mainly deal with jurisdiction, powers of enforcement and inspection, and the more detailed anti-pollution regulations are contained in six technical Annexes. The ones that are relevant for this chapter are Annex I Regulations for the Prevention of Pollution by Oil, Annex II Regulations for the Control of Pollution by Noxious Liquid Substances in Bulk, and Annex III Prevention of Pollution by Harmful Substances Carried by Sea in Packaged Form.

MARPOL 73/78 regulates the pollution from ships through both accidental and operational causes. It addresses marine pollution from ships by setting discharge standards; standards for construction, design and equipment of a vessel; and operational standards. SOLAS sets up standards with the focus on safety, whereas the standards established in MARPOL focus on the prevention of pollution. Frequently, when an amendment is made to MARPOL or SOLAS, related provisions in the other convention are also amended to keep the safety regulations in these two conventions in line with each other. Hence, this chapter focuses only on the regulations on marine pollution prevention as prescribed in MARPOL, but there is a strong parallel with the similar provisions in SOLAS.

8. E.J. Molenaar (ed.), *Coastal State Jurisdiction over Vessel-Source Pollution* (The Hague: Kluwer Law International, 1998), 20.

3.2 Oil

3.2.1 Discharge Standards in MARPOL Annex I

The discharge standards are prescribed in two aspects in Annex I: one is the positive description of the standards and the other is the establishment of 'special areas' in which more stringent discharge standards apply.

MARPOL 1973 maintained the oil discharge standards prescribed in the 1969 amendments to the OILPOL 1954 without substantial changes. The operational discharges of oil from tankers are allowed only when they meet the following conditions: first, the total quantity of oil a tanker may discharge in any ballast voyage while under way must not exceed 1/15,000 of the total cargo-carrying capacity of the vessel; second, the rate at which oil may be discharged must not exceed 60 litres per mile travelled by the ship; third, no discharge of any oil whatsoever must be made from the cargo spaces of a tanker within fifty miles of the nearest land.

Moreover, MARPOL 1973 reduced the maximum quantity of oil permitted to be discharged on a ballast voyage of new oil tankers from 1/15,000 of the cargo capacity to 30,000 of the amount of charge carried. These criteria applied equally to both persistent and non-persistent oils.

These discharge standards were amended in 1992. The 1992 amendments have considerably reduced the amount of oil allowed to discharge into the sea from ships (e.g., following the cleaning of cargo tanks or from engine room bilges). Originally, oil tankers were permitted to discharge oil or oily moistures at the rate of sixty litres per nautical mile. The 1992 amendments reduced this to thirty litres. As for non-tankers of 400 gross tonnes (GT) and above, the permitted oil content of the effluent that may be discharged into the sea is cut from 100 parts per million to fifteen parts per million.

'Special area' is a concept first established in MARPOL 73. The special areas are those vulnerable to pollution risks, and thus more stringent discharge standards apply in these areas.[9] MARPOL 73 designated the Mediterranean Sea, the Baltic Sea, the Black Sea, the Red Sea and the 'Gulfs area' as the special areas. Through the years, more special areas were designated. So far, the special areas for this purpose, in addition to those mentioned earlier, are Gulf of Aden, the Antarctic, the North West European waters, the English Channel, the Oman Sea and the Southern South Africa waters.[10] The special areas will be implemented only when the littoral

9. MARPOL 1973 Annex I Regulation 1 (10) defines 'special area' to be 'a sea area where for recognized technical reasons in relation to its oceanographical and ecological condition and to the particular character of its traffic the adoption of special mandatory methods for the prevention of sea pollution by oil is required.'
10. Gulf of Aden was included in the Annex I special area in the 1987 amendments, which were adopted in December 1987 and entered into force on 1 Apr. 1989; Antarctic became the special area through the 1990 amendments which were adopted in November 1990 and entered into force on 17 Mar. 1992; the North West European waters were designated as special area under

states concerned have provided adequate reception facilities for dirty ballast and other oily residues.

The use of special areas may be seen as a prescription of discharge standards in a qualitative manner, which means that the quality standards of the seawater in those special areas are higher than in others.

3.2.2 Construction, Design and Equipment Standards

In Annex I of MARPOL 73/78, one important regulation concerning the construction and design of tankers is the regulation requiring segregated ballast tanks (SBT).[11] The goal of this regulation was to ensure that ballast water shall not be contaminated by oil carried as cargo or fuel. Originally, in MARPOL 73, this SBT requirement applies to new tankers over 70,000 deadweight tonnes (DWT). MARPOL 78 expands the SBT requirement to all new crude oil tankers of 20,000 DWT and above and to all new product carriers of 30,000 DWT and above. Moreover, MARPOL 78 also requires SBT to be protectively located in the ship where the possibility and amount of oil outflow from cargo tanks in case of a collision or grounding can be minimized.

The so-called double hull design was not required at the international level until 1992.[12] This was because the US enacted unilaterally in OPA 90 a compulsory requirement on double hulls for the tankers visiting its ports. However, OPA 1990 does not force single hull tankers to retire but only bans them from trade to the United States. Hence, these single hull tankers that are not allowed to operate in US waters will very likely be operating in European waters and other areas of the world where they can still be allowed to operate. In light of this situation, the IMO introduced in the 1992 amendments two new regulations into Annex I of MARPOL concerning the double hull design of tankers, being Regulations 13F and 13G. Regulation 13F (prevention of oil pollution in the event of collision or stranding) introduces double hulls or equivalent alternatives to new tankers, whereas Regulation 13G (prevention of oil pollution in the event of collision or stranding – measures for existing tankers) applies to existing tankers.

Regulation 10 of Annex I in the 1997 amendments which were adopted on 23 Sep. 1997 and entered into force on 1 Feb. 1999. The waters cover the North Sea and its approaches, the Irish Sea and its approaches, the Celtic Sea, the English Channel and its approaches and part of the North East Atlantic immediately to the West of Ireland; the Oman Sea area of the Arabian Seas is designated as a special area in the 2004 amendments which were adopted on 15 Oct. 2004 and entered into force on 1 Jan. 2007; the Southern South Africa waters were designated as a special area through the 2006 amendments adopted October 2006 and entered into force on 1 Mar. 2008.

11. When a ship is sailing empty to pick up cargo or returning to its origin, some water will be taken on board to maintain stability of the ship. Such water is known as ballast water, and the tanks to hold ballast water are called ballast tanks.
12. Some studies assign the tanker design into four major categories: single hulls (either pre-MARPOL or MARPOL), double side or double bottom and double hulls. See Double-Hull Tanker Legislation, An assessment of the Oil Pollution Act of 1990, Committee on Oil Pollution Act of 1990 (s. 4115) Implementation Review, National Academy Press, Washington, DC, 1998, 66.

Regulation 13F entered into force in July 1993, and Regulation 13G entered into force on 6 July 1995.

Regulation 13F requires new tankers[13] of 5,000 DWT and above to be fitted with double hulls (double bottoms and wing tanks extending the full depth of the ship's side) separated by a space of up to two meters (and on tankers below 5,000 DWT such a space should be at least 0.76 meters). As an alternative, tankers may incorporate the mid-deck design, in which the pressure within the cargo tank does not exceed the external hydrostatic water pressures (tankers built under this design have double sides but not a double bottom). Instead, another deck is installed inside the cargo tank, with the venting arranged in such a way that there is an upward pressure on the bottom of the hull. Other accepted alternatives must ensure the same level of protection against oil pollution in the event of a collision or stranding.

Regulation 13G applies to existing crude oil tankers of 20,000 DWT and product carriers of 30,000 DWT and above. Tankers that are twenty-five years old and were not constructed according to the requirements of MARPOL 73/78 have to be fitted with double sides and double bottoms. Tankers built according to the standards of MARPOL 73/78 are exempt until they reach the age of thirty.

Existing tankers are subject to an enhanced programme of inspections during their periodical, intermediate and annual surveys. Tankers that are five years old or more must carry on board a completed file of survey reports together with a conditional evaluation report endorsed by the flag state. Tankers built in the 1970s that are at or past their twenty-fifth year must comply with Regulation 13F. If not, their owners must decide whether to convert them to the standards set out in regulation 13F or to scrap them. Regulation 13G also allowed for future acceptance of other structural or operational arrangements, such as hydrostatic balance loading (HBL), as alternatives to the protective measures in the Regulation.

Following the *Erika* incident of 1999, the European Commission proposed an accelerated schedule to phase out single hull tankers that is to a large extent similar to the schedule introduced in OPA 90. Moreover, there was concern from the European Union (EU) Member States that single hull tankers forced to withdraw from American waters would start to operate in other parts of the world, including European waters. However, financial incentives as prescribed in Regulation 2978/94 were deleted in the final adopted proposal.

The EU Member States preferred to have the issue decided at the international level rather than the EU level. Therefore, the European Parliament waited for the final decision by the IMO. In 2001, amendments were introduced to speed up the process of phasing out single hull tankers. According to the 2001 scheme, the principal cutoff date for all single hull tankers is 2015 or when the tanker reaches twenty-five years after delivery.

13. New tankers are defined as those the building contract placed on or after 6 Jul. 1993, or the keels are laid on or after 6 Jan. 1994, or delivered on or after 6 Jul. 1996; in the event of a major conversion, new tankers are those the conversion contract placed after 7 Jul. 1993, or the construction work is begun after 6 Jan. 1994, or the major conversion is completed after 6 Jul. 1996. See Regulation 13F (1) of MARPOL Annex I.

Subsequently, the EU adopted Regulation 417/2002[14] in February 2002 concerning the accelerated phasing out of single hull tankers. This Regulation reflected the changes made to the MARPOL Convention in 2001. One deviation in the Regulation is that it legitimately makes use of paragraph 8(b) to deny entry of single hull tankers to European ports beyond 2015. Article 7 of Regulation 417/2002 provides that after 2015, 'the continued operation, in accordance with paragraph 5 of revised Regulation 13G of Annex I to MARPOL 73/78, of Category (2) and Category (3) oil tankers under the flag of a Member State' and 'the entry into the ports or offshore terminals under the jurisdiction of a Member State of Category (2) and Category (3) oil tankers, irrespective of the fact that they continue to operate under the flag of a third State in accordance with paragraph 5 of the revised Regulation 13G of Annex I to MARPOL 73/78' shall no longer be allowed.

On 13 November 2002, the catastrophe caused by another single hull tanker, *Prestige*, in European waters triggered again the debate on the phasing out of single hull tankers. Immediately, in December 2002, the European Commission proposed to further accelerate the phasing out of single hull tankers to make it in line with the relevant OPA provisions. Regulation 417/2002 was quickly amended on 22 July 2003 through the adoption of Regulation 1726/2003,[15] which entered into force on 21 October 2003. Therefore, the phasing out of single hull tankers in EU States has followed the same schedule as the United States since then.

Following the adoption of Regulation 1726/2003, the EU Commission also submitted a proposal to the IMO to amend *inter alia* the phasing out timetable for single hull tankers. Urged by the EU activism, the IMO adopted in 2003 the accelerated schedule to withdraw single hull tankers from operation. The phasing-out date for Category 1 tankers (pre-MARPOL tankers) was brought forward from 2007 (in 2001 amendments) to 2005 (in 2003 amendments), and as for category 2 and 3 tankers (MARPOL tankers and smaller tankers), the phasing-out date was brought forward from 2015 (in 2001 amendments) to 2010 (in 2003 amendments). Table 2.1 is a timetable for the phasing out of single hull tankers as provided in 2003 amendments to MARPOL:

Category of tankers	*Phasing out time*
Category 1 oil tankers of 20,000 DWT and above carrying crude oil, fuel oil, heavy diesel oil or lubricating oil as cargo, and of 30,000 DWT and above	5 April 2005: for ships delivered on 5 April 1982 or earlier
	Anniversary date in 2005: for ships delivered after 5 April 1982

14. Regulation European Commission (EC) No. 417/2002 of the European Parliament and of the Council of 18 Feb. 2002 on the accelerated phasing-in of double hull or equivalent design requirement for single hull oil tankers and repealing Council Regulation (EC) No. 2978/94, entered into force on 1 Sep. 2002.
15. Regulation (EC) No. 1726/2003 of the European Parliament and of the Council of 22 Jul. 2003 amending Regulation 417/2002 on the accelerated phasing-in of double-hull or equivalent design requirement for single hull tankers.

Category of tankers	*Phasing out time*
carrying other oils, which do not comply with the requirements for protectively located segregated ballast tanks (commonly known as Pre-MARPOL tankers)	
Category 2 oil tankers of 20,000 DWT and above carrying crude oil, fuel oil, heavy diesel oil or lubricating oil as cargo, and of 30,000 DWT and above carrying other oils, which do comply with the protectively located segregated ballast tank requirements (MARPOL tankers), and Category 3 oil tankers of 5,000 DWT and above but less than the tonnage specified for Category 1 and 2 tankers	5 April 2005: for ships delivered on 5 April 1977 or earlier
	Anniversary date in 2005: for ships delivered after 5 April 1977 but before 1 January 1978
	Anniversary date in 2006: for ships delivered in 1978 and 1979
	Anniversary date in 2007: for ships delivered in 1980 and 1981
	Anniversary date in 2008: for ships delivered in 1982
	Anniversary date in 2009: for ships delivered in 1983
	Anniversary date in 2010: for ships delivered in 1984 or later

3.3 Hazardous and Noxious Substances

3.3.1 Definition of HNS

HNS broadly defined are those substances that because of their intrinsic properties, once released into sea, may endanger human life, the environment or property.[16] The HNS discussed in this chapter exclude oil because there are special legal regimes dealing with pollution caused by oil.

HNS carried by sea often present two forms – liquid or solid. Chemicals in liquid form are mostly transported in chemical tankers.[17] Chemical tankers are generally designed to carry as many different chemical cargoes as possible.[18]

16. Manual on Chemical Pollution, International Maritime Organization, 1999, 1.
17. See Annex II of MARPOL, Ch. 1, Regulation 1.16. In addition, there are also noxious liquid substances tankers.
18. K. Rawson (ed.), *The Carriage of Bulk Oil and Chemicals at Sea* (Great Britain: Institution of Chemical Engineers, 1994), 98.

Generally, chemical tankers are smaller in their average size (chemical tankers range from 5,000 to 40,000 DWT) compared with oil tankers (which range from general purpose tanker of 10,000 DWT to ultra-large crude carrier of 550,000).[19] This is mainly because of the special nature of the chemical cargoes, and moreover there are often size restrictions on chemical terminals at port terminals. The structure of chemical tankers is more sophisticated. The packaged form of HNS also can be transported in container ships.

Recent years have seen a growth in transport of HNS. The potential danger to the marine environment posed by HNS at sea is acknowledged by the industry and governments. However, it seems that pollution incidents involving HNS receive less public attention than those caused by oil. This may be because HNS incidents do not occur as frequent as oil spills and probably also because of most HNS being less visible.[20]

The prevention of pollution from HNS is a complicated issue because the HNS exist in different forms (liquid or packaged) and hence may require different measures to control the risks they pose to the environment. The prevention of pollution from HNS is addressed in Annex II and III of MARPOL.

3.3.2 Annex II

MARPOL Annex II (and SOLAS Chapter VII carriage of dangerous goods) regulates prevention for pollution from noxious liquid substances. It categorizes the liquid HNS according to the hazard posed to the marine environment. Originally, Annex II categorized the NHS into categories A to D, with their hazard decreasing. The idea is that one category of substances should be carried on board a corresponding type of tanker, the construction, design and equipment of the tanker should be related to the nature of the category of cargo. For different categories of substances, the discharge standards also vary.

The means of categorization was amended in October 2004, effective as of 1 January 2007. The new categorization distinguishes the noxious liquid substance as X, Y, Z and other substances.[21] Through the 2004 amendments, the anti-pollution measures were made more specific.[22] For example, in the original Annex II, for Category B cargo that presents hazard to the marine environment, the anti-pollution measures are 'special measures', whereas in the new Annex II, for Category Y, which also presents hazard to the marine environment, the anti-pollution measures are 'limitation on the quality and quantity of the discharge'.

This change not only makes the enforcement a lot easier but also reflects the developments in technology. Moreover, the use of new technology (for example,

19. BONN 99/3/6-E (L), Chemical spills at sea – Case Studies, presented by France, Agreement for Cooperation in Dealing with Pollution of the North Sea by Oil and Other Harmful Substances, 1983, Eleventh meeting of the contracting parties, Brest, 29 Sep.–1 Oct. 1999,
20. Manual on Chemical Pollution, *supra* n. 16.
21. Chapter 2 of MARPOL Annex II.
22. Chapter 5 of MARPOL Annex II.

efficient stripping techniques) makes it possible to discharge certain products at substantially lower levels. Therefore, the discharge standards are also adjusted accordingly in the amendments. For ships constructed on or after 1 January 2007 the maximum permitted residue in the tank and its associated piping left after discharge will be set at a maximum of seventy-five litres for products in categories X, Y and Z compared with previous limits that set a maximum of 100 or 300 litres, depending on the product category.

In addition to the discharge standards, Annex II also designated a special area where discharge is prohibited, the Antarctic area.

MARPOL in Annex II establishes discharge standards for liquid noxious substances and construction standards for chemical carriers. The prescription of discharge standards to a large extent resemble the standards for oil discharge in Annex I, being a positive prescription of quantitative standards and a qualitative prescription of special areas where higher discharge standards apply.

Annex II contains a provision that requires cargo tanks to be protectively located. This is important so that in the case of a casualty the liquid cargo shall not be discharged immediately to sea.

3.3.2 Annex III

Marine pollution caused by hazardous substances carried in packaged form is regulated by MARPOL Annex III (and Part A of SOLAS Chapter VII Carriage of dangerous goods). This Annex contains general requirements for the issuing of detailed standards on packing, marking, labelling, documentation, stowage, quantity limitations, exceptions and notifications for preventing pollution by harmful substances. The state parties are required to issue instructions at the national level, and the Chapter refers to the International Maritime Dangerous Goods (IMDG) Code developed by the IMO. The IMDG Code is constantly updated to accommodate new dangerous goods and supplement or revise existing provisions.

The IMDG Code is a uniform international code for shipping dangerous goods. It covers matters such as packing, container traffic and stowage, with particular reference to the segregation of incompatible substances. The IMDG Code includes products considered to be marine pollutants. For the purpose of Annex III, 'harmful substances' are those identified as 'marine pollutants' in the IMDG Code.

The general attitude in the MARPOL is that the transport of such substances is prohibited unless all the conditions are complied with.

3.4 Fuel

The MARPOL Convention contains rules to limit the size of oil fuel tanks on new ships and ensure that they are protectively located. This is the regulation on oil fuel tank protection that was adopted in 2004 and entered into force on 1 January 2007.

The regulation applies to all ships delivered on or after 1 August 2010 with an aggregate oil fuel capacity of 600 m^3 and above. It includes requirements for the protected location of the fuels tanks and performance standards for accidental oil fuel outflow. A maximum capacity limit of 2,500 m^3 per oil fuel tank is included in the regulation, which also requires flag state administrations to consider general safety aspects, including the need for maintenance and inspection of wing and double-bottom tanks or spaces, when approving the design and construction of ships in accordance with the regulation.

3.5 Comparative Analysis

It seems from the previous discussion that the regulations concerning pollution prevention from oil are far more intensive than those concerning HNS. One of the major reasons might be that oil spills receive much more political attention than pollution incidents caused by HNS. This is due to the less visual nature of HNS and the small size of HNS carriers. One important reason is that the activity level of HNS transportation is much lower than that of oil transport by sea, which means the chances of having HNS pollution incidents are lower than with oil. Moreover, the strict requirement for the structure of chemical tankers may contribute to prevent certain spills as well.

As far as the prevention regulations for oil and noxious liquid substances are concerned, MARPOL addresses the issue from two perspectives – the discharge standards and the construction, design and equipment standards of the ship. The discharge standards concern the requirement for operation, and the construction, design and equipment requirements are especially important so that the ship can survive a casualty and minimize pollution damage.

Almost all regulations tackling different pollutants have the requirement of certification. However, the certification for chemical carriers may present a heavier administrative burden because the conditions of each cargo tanks need to be specified. The chemical tanker is more sophisticated in its structure and operational requirements, whereas this is not the case with oil tanker.

The ship should be certified by the flag state authority. This is often the classification societies.

It has always been recognized by the IMO that the effectiveness of the international conventions depends upon the degree to which they are enforced.[23]

Generally, under the MARPOL Convention, the jurisdiction is authorized to the flag state or the state under whose authority the ship is operating, known as the 'administration'.[24] In the case of violation, it is the 'administration' to initiate

23. The website of IMO: <www.imo.org>.
24. Article 2(5) of MARPOL provides that 'Administration' means 'the Government of the State under whose authority the ship is operating. With respect to a ship entitled to fly a flag of any State, the Administration is the Government of that State.'
    Art. 3(1) provides that 'The present Convention shall apply to:
    (a) ships entitled to fly the flag of a Party to the Convention; and
    (b) ships not entitled to fly the flag of a party but which operate under the authority of a Party.'

proceedings or take other actions.[25] This may present some potential problems because violation of the international conventions (for example, illegal discharge of polluting substances into the sea) often occurs outside the flag state, and influences a coastal state. Because of the transboundary nature of marine pollution, a flag state may not have sufficient incentives to take actions. This reflects the traditional conflict between the flag state and the coastal state or between the shipping (the trade) interests and the environmental interests.

The interest of a flag state is more an economic one because it may benefit from the ships registering in its state, whereas damage caused by those ships are often incurred to another state with coastal interests. The transboundary nature in marine pollution may explain why typical flag states (especially those known as flag of convenience states or open registry states) may have little incentive to impose stringent pollution control standards and to effectively enforce those standards. Moreover, out of the fear of losing the economic benefits when ships choose to 'flag out' to another state (probably with more lenient regulations), flag states will have no incentives whatsoever to do so.

In MARPOL there are also provisions on the power of a port state authority to inspect a ship in its port. This is incorporated into the Convention via the 1994 amendments. These amendments made it possible for ships to be inspected when in the ports of other parties to the Convention, to ensure that crews are able to carry out essential shipboard procedures relating to marine pollution. The 1994 amendments entered into force on 3 March 1996 and apply to both oil (Annex I) and HNS (Annexes II and III).

The port state being empowered to carry out inspections may be useful in the sense that it can remedy some deficiencies in implementation by a flag state. The interest of a port state to some extent can be aligned with the interest of shipping nation because a port definitely has economic interests in the vessels calling there. On the other hand, a port state is a coastal state as well and is exposed to the risk for pollution incidents. Therefore, a port state may have a more balanced interest in exercising its jurisdiction. MARPOL provides that a port authority may undertake inspections with respect to certificates required by relevant conventions and discharge of harmful substances in violation of related regulations.[26] However, these rights of inspection on ships visiting its port or offshore terminals to a state are to certain extent restricted. Therefore, the provisions in MARPOL on the implementation of the convention try to strike a balance between maintaining the maritime tradition of flag state jurisdiction and protecting the marine environment.

One the other hand, all of these regulations contain provisions related to port state control.

There are in MARPOL also provisions on the power of a port state authority to inspect a ship in its port. This is incorporated into the Convention via the 1994 amendments. These amendments made it possible for ships to be inspected when in

25. Article 6 of MARPOL Detection of violations and enforcement of the Convention.
26. Articles 5 and 6 of MARPOL Convention.

the ports of other parties to the convention, to ensure that crews are able to carry out essential shipboard procedures relating to marine pollution. The 1994 amendments entered into force on 3 March 1996 and apply to both oil (Annex I) and HNS (Annexes II and III).

MARPOL provides that a port authority may undertake inspections with respect to certificates required by relevant conventions and discharge of harmful substances in violation of related regulations.[27] However, these rights of inspection on ships visiting its port or offshore terminals to a state are to certain extent restricted. If, and only if, there are clear grounds to believe that the condition of the ship or its equipment does not correspond substantially with the particulars of the certificate, or if the ship does not carry a valid certificate, the port authority carrying out the inspection may detain the ship until it is satisfied that the ship can proceed to sea without presenting unreasonable threat of harm to the marine environment.

On the other hand, the port state may conduct examination to verify the ship's compliance with international standards and is entitled to detain the ship if it finds significant non-compliance; the port state should nevertheless refer the case to flag state, which shall decide whether to initiate a legal proceeding. Studies have shown that in such cases, the response rate from flag states is poor.[28] Moreover, the only condition for a port state to detain the ship is when there is evidence to show significant non-compliance.

Therefore, the provisions in MARPOL on the implementation of the convention try to strike a balance between maintaining the maritime tradition of flag state jurisdiction and protecting the marine environment.

In 1994, amendments were made to improve the implementation of MARPOL.[29] They extend port state control to operational requirements.

4 CIVIL LIABILITY

4.1 Oil

The civil liability regime for marine oil pollution is established through the CLC 1969 and the Fund Convention 1971. CLC imposes strict liability exclusively on the ship owner (known as the channelling provision because it channels the liability solely to the ship owner) and has requirements concerning compulsory insurance. The Fund Convention establishes an International Oil Pollution Compensation Fund (referred to as the Fund or the IOPC Fund) that is financed by a levy on the

27. Articles 5 and 6 of MARPOL Convention.
28. Marine Pollution, Report to Congressional Request, GAO/RCED-00-48, United States General Accounting Office (GAO), February 2000, 19–21.
29. The 1994 amendments were adopted on 13 November and entered into force on 3 Mar. 1996. They are contained in Resolution A. 742 (18) which was adopted by the IMO Assembly in November 1993.

oil imported. In 1992, these two conventions were amended through the 1992 Protocols, whereby the scope of compensation is expanded, and the amount of compensation is substantially increased. However, the basic principles established in the CLC 1969 and Fund Convention 1971 remain the same, being strict liability, channelling of liability and compulsory insurance.

These conventions were amended again in 2000, whereby the limitation of liability is again increased by 50%. On 16 May 2003, another Protocol on the Establishment of a Supplementary Fund for Oil Pollution Damage was adopted. It established a Supplementary Fund of Special Drawing Right (SDR) 750 million SDR (at the time of adoption, this corresponded to approximately EUR 920 million or USD 1,000 million). The Supplementary Fund does not replace the existing Fund, but only make additional compensation available to victims in the states that have ratified the 2003 Protocol. The creation of this new fund is aimed at improving the compensation to pollution victims.

4.2 HNS Convention

Probably less known than the oil spill incidents, pollution incidents from hazardous and noxious substances other than oil do occur on a regular basis. To name a few recent ones, on 23 August 1999, the Panamanian registered container vessel *Ever Decent* collided with the cruise ship *Norwegian Dream* twenty miles east of Margate, southeast England. *Ever Decent* lost some containers and caught on fire. It carried on board HNS (including almost all IMO classes except Class 1, and cyanide, which poses a significant toxic hazard). Thanks to the quick response from the UK authorities, there was no catastrophic damage. However, the UK Government has incurred significant costs for cleanup and preventive measures.

There have been early efforts to establish civil liability for HNS pollution. When the liability system for oil pollution damage was established in 1969, HNS was specifically excluded because this would have complicated the issue. Later, when the oil pollution compensation regime was amended in 1984, the issue was raised as to whether HNS should be included in the system as well. However, these efforts were not particularly successful.

The HNS Convention 1996 adopted a two-tier system modelled on the CLC/Fund Convention (being a strict liability on the shipowner, with a fund on top), and it also adopts strict liability of the shipowner and has a requirement of compulsory insurance. The HNS Convention goes further in that it covers not only pollution damage but also the risks for fire and explosion, including loss of life or personal injury as well as loss of or damage to property. The compensation provided to the victims of accidents involving HNS (such as chemicals) is also up to a certain amount, as in the oil pollution compensation regime.[30] The HNS is defined in the

30. For a discussion of the HNS Convention, see An Overview of the HNS Convention, prepared by a Correspondence Group under the leadership of the United Kingdom and approved by the

convention by reference to lists of substances included in various Annexes of the MARPOL Convention. However, the HNS Convention has not entered into force thus far.

Major difficulties impeding the ratification process of HNS are considered to include mainly three aspects: first, as for packaged goods, it is difficult to set up the reporting system; second, the concept of contributor is difficult to identify; and third, the submission of reports on contributing cargo is rarely carried out.

In order to bring HNS Convention into effect, the Legal Committee of the IMO approved in April 2009 a draft Protocol aiming at solving these difficulties. In this Protocol, first, packaged HNS are excluded from the definition of contributing cargo. Therefore, the receivers of packaged goods will not contribute to the HNS Fund. However, it is also realized that incidents involving packaged goods will remain eligible for compensation and the shipowner's liability limits for incidents involving these substances shall be increased. The precise level of increase though remains to be determined.

Second, under the current HNS Convention, the person liable for Liquefied Natural Gas (LNG) contributions is the person who holds title to an LNG cargo immediately before its discharge. In the case of other accounts, the person liable is the receiver. The receiver must be subject to the jurisdiction of a state party to the Convention, whereas the title holder need not be. It seems impossible to enforce payment of contribution to the LNG account by title holders in a non-state party. In the draft protocol, the receiver, as defined in Article 1.4 will be liable for annual contributions to the LNG account, except in the limited situation in which the title holder pays them, following an agreement to this effect with the receiver and the receiver has informed the state party that such an agreement exists.

Third, very few states will submit reports on contributing oil when ratifying the convention, although this is an obligation to do so. This is an important issue, especially given the experience of the IOPC Fund. The draft protocol tries to solve this problem by withholding compensation in respect to the state that fails this obligation.

There might be another reason for the slow ratification process. The compensation for oil pollution establishes a compensation fund in a separate convention. Countries such as China may ratify only the CLC but not the Fund Convention. There are two oil majors dominating the Chinese oil market, and they are state owned. It seems these two oil majors have great influence in lobbying the Chinese government not to ratify the Fund Convention. As for the case of HNS, which tries to establish the two tiers of liability in a single convention, this might discourage countries such as China from ratifying the convention. Indeed, the technical difficulties considered now are all related to the HNS fund.

Legal Committee of the IMO at its 84th session held in April 2002. See the website: <www.LED\INF\FACTS\LIABILITY AND COMPENSATION LEGAL ISSUES\HNS conventions overview.doc>.

4.3 Bunker Convention

More than 80% of the world merchandise trade by volume is carried by sea.[31] Maritime transport relies on oil for its propulsion. Therefore, controlling the risks posed by fuel oil to the marine environment is crucial.

The issue of an international liability and compensation regime for pollution damage caused by bunker has been raised several times during the discussions on the liability regimes for oil and HNS, but these efforts were abandoned, and it was only in 2001 that such an international liability regime was established.

At the international conference preceding the conclusion of CLC 69, bunker oil was specifically excluded because most delegations were afraid the inclusion of bunker would complicate the discussion on a yet-to-decide issue.[32] It was not decided at the time whether the shipping or the cargo interest should bear the responsibility for oil pollution damage; thus the delegations were concerned that if the cargo liability were adopted, it would have been unreasonable for the cargo owner to be held liable for the pollution caused by bunker.

The issue of bunker pollution compensation was raised again at the 1992 conference (which led to adoption of the 1992 Protocols to CLC 69 and the Fund convention 71). However, there was the fear that to include bunkers in the oil pollution compensation regime would complicate matters because there is a clear difference between oil carried as cargo and bunker as fuel oil.[33]

Later, in 1994, Australia submitted a proposal to the IMO to establish a regime for liability and compensation for pollution damage caused by ships' bunkers.[34] At the Legal Committee debate on the HNS Convention in 1996, this issue was raised again, but it was finally decided that bunkers were left out of the HNS Convention. Nevertheless, there was a firm understanding by several delegations that a bunker convention would be developed at the earliest possible opportunity thereafter.

The need for a bunkers convention was highlighted in 1996 in a joint submission to the seventy-fifth session of the Legal Committee by Australia, Canada, Finland, Norway, South Africa, Sweden and the United Kingdom. It referred to the UK Protection and Indemnity (P&I) Club's Analysis of Major Claims 1993, which

31. Report by the United Nations Conference on Trade and Development (UNCTAD) secretariat *Review of Maritime Transport* (2008): 25.
32. See LEG/CONF/C.2/SR.5, Official Records of the International Legal Conference on Marine Pollution Damage (1969) International Maritime Organization 637–640.
33. Before, there was the opinion that it might be realistic to expect cargo interest to contribute to losses caused by oil cargo and bunker oil when the tanker was laden, but when the ship is in ballast, it is considered to be used for the exclusive use of ship; therefore, the cargo interest should not be involved in compensation. However, the CLC 92 has changed the definition of a 'ship' to include a ship in ballast voyage as well. See Art. I. 1 of CLC 92; however, bunkers from non-tankers still fall outside the scope of the CLC.
34. In 1994, Australia submitted to the 38th session of IMO's Marine Environment Protection Committee (MEPC), the MEPC requested the Legal Committee to give the matter due consideration and the Legal Committee's 73rd session in 1995 agreed to include the matter in its work programme.

stated that 'half of the total number of pollution claims arose from incidents involving ships not carrying oil cargo'. Furthermore, Oil Spill Intelligence Report data confirmed that even for larger spills the number of non-tanker vessel spills was significantly greater than the number of tanker spills. Dealing with bunker spills from non-tankers was made more difficult by the lack of a liability and compensation regime, while the nature of fuel oil itself made spills of such oils more difficult – and more costly – to clean up. It has been estimated that, on average, the amount of bunkers carried in non-tankers is around 14 million tonnes at any given time – compared with approximately 130 million tonnes of oil which is carried as cargo on the world's seas. Some bulk carriers and container ships carry more oil as bunkers than coastal tankers carry as cargo.

It was considered to be a gap in the pollution liability and compensation regime that bunkers of non-tankers are not covered by any compensation regime. Hence, at a diplomatic conference in March 2001, the International Convention on Civil Liability for Bunker Oil Pollution Damage 2001 (Bunker Convention 2001) was adopted to address this gap. The Bunker Convention entered into force on 21 November 2008. It to a large extent resembles the liability system for oil pollution damage, but with some differences.[35] Similar to the CLC, the Bunker Convention 2001 also imposes strict liability on the shipowner, up to a certain amount, and has the requirement of compulsory insurance, but with a major deviation from the CLC that the so-called channelling provision in the CLC (whereby the liability of other potential liable parties is specifically excluded) is missing in the Bunker Convention 2001.

Ships over 1,000 GT registered in a State Party to the Convention are required to carry on board a certificate certifying that the ship has insurance or other financial security, such as the guarantee of a bank or similar financial institution, to cover the liability of the registered owner for pollution damage in an amount equal to the limits of liability under the applicable national or international limitation regime. In all cases, this amount should not exceed an amount calculated in accordance with the Convention on Limitation of Liability for Maritime Claims 1976, as amended.

The Convention will make the shipowner, defined broadly so as to include the owner, registered owner, bareboat charterer, manager and operator of a ship, liable to pay compensation for pollution damage (including the costs of preventative measures) caused in the territory, including the territorial sea of a State Party, as well as in its exclusive economic zone, or if a State Party has not established one, in an equivalent area.

During the discussions in the Legal Committee, some states were keen to see a separate fund contributed by shipowners to satisfy the bunker pollution claims. However, there was strong opposition from the shipowning interest and the insurance

35. See P. Griggs, International Convention on Civil Liability for Bunker Oil Pollution Damage 2001, at <www.bmla.org.uk/documents/imo-bunker-convention.doc>, and also W. Chao, 'Liability and Compensation for Bunker Pollution', *Journal of Maritime Law and Commerce* 33 (2002): 553–567.

industry. Thus, the finally adopted Bunker Convention provides that bunker pollution claims are subject to existing laws concerning the limitation of liability, which was the Limitation of Liability for Maritime Claims 1976 at the time. The 2001 conference also adopted a resolution to urge states to ratify or accede to the 1996 Protocol to the Limitation of Liability for Maritime Claims (LLMC) 1996, whereby the compensation limits for bunker pollution claims could be increased.[36]

The European Council adopted in September 2002 a decision to urge the Member States to ratify the Bunker Convention.[37] Thanks to the EU activism, the entry into force conditions were satisfied quickly and the Bunker Convention entered into force on 21 November 2008.[38]

Another issue worth mentioning is a resolution adopted at the conference on protection for persons taking preventive measures.[39] The resolution urges states when implementing the convention to consider the need to introduce in domestic legislation provisions to protect persons taking measures to prevent or minimize the effects of bunker oil pollution. It recommends that such persons taking reasonable measures to prevent or minimize the effects of oil pollution should be exempt from liability unless the liability in question resulted from their personal act or omission, committed with the intent to cause damage, or recklessly and with knowledge that such damage would probably result.

This resolution came as a compromise to the request of some industry groups (including International Tanker Owners Pollution Federation Limited (ITOPF), Comité Maritime International (CMI), International Association of Independent Tanker Owners (INTERTANKO), International Association of Ports and Harbours (IAPH), International Chamber of Shipping (ICS), International Group of Protection and Indemnity Clubs (IG) International Salvage Union (ISU) Oil Companies International Marine Forum (OCIMF) and Baltic and International Maritime Council (BIMCO)) to include similar provision into the convention.[40] The idea is that such legal protection would encourage prompt and effective response in the event of a pollution incident.

---

36. Annex I: Conference Resolution on Limitation of Liability.
37. COUNCIL DECISION of 19 Sep. 2002 authorizing the Member States, in the interest of the Community, to sign, ratify or accede to the International Convention on civil liability for Bunker Oil Pollution Damage, 2001 (the Bunkers Convention), 2002/762/EC, OJ L256, 25 Sep. 2002, 7–16.
38. The Bunker Convention requires eighteen states to adopt it. These states are Bahamas, Bulgaria, Croatia, Cyprus, Estonia, Germany, Greece, Hungary, Iceland, Jamaica, Latvia, Lithuania, Luxembourg, Marshall Islands, Norway, Poland, Samoa, Singapore, Slovenia, Spain, Tonga, United Kingdom and Sierra Leone. Among the first eighteen states that have ratified the Bunker Convention, thirteen of them are the EU Member States. Thus, the role of EU in facilitating the entry into force of the Bunker Convention should not be denied.
39. Resolution on protection for persons taking measures to prevent or minimize the effects of oil pollution.
40. LEG/CONF.12/8, 12 Jan. 2001. This proposal to include specific wording into the Convention was rejected. However, a compromise proposal was put forward by Australia, Denmark, Indonesia, Ireland, the Netherlands, Switzerland, the United Kingdom and Hong Kong to attach a resolution to the Convention. See LEG/CONF.12/11, 16 Feb. 2001.

4.4 Comparative Analysis

All three liability regimes examined in this chapter set up a strict liability on the shipowner. Strict liability of the polluter has been widely used in conventions on environmental liability not only in the case of marine pollution but in other domains of environmental law as well.

Another common feature of these three liability systems is the compulsory insurance requirement. Compulsory insurance is a good remedy in the event of insolvency of the shipowner. Some experience shows that when compulsory insurance is required, 80% of the oil pollution damage is compensated, but where there is no guarantee of compulsory insurance, the compensation rate is much lower, around only 20%.

A third general element in these liability regimes is the limitation of liability of the shipowner. This is a mechanism that has long tradition in maritime law. The adoption of limitation of shipowner's liability for oil pollution regime came almost as natural pattern. The liability limits of the regimes are based on the tonnage of the ship involved, although one may doubt whether the tonnage of the ship is the only deciding factor on the damage a vessel may cause.

One important aspect of the compensation regime is the channelling of liability. In both the CLC and HNS Convention, liability is exclusively imposed on the shipowner, whereas the liability of other parties who may potentially influence accident risk is specifically excluded. In the Bunker Convention, the so-called channelling provision is abolished.

Currently, only two conventions are in force. The CLC covers oil spills but does not include bunker oil spills from vessels other than tankers; the Bunker Convention covers liability and compensation for pollution damage caused by spills of oil when carried as fuel in ships' bunkers.

Bunker oil pollution from tankers is still covered by the CLC, whereas the Bunker Convention applies to bunker pollution from other ships. Questions may arise when bunker oil has caused pollution, and different compensation limits will apply depending upon the types of ships from which bunker is discharged into the sea.

One distinguishing characteristic of the Bunker Convention is that it does not have a channelling provision as in the CLC. This deviates from the CLC regime. However, the obligation to obtain the required insurance under the Bunker Convention rests on the registered owner alone,[41] not on the other liable parties under the convention.[42] One possible explanation for this seemingly anomalous provision is that it might be difficult for all of the parties, such as a bareboat charterer, to arrange the insurance.[43]

41. Article 7(1) of Bunker Convention 2001.
42. Article 1(3) defines shipowner to be 'the owner, including the registered owner, bareboat charterer, manager and operator of the ship'.
43. Griggs, *supra* n. 35.

Pollution damage caused by oil is covered under the CLC/Fund Convention regime, but non-contamination risks, such as damage resulting from an explosion, would be covered under the HNS Convention.

All three of these liability regimes have the requirement of compulsory insurance and hence refer to the obligatory insurance certificate. It is the flag state's responsibility to issue such a certificate confirming that appropriate insurance or financial security is in place. This places an administrative burden on the flag state. In practice, the liability insurance is often placed with the P&I Clubs. Article 7 of the Bunker Convention provides that a state party may authorize another institution or organization to issue the certificate. How this power of delegation is used in practice remains an interesting question.

5 A CRITICAL ANALYSIS

5.1 Regulation versus Civil Liability

Despite the existence of a vast number of international conventions addressing pollution prevention, pollution incidents may nevertheless happen, and thus it will be necessary to have a mechanism to compensate the pollution victims. However, currently only part of the designed compensation regimes are in force. When oil is discharged into the sea either as cargo or as bunker, there is a compensation regime available. The compensation for HNS pollution is not in force yet, but hopefully with the efforts of IMO, the technical difficulties will be overcome soon.

5.2 Regulation

The IMO is the institution that organizes international diplomatic conferences; it is under the auspices of the IMO that the international conventions on maritime safety and marine pollution have been concluded. The IMO has been recognized worldwide as the empowered organization to legislate on maritime shipping–related issues.[44] The implementation of IMO regulations depend on the Member States ratifying the conventions. If the state ratifies the convention but does not implement it in its national law, there is no legal remedy at the international level, unless there is an incident that leads to pollution, and only then may the victims may claim compensation under the civil liability regime, but there will still be no sanctions imposed on the state that fails to implement relevant IMO conventions.

In this respect, it might be interesting to look at the institutions of the EU. The European Commission is the institution to implement the EU law. When a Member State has not implemented the Community law by the required date, the

44. For a detailed discussion on the competence of the IMO, see C. Henry (ed.), *The Carriage of Dangerous Goods by Sea, the Role of the International Maritime Organization in International Legislation* (London Frances Pinter, 1985), 37–57.

European Commission is entitled to initiate actions against the Member State for failure to do so. Moreover, the legal action launched may ultimately lead to fines imposed on Member States if they consistently fail to implement the Community law. For example, in 1995, the European Council adopted a Directive regarding port state control[45] to set up strengthened and harmonized inspection rules and criteria. Some Member States did not inspect the required number of ships to the required standards. In 2002, the European Commission brought France and Ireland to the European Court of Justice for not observing the annual threshold of 25% of ships inspected by the port state as provided for by the previously mentioned Directive.[46] It seems the European Community has a more effective mechanism to implement Community law, and the lack of such an institution makes the implementation of IMO conventions not always optimal.

Another issue concerning the implementation of MARPOL is the specific measures such as economic incentives for enforcement of the Convention. In this respect, one European legislative instrument may provide some useful insight. This is the Regulation 2978/94[47] adopted in 1994 to address the issue of double hull tankers in the European waters. The goal of this regulation is to promote the use of environmental friendly tankers, including double hulls. In this legislation, port fees were reduced for tankers with segregated ballast oil tanks and double hull tankers. Such an economic incentive may indeed contribute to an effective implementation of the Regulation. However, in the later political debate preceding the amendments of MARPOL, similar provisions on such economic incentives were deleted.

The MARPOL 73/78 contains regulations for the survey and certification of ships. When it comes to implementation of these rules, it is again the competent authority in the flag state that should certify the ship's compliance with relevant MARPOL pollution prevention standards. These competent authorities are often classification societies.[48] Although the classification societies are supposed to act on behalf of the flag state, in practice they are often designated by the shipowners on a commercial basis. One may doubt whether the shipowner can exert some influence on the societies' decision. Moreover, as a result of economic pressure, the confidence from the maritime industry and the public in the classification societies has recently declined.[49] The implementation of required surveys and

45. 95/21/EC, Council Directive of 19 Jun. 1995 concerning the Enforcement, in respect of Shipping Using community Ports and sailing in the waters under the jurisdiction of the Member States, of international standards for ship safety, pollution prevention and shipboard living and working conditions (port state control), OJ L157, 7 Jul. 1995.
46. IP/02/931.
47. Council Regulation (EC) No. 2978/94 of 21 Nov. 1994 on the implementation of IMO Resolution A.747 (18) on the application of tonnage measurement of ballast spaces in segregated ballast oil tankers.
48. For a more detailed discussion on the role of classification societies, see M. Ferrer, *La Responsabilité des Societies de Classification* (Marseille: Aix-Marseille University Press, 2004).
49. For instance, the EU expressed its doubts concerning the reliability of classification societies in its Council Directive 94/57/EC of 21 Nov. 1994 on common rules and standards for ship

certification through the classification societies may in turn lead to the doubt on the effectiveness of such a system.

On the other hand, the port state may conduct examination to verify the ship's compliance with international standards and is entitled to detain the ship if it finds significant non-compliance, the port state should nevertheless refer the case to the flag state who shall decide on whether to initiate a legal proceeding or not. There have been studies showing that in such cases, the response rate from flag state is poor.[50] Moreover, the condition for a port state to detain the ship is only when there is evidence to show significant non-compliance.

In 1994, some amendments were made to improve the implementation of MARPOL.[51] They extend the port state control to operational requirements.

Article 8 of MARPOL (which refers to Protocol I to MARPOL)[52] provides that the master or any other person in charge of the ship should report the discharge or probable discharge of oil or HNS to the nearest coastal state.

The major tools for marine pollution prevention are the regulations concerning discharge standards, construction, design and equipment standards of the ships to with the potentially polluting cargo they carry on board. This way of regulation may be useful; however, these regulations depend very much on implementation, which is rather weak at the international level. There may be several reasons for this: first, there is no international institution like the Commission in the EU that can ensure enforcement of relevant regulations. Second, as discussed earlier, the international convention on marine pollution (particularly MARPOL) relies on the flag state to implement the regulations. However, in reality such enforcement by flag state is rather weak given the transboundary nature of shipping. Port state control is thus designed to remedy the ineffective implementation by flag states. However, a port state can exercise such control only when a ship enters its port and the contents of such control are not unlimited.

5.3 Civil Liability

A civil liability regime does not only work *ex post*; the fact that it might be exposed to potential liability makes a rational decision maker take preventive measures

---

inspection and survey organizations and for the relevant activities of maritime administrations, OJ L319, 12 Dec. 1994, 20–27. It reads: 'worldwide a large number of the existing classification societies do not ensure either adequate implementation of the rules or reliability when acting on behalf of national administrations as they do not have adequate structures and experience to be relied upon and to enable them to carry out their duties in a highly professional manner.'

50. Marine Pollution, Report to Congressional Request, GAO/RCED-00-48, United States General Accounting Office (GAO), February 2000, 19–21.
51. The 1994 amendments were adopted on 13 November and entered into force on 3 Mar. 1996. They are contained in Resolution A. 742 (18), which was adopted by the IMO Assembly in November 1993.
52. In 1996, some amendments were adopted to add on the requirement to report incidents involving harmful substances. These amendments were adopted on 10 Jul. 1996 and entered into force on 1 Jan. 1998.

according to the exposure to liability when making an economic decision. An optimal liability system should thus be designed to make potential liable parties exposed to the liability to the extent of damage that it could cause so that its prevention level will be efficient.

5.4 Empirics

As far as oil pollution is concerned, statistics show a trend of decreasing incidents since the 1990s. It is nevertheless difficult to judge whether this is the effect of prevention regulations alone or in combination with the civil liability regime.

According to the Fund, the majority of claims are fully covered; only a few catastrophic incidents are not fully covered. It is these major incidents that attract the most public attention, and thus the political need arises for changes in legal regimes.

As for HNS, thirteen major members of the International Group of P&I Clubs have collected data relating to incidents involving HNS by vessels entered in their Club between 2002 and 2007.[53] According to these data, there have been 126 incidents during this period that would have been governed by the HNS Convention, if it were in force at the time of the incident. The total costs of claims of these incidents were approximately SDR 137 million (USD 212 million, as per the exchange rate in January 2008).[54] Of the 126 incidents, only two caused damage that would exceed the shipowner's liability limits under the HNS Convention. One of these two incidents arose from an explosion on board a chemical tanker in Brazil. The tanker is of 11,636 GT; correspondingly, the liability limit is SDR 24.5 million under the HNS Convention, whereas the damage was SDR 35.7 million SDR (USD 55 million). The other incident occurred as a result of gasoline discharge at a terminal in the United States. It incurred a claim of approximately SDR 22.6 million, whereas the liability limit under the HNS Convention would have been SDR 16.6 million (the vessel is of 6,411 GT). Moreover, from the total damage of SDR 137 millions in respect to 126 incidents, approximately SDR 120 million would have been met by the shipowner and only SDR 17 million by the HNS Fund.[55]

As for HNS in packaged form, it seems difficult to identify these goods in the claims data. However, the incidents caused by packaged goods would be at most 23 of 126, if all of the unknown or unclassified good are included. It is also notable that the earlier-mentioned two large accidents did not involve HNS in packaged form. In data from the International Group, the highest claim incurred involving

53. This was at the request from the UK government on 14 Dec. 2007, on behalf of a number of states, for 'information and data in respect to claims history for incidents that would have been caught by the HNS Convention had it been in force at the time of the incident occurring'. See document 92FUND/WGR.5/5, 18 Jan. 2008.
54. See document 92FUND/WGR.5/5, *supra* n. 54, 2.
55. *Ibid.*

HNS in packaged form totalled approximately SDR 18 million (USD 28.2 million). This was caused by a vessel of 16,800 GT, and hence the liability limit of the shipowner is capped at 32.2 million SDR. Therefore, the damage could be fully recovered. The second highest claim according to the data was SDR 5.2 million, whereas the liability in that case would have been SDR 10 million.

A further breakdown of the incident claims shows that of the 126 incidents, only five of the reported incidents caused damage above SDR 10 million (which is the shipowner's minimum limit of liability). The size of these five vessels ranged from 6,400 to 38,600 GT. The SDR 10 million liability limit corresponds with 2,000 GT. It may be odd to notice that there are sixty-three of the 126 incidents incurring no claims for damage under the HNS Convention at all, even if the Convention were in force.

The International Group thus reached the conclusion that the current limits of liability of the shipowner under the HNS Convention are considered sufficient to cover most incidents involving HNS, including both liquid and packaged forms. Based on empirical data, there is no need to increase the compensation amount under the HNS Convention.[56]

6 CONCLUDING REMARKS

Shipping remains vital today for world trade just as it was centuries ago, and because of the uneven division of raw material and resources around the world, transportation by sea continues and the risk for marine pollution still exists.

The legal regime for marine pollution caused by oil was established earlier than the regimes for other polluting substances (such as HNS or bunker) probably because oil is the most transported product by sea and the world has witnessed many catastrophic oil spill incidents. The compensation regime for HNS is still not yet established, probably because there have not been so many pollution incidents caused by HNS as such or they are not on such a scale that the world will pay enough attention to the pollution caused by HNS. Therefore, there seems to be no political demand for such a regime yet. On the other hand, this may lead to the conclusion that the prevention regime for HNS is adequate at least at the current stage. However, this is too hasty a remark to make, and it may take another pollution incident to prove the insufficiency of the regime.

Prevention and compensation complement each other. The international regime on prevention and compensation for marine pollution has been developed since the late 1960s, mainly as a result of reactions to certain major incidents. The legal regime concerning oil pollution prevention and compensation seems to be the first to be developed and the most complete in the sense that it covers both sides – prevention and compensation. The legal regime concerning HNS is not yet complete because the compensation regime has not yet been effective. Bunker has

56. See document 92FUND/WGR.5/5, *supra* n. 54, 4.

become a debated issue only in very recent years, but policy in regard to it was quickly completed.

This may be explained by the fact that there have been many oil pollution incidents that have trigger the changes of the international regime, whereas the incidents caused by chemicals and other harmful substances have not occurred at such a frequency or scale that the world had paid attention to them.

Implementation of international conventions relying on flag states is not sufficient, and the use of port state control may remedy some of the inefficiencies in this respect.

Regional actions contribute to the development of the international regimes, especially the EU and United States, who have been victims of major pollution incidents in recent years. This may explain their activism in promoting, for example, double hull tanker design to prevent future incidents. The threat of unilateral action from the United States and EU leads to change at international level.

References

Alcock, T. 'Ecology Tankers and the Oil Pollution Act of 1990: A History of Efforts to Require Double Hulls on Oil Tankers'. *Ecology Law Quarterly* 19 (1992): 127.

Ferrer, M. *La Responsabilité des Sociétés de Classification.* Marseille: Aix-Marseille University Press, 2004.

Griggs, P. International Convention on Civil Liability for Bunker Oil Pollution Damage 2001. <www.bmla.org.uk/documents/imo-bunker-convention.doc>.

Henry, C. (ed.). *The Carriage of Dangerous Goods by Sea, the Role of the International Maritime Organization in International Legislation.* London: Frances Pinter, 1985.

Molenaar, E.J. (ed.). *Coastal State Jurisdiction over Vessel-Source Pollution.* The Hague: Kluwer Law International, 1998.

Rawson, K. (ed.). *The Carriage of Bulk Oil and Chemicals at Sea.* London: Institution of Chemical Engineers, 1994.

Wu, C. 'Liability and Compensation for Bunker Pollution'. *Journal of Maritime Law and Commerce* 33 (2003): 553–567.

Chapter 3

Marine Pollution and the Right of Entry in Maritime Ports for Ships in Distress

*Eddy Somers**

1 INTRODUCTION

Vessels in distress at sea carry an important risk for marine pollution. Particularly in the case of (oil) tankers and ships transporting dangerous or hazardous substances. Unfortunately, it has at regular intervals been demonstrated in the past that there is a real danger for serious pollution whenever ships in distress cannot find a haven that protects them from further deterioration and subsequent disaster. It must be obvious, however, that it is not always possible to offer such a safe place or that there might not be other solutions available at sea that can save a vessel from tragedy. In many instances the salvage industry can offer acceptable solutions without endangering the marine environment. Of course, finding shelter against the sometimes devastating effects of nature will in most cases be the answer. Hence, it is a legitimate question to ask whether ships in distress have a right of entry in maritime ports, roadsteads or other places of refuge. This question fits within the more general legal right of entry in internal waters and maritime ports.

* Ghent University, Belgium.

Michael G. Faure, Han Lixin & Shan Hongjun, *Maritime Pollution Liability and Policy: China, Europe and the US*, 41–53.

2 MARITIME PORT CONCEPT

Within the framework of the subject it will be useful to clarify the legal meaning of the concept 'port', in particular maritime port. International conventions, however, are not particularly helpful because they do not necessarily reflect the actual reality of the maritime industry. The 1923 Geneva Convention on Ports (Port Statute) states that all ports normally frequented by sea-going vessels and used for foreign trade shall be deemed maritime ports within the meaning of the present statute (Article 1 of the Statute). This is a rather restrictive concept that excludes some ports, for example, ports solely used for fishing vessels, those used only for war ships or where non-sea-going vessels dock. Moreover, ports used only for internal trade (cabotage) are left out of the application of the 1923 Geneva Port Statute. As a consequence, they will not be subjected to the reciprocity rule introduced by the Convention. A broader definition might refer to places where ships commonly call at for loading, discharging or anchoring.

3 THE RIGHT OF ACCESS

Ports belong to the internal waters of a coastal state, and thus they will fall under its sovereignty and will in principle be subjected to its legislation[1] The crux of the matter of the right of access to internal waters is whether coastal states can deny such access for ships flying a foreign flag. As far as inland maritime ports are concerned, it will be the legal regime of the connecting waterway that will be basically decisive. The principle of freedom of navigation such as we can find in international river conventions (e.g., Rhine, Scheldt, etc.) or still in the Barcelona International Waterway Statute of 1921 implies a primary right of access without which the freedom of navigation would be defective.[2] With respect to coastal maritime ports, it was maintained in the Aramco arbitration that:

> According to a great principle of public international law the ports of every State must be open to foreign merchant vessels and can only be closed when the vital interests of the State so require.[3]

1. Notwithstanding the internationally agreed restrictions of sovereignty through the conclusion of treaties, for example, the capitularian treaties from the eighteenth to nineteenth centuries and from the late medieval period. Also, 'free ports' such as Danzig and Trieste enjoyed a specific status not entirely covered by the coastal state's sovereignty.
2. Whenever the waterway connecting the port with the sea does not come under any international regulation, the internal legal regime of the riparian state will be applicable. Thus, when the occasion arises a right of access to an inland maritime port could be denied.
3. Aramco, Arbitration Tribunal 23 Aug. 1958, *International Law Reports* (1963): 212; A.V. Lowe, 'The Right of Entry into Maritime Ports in International Law', *San Diego Law Review* (1977): 598–600.

This seemingly bold statement, however, cannot withstand any serious critical approach[4] There is definitely no general rule of international customary law that can uphold such a view. At the most, there is a presumption that coastal states have opened up their maritime ports for foreign vessels because their economic interests so require. This presumption, however, has not attained the status of 'law' according to international customary law. Therefore, so long as a right of access has not been agreed in a treaty, coastal states can regulate that themselves, which corresponds to the (unrestricted) sovereignty of the coastal states over their internal waters.

4 A RIGHT OF ACCESS TO MARITIME PORTS IS GENERALLY BASED ON CONVENTIONAL AGREEMENT (OR NATIONAL LEGISLATION OR REGULATIONS)

It cannot be disputed that the Geneva International Port Statute of 1923 did not introduce the right of access to maritime ports in the law of the sea. Access is subject to the reciprocity principle, and as long that there is no reciprocity agreed upon, the advantages of the Port Statute cannot be invoked. Rather than introducing the right of access, the Port Statute regulates the conditions for access aiming at equal treatment of all foreign vessels in a port. Moreover, this prohibition of flag discrimination by the coastal state is not absolute because contracting parties can conclude agreements wherein they grant each other more facilities than provided in the Statute itself. This of course also explains the great number of bilateral agreements concluded concerning maritime transport. These bilateral agreements confirm the lack of a generally recognized right of access because they precisely regulate access and equal treatment between the parties involved. Besides, frequently such treaties will mention which ports will be open for trade and navigation. Stipulating which ports will be open for navigation is the prerogative of the state, which implies that states can also decide which ports will remain closed (in particular in times of armed conflict or for reasons of self-preservation). Closing of ports that are normally open must be announced in a timely manner because of the risk for the closing state being held liable in the case of damage. Various multilateral international conventions, moreover, confirm explicitly or implicitly that coastal states can exclude foreign ships from their internal waters. For instance, the International Convention for the Safety of Life at Sea (SOLAS) convention (Chapter I), the 1954 Oil Pollution (OILPOL) convention (Article VI*bis*, 4, 1971 amendments), the International Convention for the Prevention of Pollution

4. See L. De La Fayette, 'Access to Ports in International Law', *The International Journal of Marine and Coastal Law* 15–16; Lowe, *supra* n. 3, 600–618; R. Churchill & A.V. Lowe, *The Law of the Sea* (Manchester: Manchester University Press, 1999), 61.

from Ships (MARPOL) convention (Article 5, paragraph 3) and the European directives on port state control are examples of such practice.[5]

State practice shows that states deny access to their ports for various reasons such as the protection of public health and public order or with the aim of protecting the marine environment. Also, specific rules can apply with respect to substandard ships or ships containing special risks (e.g., nuclear vessels). Finally, international legal doctrine offers a widespread opinion of a vast majority of legal scholars allowing for coastal states to regulate access (including denial) to their internal waters.[6]

5 RIGHT[7] OF ACCESS FOR SHIPS IN DISTRESS

A large part of the legal doctrine is undauntedly of the opinion that ships in distress have always or under certain conditions a right of entry into maritime ports. Such a standpoint, however, needs to be examined more carefully on the basis of the sources of international law such as treaties and customary international law as well as state practice.

5.1 Multilateral Treaties

Is there a right of entry for ships in distress to be found in multilateral treaties? It is quite clear that such a right cannot be found in any multilateral convention, not

5. Also Art. 23, para. 3 of the 1995 UN Straddling Fish Stocks Convention. The fact that the 1958 Geneva Convention on the high seas (as well as Part X of the 1982 LOSC) provides that land-locked states must be given access to the sea does not imply the existence of a general right of access to ports because this is exclusively limited to land-locked states.
6. See the report of the UNCTAD Secretariat: Economic Cooperation in Merchant Shipping. Treatment of Foreign Merchant Vessels in Ports, UN DOC.TD/BC.4 9 Sep. 1975. Generally see Churchill & Lowe, *supra* n. 4, 61–65; furthermore: G. Gidel, *Le droit international public de la mer* (Paris: Chateauroux, 1932–1934), 50; J.P.A. François, *Handboek van het volkenrecht* (Zwolle: Tjeenk Willink, 1949), 906; H.W. Briggs, *The Law of Nations* (New York: Appleton Century Crofts, 1952), 348; L.N. Hydeman & W.H. Berman, *International Control of Nuclear Maritime Activities* (Ann Arbor: University of Michigan, 1960), 130–142; M.S. McDougal, & W.T. Burke, *The Public Order of the Oceans* (New Haven and London: Yale University Press, 1962), 93; P. Reuter, *Droit international public* (Paris: Presses Universitaire de France, 1976), 312; D.P. O'Connell, *The International Law of the Sea*, vol. 2 (Oxford: Clarendon Press, 1984), 848; R.J. Dupuy, & D. Vignes, *Traité du Nouveau Droit de La Mer* (Paris/Brussels: Economica/Bruylant, 1985), 780; V.D. Degan, 'Internal Waters' *Netherlands Yearbook of International Law* (1986): 3; N.O. Dinh, P. Dallier & A. Pellet, *Droit International Public* (Paris: LGDJ, 1987), 970; R.J. Dupuy & D. Vignes, *A Handbook of the New Law of the Sea*, vol. 2 (Dordrecht: Martinus Nijhoff, 1991), 940; R. Bernhardt, *Encyclopaedia of Public International Law*, vol. 2 (Amsterdam/New York/Oxford: North-Holland Publishing Company, 1987), 264; J. Combacu & S. Sur, *Droit International Public* (Paris: Montchrestien, 2001), 461.

   Contra: C. Rousseau, *Droit International Public IV* (Paris: Sirey, 1980), 342–343; C.J. Colombos, *International Law of the Sea* (London: Longmans, 1967).
7. 'Right' here is to be understood as something that eventually can be claimed from the judiciary if it is being denied.

even in the 1923 Geneva Port Statute[8] Besides, restrictions of sovereignty, including access to internal waters and ports also for ships in distress, cannot be presumed. Such restrictions must show undisputedly through accepted national or international norms, which are non-existent in this case.[9] On the contrary, the 1989 London Salvage Convention mentions that contracting parties can regulate and decide upon matters relating to salvage operations such as admittance to ports for ships in distress (Article 11).[10] Nothing in the Convention affects the right of the coastal state to take measures in accordance with generally recognized principles of international law to protect its coastline or related interests from pollution (including giving directions for salvage operations) (Article 9). These are unambiguous, positive law norms allowing the coastal state to regulate and eventually deny the right of access for ships in distress.[11] If ever states can be convinced to

---

8. According to some scholars, the Contracting Parties to the 1923 Geneva Convention were firmly convinced that such a right was absolutely evident and that therefore they did not consider it necessary to include such a right in the convention itself. The basis for this assumption were two declarations of Belgian and Dutch delegates. There is, however, not a shred of evidence that such was the opinion of the majority of all the delegations anyway, which makes this position flawed. See: E. Van Hooydonk, 'Some Remarks on Financial Securities Imposed by Public Authorities on Casualty Ships as a Condition for Entry into Ports', in *Marine Insurance at the Turn of the Millennium*, ed. M. Huybrechts, E. Van Hooydonk & C. Dieryck, vol. 2 (Intersentia, Antwerp, 2000), 127; also L. Lucchini & M. Voelckel, *Droit de la mer II* (Paris: Pédone, 1996), 296.
9. The existence of national norms regulating access for ships in distress confirm moreover that there is no general international norm available upon which the right of access for ships in distress can be based; see, e.g., the Belgian law of 20 Jan. 1999 for the protection of the marine environment in maritime areas under Belgian jurisdiction, *Belgisch Staatsblad* 12 Mar. 1999 (ed. 2). Art. 23§ 1 does not allow the master of a ship in distress to enter a Belgian port unconditionally; more: Arts 3 and 9 of the French decree of 22 Jul. 1977 and Art. 5 of the French decree of 6 Feb. 1985, cf. J.P. Queneudec, 'Chronique du droit de la mer', *Annuaire français de droit international* (1985): 783–789; the Spanish Royal Decree 210/2004 on the monitoring and information of maritime traffic, mentioned by R. Roman, 'Port Perspectives and Environmental Management Considerations', in *Places of Refuge for Ships: Emerging Environmental Concerns of a Maritime Custom*, ed. A. Chircop & O. Linden (Leiden & Boston: Martinus Nijhoff Publishers, 2006), 119–143, at 132. For an overview of national legislation see *CMI Yearbook* 2002, 123–125, which shows clearly that states reserve the right to decide whether or not access will be granted and under what conditions eventually. A 'Schedule of casualty experience' framed by the CMI equally shows that coastal states decide for themselves in this matter. As such it can be concluded that there is no *de lege lata* (enforceable) right of access for ships in distress, *CMI Yearbook* (2002): 139–142. Also A. Chircop & O. Linden (eds), *Places of Refuge for Ships: Emerging Environmental Concerns of a Maritime* Custom (Leiden & Boston: Martinus Nijhoff Publishers, 2006).
10. The requirement to develop an oil pollution response contingency plan in the OPRC 1990 does not imply by itself that ships in distress must be given access to internal waters. Moreover, only Australia, Germany and New Zealand include in their contingency plans provisions concerning admission of ships in distress in case of a threat of pollution; however, the final decision on admittance remains with the coastal state, *CMI Yearbook* (2002): 121.
11. Similarly, see the British Merchant Shipping Act (1995), Art. 137, in which the empowered Special Representative of the Secretary of State (SOSREP) is given extensive jurisdiction in the case of marine pollution or threat of such pollution. See F. Rose, *Kennedy and Rose: The Law of Salvage* (London: Sweet & Maxwell, 2002), 40–41; Art. 20 of the Directive 2002/59/EC establishing a European monitoring, controlling and information system provides that Member

allow a right of entry for ships in distress into maritime ports, this should be arranged using the vehicle of a multilateral treaty.[12] Without such a treaty there is *de lege lata* no right of entry for ships in distress, at least not on a conventional basis.

5.2 Customary International Law

So is there perhaps then a general customary legal rule providing for ships in distress a right of access to maritime ports? A great number of legal scholars are convinced that such is the case. However, this is seemingly rather a position *de lege ferenda*, because there are no arguments de jure to sustain it. The fact that scholars repeat each other[13] is as such not evidence of a customary legal norm; neither is the

States have to draw up plans to accommodate ships in distress in waters under their jurisdiction. Such plans shall contain the necessary arrangements and procedures taking into account operational and environmental constraints, to ensure that ships in distress may immediately go to a place of refuge subject to authorization by the competent authority. Again this confirms that the coastal state has the final word and is allowed eventually to deny access; generally see A. Bellayer-Roille, *Le transport maritime et les politiques de sécurité de l'Union Européenne* (Rennes: Ed. Apogée, 2000). The provisions of the Erika III package draft directive on the amending of the Directive 2002/59/EC is making an opening for a more balanced approach. Requests for entry of ships in distress will be taken into consideration by a specific competent authority that can make an independent decision whether to allow or refuse entry. Allowing entry can be further regulated by such authority; for example, financial security can be asked for, the lack of which, however, cannot be solely the reason for denying access. Although the amendments to the monitoring directive allow for more transparency in the process of considering entry requests, it does not by itself introduce a right of entry as such. it merely clarifies the conditions that can be imposed with respect to entry. Furthermore, it makes clear that states need to scrutinize and assess any request seriously and that the final decision lies with an independent authority designated by the Member States. Cf. the text accepted by the Council and the Parliament; Report A6-0100/2009 and draft text C6-0045/2009.

12. See N. Gaskell, 'The 1989 Salvage Convention and the Lloyd's Open form (LOF) Salvage Agreement 1990', *Tulane Maritime Law Journal* (1991): 20; Van Hooydonk, *supra* n. 8, 132. Also see the CMI proposal draft instrument on places of refuge, *CMI Yearbook* 2007–2008, 128. The idea of introducing a new convention, however, is not welcomed unanimously, not to say that there is considerable opposition, even when the CMI draft does not yet impose a right of access but merely proposes to clarify the conditions of such right when granted. On the work of the CMI see also R. Shaw, 'Places of Refuge: The Debate Moves On', *Journal of International Maritime Law* (2004): 174–187; and R. Shaw, 'CMI Draft Instrument', *Journal of International Maritime Law* (2008): 434.
13. McDougal & Burke, *supra* n. 6, 110; Dinh et al., *supra* n. 6, 1106; A.V. Lowe, 'The United Kingdom and the Law of the Sea', in *The Law of the Sea. The European Union and Its Member State*, ed. T. Treves (The Hague: Martinus Nijhoff, 1997), 524; E. Beckert & G. Breuer, *Öffentliches Seerecht* (Berlin: W. De Gruyter, 1991), 138; De La Fayette, *supra* n. 4, 1; Churchill & Lowe, *supra* n. 4, 63 who put things in perspective, however; O'Connell, *supra* n. 6, 627–629; L. Hydeman & W. Berman, *supra* n. 6, 153–162; K. Hakapaa, *Marine Pollution in International Law* (Helsinki: Suomalainen Taideakatemia, 1981), 167; L. Lucchini, & M. Voelckel, *supra* n. 8, 295–299, who are of the opinion that we are dealing with 'un droit fort ancien' but give themselves a number of examples from state practice to the contrary;

verdict of a national judge or ad hoc arbitration.[14] It is generally accepted that in the nineteenth and early twentieth century a customary legal basis existed to allow ships in distress a right of entry into maritime ports.[15] That is no longer the case, however, in contemporary public international law. In earlier centuries the customary rule was understandable, and given the humanitarian concern such a general principle would have been agreed to. Ships were much smaller, did not carry inherently dangerous substances and did not constitute a threat for the marine environment. The potential risk for the coastal states was negligible, and ultimately coastal states could deal with the risk themselves more easily.[16] Apparently, developments after the Second World War in particular drastically changed the situation. Technological evolution produced a marked increase in tonnage worldwide, as well as the launching of several new types of vessels steadily growing also in size. Both the cargo transported – in ever-increasing quantities – and the heavy fuel used for propulsion contain inherently serious risks for marine pollution. As a rule, coastal states will no longer be solely capable of dealing with the danger resulting from a distress situation at sea. Possible catastrophic damage to the marine environment can be caused by oil tankers and by ships carrying dangerous goods, including nuclear material, toxic waste or chemicals in bulk or packaged form. Taking such circumstances into consideration, it is not reasonable to expect that a ship in distress will be allowed entrance into a port as a right. In order to deter a grave and immediate danger for the coastal state's marine and terrestrial environment, states can regulate and even prohibit entry into their ports.[17] Even those scholars who are fervently in favour of a right of entry for ships in distress recognize that it is not of an absolute character and that an appraisal of the interests of the ship and of the coastal state is necessary. Whenever the interests of the ship are subordinate to those of the coastal state, the latter can deny access.[18] The theory of

---

G. Kasoulides, 'Vessels in Distress: "Safe Havens" for Crippled Tankers', *Marine Policy* (1987): 184.

14. Particularly in the Anglo-American doctrine the *Rebecca* case (1929) is considered to be evidence for the existence of 'the long-established right of entry into a foreign port by vessels in distress', E.D. Brown, *The International Law of the Sea*, vol. 1 (Dartmouth: Aldershot, 1994), 39. Concerning the *Rebecca* case see *R.I.A.A.*, IV, 444 (2 Apr. 1929). However, this case concerns a decision of the Mexican-American General Claims Commission dealing with the right of immunity of a vessel in distress that entered a foreign port; also see *Annual Digest* 1929–1930, 129.
15. A. Chircop, 'The Customary Law of Refuge for Ships in Distress', in *Places of Refuge for Ships: Emerging Environmental Concerns of a Maritime Custom*, ed. A. Chircop & O. Linden (Leiden & Boston: Martinus Nijhoff Publishers, 2006), 163–229.
16. See A. Chircop, O. Linden & D. Nielsen, 'Characterising the Problem of Places of Refuge for Ships', in *Places of Refuge for Ships: Emerging Environmental Concerns of a Maritime Custom*, ed. A. Chircop & O. Linden (Leiden & Boston: Martinus Nijhoff Publishers, 2006), 13–16.
17. Based on the principle of self-help or even self-preservation and the status of necessity. See D. Devine, 'Ships in Distress. A Judicial Contribution from the South-Atlantic', *Marine Policy* (1996): 230; E. Van Bogaert, *Volkenrecht* (Antwerp: Kluwer Rechtswetenschappen, 1982), 158.
18. Hydeman & Berman, *supra* n. 6, 157; McDougal & Burke, *supra* n. 6, 110; A. De Zayas, 'Ships in Distress', in *Encyclopaedia of Public International Law IV* (Amsterdam: Elsevier, 2000),

a balance of interest clearly brings the problem into perspective. However, it does not lead to the conclusion that there is a general right of access for ships in distress. Therefore, upon this consideration it cannot convincingly be decided that ships in distress must be granted a general right of access.[19]

Lacking positive legal arguments, also of a customary legal nature, necessarily leads to the conclusion that *de lege lata* there is no international legal norm that allows a general right of access for ships in distress. Moreover, state practice produces examples in which coastal states for various reasons have denied access to internal waters and ports even for ships in distress.[20] Striking cases are the *Erika*,[21] *Tampa*[22] and *Prestige* cases.

It will be clear that the decision to allow or refuse entrance for a ship in distress is mainly one of a policy nature, taking into account the different interests concerned. The decision will be unilaterally taken or on a regional basis and after consultations.[23]

5.3 No General Right of Access

Finally, it can be stated that there are no legal norms in international law – be them customary, as a general principle or conventionally – to conclude that at present there is a general right of access to maritime ports for ships in distress. It is accepted that such a right has existed in previous times, but states today will not accept a

---

399; H. Meijers, Comments Hoge Raad 7 Feb. 1986 (ms Attican Unity), *Nederlandse Jurisprudentie* (1986): 1838; Churchill & Lowe, *supra* n. 4, 63 who indicate that measures must be taken to save lives before access can be denied; Hakapaa, *supra* n. 13, 168; also Raad van State (Ned.) 10 Apr. 1995 (ms Long Lin), (1995) 96 *Schip en Schade*; Ch. 26 Bonn Agreement Manual within the framework of the Bonn Agreement of 13 Sep. 1983.

Contra Van Hooydonk, *supra* n. 8, 128: 'To close off all seas and ports as soon as one smells danger, thereby implementing the result of a so-called balance of interests, clearly is not what international law on ships in distress intends.'

19. See the Guide Lines 2003 of the IMO concerning 'places of refuge' where it is stated that whenever permission is asked for access to a 'place of refuge' 'there is no obligation for the coastal State to grant it, but the coastal State should weigh all the factors and risks in a balance and give shelter whenever reasonably possible' and further: 'In the light of the outcome of the assessment provided for above, the coastal State should decide to allow or refuse admittance, coupled where necessary, with practical requirements' (IMO Doc. A23/17/Add.1 para. 3.12 & 3.13). The final decision therefore remains with the coastal state, implying that there is no right of access.

    Also A. Chircop, 'The IMO Guidelines on Places of Refuge for Ships in Need of Assistance', in *Places of Refuge for Ships: Emerging Environmental Concerns of a Maritime Custom*, ed. A. Chircop & O. Linden (Leiden & Boston: Martinus Nijhoff Publishers, 2006), 35–60.
20. See G. Kasoulides, *Marine Policy* (1987): 185–187; also 'schedule of casualty experience' in *CMI Yearbook* 2002, 139–142. Also Chircop et al., *supra* n. 16, 3–6.
21. See J. Le Corre, *Chronique d'un désastre annoncé* (Rennes: Ed. Apogée, 2000); J. Le Corre, *Mers noires* (Rennes: Ed. Apogée, 2001).
22. *Fairplay Shipping Weekly* 8 Aug. 2002; M. White, 'M.V. Tampa and Christmas Island Incident', August 2001, *BIMCO Review*, October 2001.
23. The Bonn Agreement 1983 recognized the importance of a safe haven for ships in distress.

restriction of their sovereign prerogatives failing a conventional ruling.[24] State practice, moreover, provides contraindications for the existence of an international customary norm.

6 DISTRESS, FORCE MAJEURE AND IMMUNITY

Apparently a great deal of the confusion created in legal doctrine seems to be the result of the existence of two separate problems, that is, the right of access for ships in distress and the immunity granted to such ships once they have been allowed to enter. Indeed, ships entering the internal waters of a foreign state because of distress or force majeure (and not denied entrance) are exempt from local jurisdiction. Therefore, they enjoy a privileged position and exemption of legal rules that normally would be applicable if the ship entered voluntarily. These rules concern penalization, arrest, confiscation, import dues, taxes and so forth. Definitely, such exemption is not of a general nature, because navigation rules and regulations remain applicable.[25] It is of course required that the case be of a genuine nature and that the ship is confronted with a bona fide case of distress or force majeure.[26] Evidence of a 'real distress' is a matter of fact the burden for which lies with those that invoke it.[27] Distress and force majeure have been defined in the comments on Article 17 of the Harvard Research Draft on Territorial Waters (1929). According to this document 'distress' comprises amongst others damage to the hull or the propulsion system or a shortage of fuel and bunkers provided that due diligence has been observed at the beginning of the voyage.[28] Nevertheless, coastal states can deny exemption from local jurisdiction even in cases of genuine distress or force majeure, whenever the ship is involved in illegal activities such as

---

24. Moreover, the right of denial still implies the possibility for allowance whilst the opposite is not true.
25. Generally see: P.C. Jessup, *The Law of Territorial Waters and Maritime Jurisdiction* (New York: Jennings Co., 1929), 194–208; Colombos, *supra* n. 6, 329; J. Castel, *International Law: Chiefly as Interpreted and Applied in Canada* (Toronto: Butterworth, 1976), 574; *RGDIP* 1986, 667. Also see Art. 4, para. 2 of the Council Regulation (EC) No. 1005/2008 establishing a Community system to prevent, deter and eliminate illegal, unreported and unregulated fishing.
26. Force majeure includes acts of mutiny and piracy; *American Journal of International Law* (1929): Suppl., 299–300. Also US Coast Guard *Marine Safety Manual,* vol. VI, Ch. 1, in *CMI Yearbook* (2002): 146.
27. Devine, *supra* n. 17, 234 following a discussion of the case *Merk and Djakimah v. the Queen* (1992), Court of Appeal for St Helena, South Africa.
28. 'Where a ship is the cause of its own distress it need not expect any jurisdictional immunities', Devine, *supra* n. 17, 234; the fact that a ship enters a port by itself and without help is not necessarily evidence that it is not in genuine distress, cf. The *Rebecca* (*supra*). A similar position can be found in various decisions by the judiciary, see: *Rex v. Flahaut, Ann. Dig.* 1938–1940, 165. Furthermore, *May v. the King, Ann. Dig.* 1931–32, 154–156; *US v. Sullivan et al., Ann. Dig.* 1919–1942 (dl. 11), 127–130; The Eleanor (1809) in K. Simmonds, *Cases on the Law of the Sea* (New York: Oceana), I, 125; also Castel, 574–585; G. Knight, H. Chiu & L. Sohn, *The International Law of the Sea. Cases, Documents and Readings* (London: 1991), 273–276.

piracy, slave trade,[29] human trafficking, illegal arms trade, drug trafficking or smuggling.[30]

7 CONCLUSION

Without dealing with all of the aspects involved in the right of entry into maritime ports for ships in distress, for example, the issue of liability in the case of denial or subsequent damage in the case of entry,[31] it should be clear by now that a general enforceable right of entry as such does not exist *de lege lata* in international law. The ongoing discussions within the International Maritime Organization the European Union (EU) and private *fora* such as the Comité Maritime International mostly and substantially focus on the position of the ship that finds itself in a distress situation and needs assistance. This is, however, only one side of the coin. The interests of the coastal states involved should not be overlooked. After all, an entire community's livelihood can be at stake in some extreme cases and the risks involved can be threatening for the (local) economic activities. Although the interests of the coastal states overwhelmingly justify that the final decision remains on land, we should not turn a blind eye for the needs of unfortunate ships looking for help. Any denial of entry should therefore seriously and considerately be looked into beforehand, and only when the situation is of such a compelling and overriding urgency for the coastal state's interests, such a decision should be accepted. Again this position is not substantiated by the status of present international law. States have the right to protect and preserve their own environment within the contours of international law. Failing a positive rule of (international) law, they cannot be held responsible for the consequences of a previously well-considered refusal of access. The amended monitoring directive of the EU obliges Member States to entrust the decision on refusal or entry to an independent and competent authority. The question remains, however, whether such authority can really be independent, because it will be appointed by the Member States themselves. If such an authority would be of a Community stature, that might be much more preferential and credible, but its creation does not seem under the present conditions to be very realistic. Moreover, appraisal of the distress situation by an independent authority

29. In a number of cases dealt with by the 'Mixed Claims Commission' established according to the British-American convention of 8 Feb. 1853 concerning ships implied in slave trade, the United States maintained exemption of the application of local legislation with respect to liberation of slaves. See the cases of the *Comet*, the *Encounter*, the *Enterprise*, the *Hermosa* and the *Creole* (1855), J.B. Moore, *A Digest of International Law*, vol. 2 (Washington, D.C.: Government Printing Press, 1906), 350–361.
30. See, e.g., the case of the *Silver Sea* (20 Feb. 1986). This ship was involved in arms trafficking when it needed to call at the port of Brest in France due to bad weather, *RGDIP* (1986): 667–668.
31. See H. Ringbom, 'Places of Refuge and Environmental Liability and Compensation, with Particular Reference to the EU', *CMI Yearbook* (2004): 208–233.

does not necessarily guarantee a quick response, but neither is such a response guaranteed when left to the Member States' administrations or services.

The call for an international convention regulating entry into maritime ports, as well as the conditions thereof, should not be dismissed without careful thought. Such a convention would at least have the advantage of establishing clarity. Moreover, coastal states would definitely be given a conventionally agreed framework to work with and to substantiate their decisions. The shipping industry (and the salvage industry as well) on the other hand would know what to expect so that they eventually can anticipate when necessity arises. Failing a global international convention, measures could be taken on a regional basis. The amended EU monitoring directive still focuses strongly on the Member States' willingness to take the necessary action, which is not inevitably the most efficient way to respond to distress situations. Bringing decision making in this respect into the hands of a truly Community machinery might avoid narrow-minded nationalistic considerations.

References

Beckert, E. & G. Breuer. *Öffentliches Seerecht.* Berlin: W. De Gruyter, 1991.

Bernhardt, R. *Encyclopaedia of Public International Law.* Vol. 2. Amsterdam/New York/Oxford: North-Holland Publishing Company, 1987.

Briggs, H.W. *The Law of Nations.* New York: Appleton Century Crofts, 1952.

Brown, E.D. *The International Law of the Sea.* Vol. 1. Dartmouth: Aldershot, 1994.

Castel, J. *International Law: Chiefly as Interpreted and Applied in Canada.* Toronto, Butterworth, 1976.

Chircop, A. 'The Customary Law of Refuge for Ships in Distress'. In *Places of Refuge for Ships: Emerging Environmental Concerns of a Maritime Custom,* edited by A. Chircop & O. Linden. Leiden/Boston: Martinus Nijhoff Publishers, 2006.

Chircop, A. 'The IMO Guidelines on Places of Refuge for Ships in Need of Assistance'. In *Places of Refuge for Ships: Emerging Environmental Concerns of a Maritime Custom,* edited by A. Chircop & O. Linden. Leiden/Boston: Martinus Nijhoff Publishers, 2006.

Chircop, A. & O. Linden (eds). *Places of Refuge for Ships: Emerging Environmental Concerns of a Maritime Custom.* Leiden/Boston: Martinus Nijhoff Publishers, 2006.

Chircop, A., O. Linden & D. Nielsen. 'Characterising the Problem of Places of Refuge for Ships'. In *Places of Refuge for Ships: Emerging Environmental Concerns of a Maritime Custom,* edited by A. Chircop & O. Linden. Leiden & Boston: Martinus Nijhoff Publishers, 2006.

Churchill, R. & A. Lowe. *The Law of the Sea.* Manchester: Manchester University Press, 1999.

Colombos, C.J. *International Law of the Sea.* London: Longmans, 1967.

Combacu, J. & S. Sur. *Droit International Public.* Paris: Montchrestien, 2001.

De La Fayette, L. 'Access to Ports in International Law'. *The International Journal of Marine and Coastal Law* (1996): 15–16.
De Zayas, A. 'Ships in Distress'. in *Encyclopaedia of Public International Law IV*. Amsterdam: Elsevier, 2000.
Degan, V.D. 'Internal Waters'. *Netherlands Yearbook of International Law* 17 (1986): 3–44.
Devine, D. 'Ships in Distress: A Judicial Contribution from the South-Atlantic'. *Marine Policy* 20 (1996): 229–234.
Dinh, N.O., P. Dallier & A. Pellet. *Droit International Public*. Paris: LGDJ, 1987.
Dupuy, R.J. & D. Vignes. *Traité du Nouveau Droit de La Mer*. Paris/Brussels: Economica/Bruylant, 1985.
Dupuy, R.J. & D. Vignes. *A Handbook of the New Law of the Sea*. Vol. 2. Dordrecht: Martinus Nijhoff, 1991.
François, J.P.A. *Handboek van het Volkenrecht*. Zwolle: Tjeenk Willink, 1949.
Gaskell, N. 'The 1989 Salvage Convention and the Lloyd's Open form (LOF) Salvage Agreement 1990'. *Tulane Maritime Law Journal* 16 (1999): 1–106.
Gidel, G. *Le Droit International Public de la Mer*. Paris: Chateauroux, 1932–1934.
Hakapää, K. *Marine Pollution in International Law*. Helsinki: Suomalainen Taideakatemia, 1981.
Hydeman, L.N. & W.H. Berman. *International Control of Nuclear Maritime Activities*. Ann Arbor: University of Michigan, 1960.
Jessup, P.C. *The Law of Territorial Waters and Maritime Jurisdiction*. New York: Jennings Co., 1929.
Kasoulides, G. 'Vessels in Distress: "Safe Havens" for Crippled Tankers'. *Marine Policy* 11 (1987): 184.
Knight, G., H. Chiu & L. Sohn. *The International Law of the Sea: Cases, Documents and Readings*. London: Elsevier Applied Science, 1991.
Le Corre, J. *Chronique d'un désastre annoncé*. Rennes: Ed. Apogée, 2000.
Le Corre, J. *Mers noires*. Rennes: Ed. Apogée, 2001.
Lowe, A.V. 'The Right of Entry into Maritime Ports in International Law'. *San Diego Law Review* 14 (1976–1977): 598–600.
Lowe, A.V. 'The United Kingdom and the Law of the Sea'. In *The Law of the Sea. The European Union and Its Member State*, edited by T. Treves. The Hague: Martinus Nijhoff, 1997.
Lucchini, L. & M. Voelckel. *Droit de la mer Vol. I, Part II*. Paris: Pédone, 1996.
McDougal, M.S. & W.T. Burke. *The Public Order of the Oceans*. New Haven/ London: Yale University Press, 1962.
Meijers, H. 'Comments Hoge Raad 7 February 1986 (ms Attican Unity)'. *Nederlandse Jurisprudentie* (1986): 1838.
Moore, J.B. *A Digest of International Law*. Vol. 2. Washington, D.C.: Government Printing Press, 1906.
O'Connell, D.P. *The International Law of the Sea*. Vol. 2. Oxford: Clarendon Press, 1984.
Queneudec, J.P. 'Chronique du droit de la mer'. *Annuaire français de droit international* (1985): 783–789.

Reuter, P. *Droit International Public*. Paris: Presses Universitaire de France, 1976.

Ringbom, H. 'Places of Refuge and Environmental Liability and Compensation, with Particular Reference to the EU'. *CMI Yearbook* (2004): 208–233.

Roman, R. 'Port Perspectives and Environmental Management Considerations'. In *Places of Refuge for Ships: Emerging Environmental Concerns of a Maritime Custom,* edited by A. Chircop & O. Linden. Leiden/Boston: Martinus Nijhoff Publishers, 2006.

Rousseau, C. *Droit international public IV.* Paris: Sirey, 1980.

Shaw, R. 'Places of Refuge: The Debate Moves On'. *Journal of International Maritime Law* 10 (2004): 174–187.

Shaw, R. 'CMI Draft Instrument – Athens 12–17 October 2008'. *Journal of International Maritime Law* 14 (2008): 434.

Van Bogaert, E. *Volkenrecht.* Antwerp: Kluwer Rechtswetenschappen, 1982.

Van Hooydonk, E. 'Some Remarks on Financial Securities Imposed by Public Authorities on Casualty Ships as a Condition for Entry into Ports', in *Marine Insurance at the Turn of the Millennium*, edited by M. Huybrechts, E. Van Hooydonk & C. Dieryck. Vol. 2. Antwerp: Intersentia, 2000.

White, M. 'M.V. Tampa and Christmas Island Incident', August 2001. *BIMCO Review*, October 2001, <www.ila.org.au/pdfs/ex_The%20M.V.%20Tampa%20and%20the%20Christmas%20Island%20Incident%20%28Updated%29.PDF>.

Chapter 4

Lost Ecosystem Goods and Services as a Measure of Marine Damages

*James Boyd**

1 INTRODUCTION

Marine vessel, terminal and harbour operations can generate a variety of legal damages arising from liability for response and cleanup costs, damages to private property and damages to public natural resources. This paper addresses the latter – damages to public environmental resources. Public resources that can be damaged by marine incidents include water quality, beach and other coastal recreational resources, coral reefs, commercial and recreational fisheries, sea grass beds and habitats for bird and other animal populations. These resources, while public and thus not owned and traded, are nevertheless economically and socially valuable.

Liability for lost public goods and services is an established and operative legal principle. In US waters, for example, owners and operators are liable for natural resource damages (NRDs). NRD liability services an important deterrence and compensation function by forcing owner-operators to internalize the costs of damages to valuable public resources. The goal of US NRD liability is to 'make the environment and public whole' following a pollution event.[1] In economic terms

---

* Senior Fellow, Resources for the Future, Washington DC; e-mail boyd@rff.org.

1. 15 Code of Federal Regulations (CFR) 990.53.

Michael G. Faure, Han Lixin & Shan Hongjun, *Maritime Pollution Liability and Policy: China, Europe and the US*, 55–74.

this means calculating damages equivalent to the social benefit lost as a result of a release, grounding or other marine event.

At a conceptual level, NRDs require us to measure lost ecological wealth. To measure lost ecological wealth it is necessary to know two things: how natural systems produce valuable biophysical goods and services and the value of those goods and services.

Within ecology and economics assessment of ecosystem goods and services is a growing area of inquiry. Broadly, 'ecosystem services' refers to the dependence of economic wealth and human well-being on natural systems. Although the promise of a cohesive framework for assessing all types of damages is not yet operational, many are working toward this goal through more rigorous conceptualization and communication of the links between changes in natural systems and effects on human welfare.[2]

The relevance of this perspective to NRDs (and marine damages specifically) is potentially powerful. In fact, lost ecosystem goods and services are – as will be argued – the appropriate measure of damages if damages are designed to internalize social costs and make the public whole following a marine pollution or damage incident. To calculate NRDs, we should be measuring damaged or lost ecosystem goods and services and then calculating the social cost of those losses.[3] Given this equivalence between damages and lost goods and services, the calculation of marine damages can and will hinge on the degree to which ecosystem goods and services can be understood and valued.

This chapter describes the ecological and economic underpinnings of ecosystem services assessment and reflects on the concept as a guide to damage calculations. One of the reasons NRD assessment has been so difficult and controversial in the United States is that there is widespread confusion over how to account for ecosystem goods services that are lost or gained. This paper illustrates the ways in which ecosystem goods and services are measured and integrated with economic assessment. Shifting to lost ecosystem services as a measure of damages is no

2. G. Daily, *Nature's Services: Societal Dependence on Natural Ecosystems* (Washington, DC: Island Press, 1997); Taylor H. Ricketts et al. 'Economic Value of Tropical Forest to Coffee Production', *Proceedings of the National Academy of Sciences* 101 (2004): 12579–12582; S. Polasky et al., 'Conserving Species in a Working Landscape: Land Use with Biological and Economic Objectives', *Ecological Applications* 15, no. 4 (2005): 1387–1401, L.A. Wainger et al., 'Wetland Value Indicators for Scoring Mitigation Trades', *Stanford Environmental Law Journal* 20, no. 2 (2001): 413–478; J. Boyd, 'The Nonmarket Benefits of Nature: What Should Be Counted in Green GDP?', *Ecological Economics* 61 (2006): 716–723; S.R. Carpenter et al., 'Millennium Ecosystem Assessment: Research Needs', *Science* 314 (2006): 257–258; J. Boyd & S. Banzhaf, 'What Are Ecosystem Services?', *Ecological Economics* 63, nos 2–3 (2007): 616–626; E.B. Barbier et al., 'Coastal Ecosystem-Based Management with Nonlinear Ecological Functions and Values', *Science* 319 (2008): 321–323.
3. It is worth emphasizing that economics and the 'goods and services' mindset place no limits on what people think is important or valuable. The desire for beauty, awe, naturalness and protection of species other than our own may be hard to measure, but they are all benefits known to economics and accepted as important social realities that should factor in economic damage assessment.

panacea. A great deal of work in the ecological and economic sciences remains necessary to calculate damages with accuracy, but the discipline of thinking of natural systems as sources of wealth provides a guide to the kinds of information and analysis necessary to establish ecologically and economically defensible damages.

2 LIABILITY FOR DAMAGES TO PUBLIC TRUST NATURAL RESOURCES

NRDs are damages to land, fish, wildlife, biota, air, water, groundwater and other resources.[4] Physical injuries can take a variety of forms but typically relate to adverse changes in the health of a habitat or species population and in the underlying ecological processes on which they rely.[5] The analytical challenge is to convert physical damages to a resource into the economic consequences of that damage. To do so requires understanding of the larger biophysical system of which the damaged resource is a part.

Liability for events that damage resources is established in the United States under the Comprehensive Environmental Response, Compensation, and Liability Act (CERCLA), the Oil Pollution Act (OPA) and the National Marine Sanctuaries Act (NMSA).[6] Earlier, the Deepwater Port Act (DPA) of 1974 and the Clean Water Act (CWA) amendments of 1977 introduced NRD liability to US federal law.[7] The statutes create a compensable monetary liability for damage, which in turn requires calculation of the monetary value of the damage. The principle behind NRDs is well-established in US law.[8]

By their nature, NRDs acknowledge that natural resources produce a collective social benefit. This means that the value of NRD-related ecological benefits is the value arising from public goods, not goods traded and priced by markets.

4. OPA 33 USC § 2701(20); CERCLA 42 USC § 9601(16).
5. 15 CFR 990.52. 'Potential categories of injury include, but are not limited to, adverse changes in: survival, growth, and reproduction; health, physiology, and biological condition; behaviour; community composition; ecological processes and functions; physical and chemical habitat quality or structure; and public services.'
6. Respectively, 42 USC § 9607(f) (1); 33 USC § 2702(b) (2) (A); 16 USC § 1432, 1443.
7. For the CWA, see 33 USC §§ 1251 et seq. s. 311 of the CWA regulates the discharge of oil and other hazardous substances into navigable waters, allows the government to remove the substance and holds the responsible parties liable for that removal. The removal cost is defined to include 'costs for restoration or replacement of natural resources damaged or destroyed.' 33 USC § 1321. The Deepwater Port Act of 1974, which preceded the CWA, established liability for damages to natural resources to be recovered by a federal trustee and used for restoration. 33 USC §§ 1501–1524, 1982.
8. The public trust doctrine is the common law foundation upon which liability for NRDs is based. Well before the passage of the relevant federal statutes, natural resource damages were collected under common law.

Instead, government trustees must somehow calculate the lost social value in the absence privately born, market-based data.[9]

Restoration, assessment and settlement of NRD damage claims are undertaken by federal, state and tribal trustees. Only governmental trustees can seek NRDs, though private plaintiffs – if they can show a concrete harm to a legally protected, collective interest in a resource – can compel action on the part of these 'trustees'. Injury to a natural resource alone is an insufficient reason to establish liability.

Under the US Oil Pollution Act, the National Oceanic and Atmospheric Administration (NOAA) is the federal trustee for claims arising from marine injuries (the US Department of the Interior is responsible for claims arising under CERCLA). Rules guide the agencies' respective NRD assessment procedures and act as a blueprint for the determination of appropriate restoration actions and damages.[10]

In practice, the calculation of NRDs has proved difficult and controversial. When economic value is lost in a market, setting damages can be based on production, inputs, inventories, sales and price data. Pertinent economic data already collected by both private firms and governments is available as a basis for the damage calculation. NRDs, by definition, are damages to public goods for which 'market data' are not available. Thus, damage calculation is complicated by the need to assess economic damages in the absence of economic data.

A further, and more serious, complication is the need to understand how physical damages to a given resource damage other parts of the biophysical system. For example, groundings that damage sea grass beds do not just damage the sea grass, they also damage the species reliant on sea grass as habitat. Similarly, oil spills do not just create oily beaches, they can disrupt a broader range of ecological processes that ultimately can affect commercial fisheries, recreation and species abundance for years to come. Consider just one aspect of this problem: an oil spill that damages coastal wetlands. What are the public resource damages associated with damaged wetlands? The coastal wetlands likely provide necessary habitat to marine species that are commercially, recreationally or ethically valuable. What is the relationship of wetland damages to the population dynamics (breeding, foraging, migration) of these species? The wetlands may also buffer inland groundwater supplies from the intrusion of salt water and thus help provide potable water to coastal communities. What is the relationship of the wetlands to coastal hydrology? The wetlands may also protect against storm surges and flooding. If damaged, they may no longer provide this buffering and communities may be exposed to flood damages.

As these examples suggest, the damage assessment problem is two-fold: First, what are the broader biophysical consequences arising from the incident (e.g., the

---

9. There is no direct private right of action to recover NRDs under federal law. See G. Spyridon & S. LeBanc, 'The Overriding Public Interest in Privately Owned Natural Resources: Fashioning a Cause of Action', *Tulane Environmental Law Journal* (1993): 287, citing *Artesian Water Co. v. New Castle County*, 659 F. Supp. 1269, 1288 (D.Del. 1987), aff'd, 851 F.2d 643 (3rd Cir. 1988).
10. 15 CFR 990 (the NOAA regulations); 43 CFR 11 (the DOI regulations).

spill, grounding or other accident)? Second, what is the economic loss associated with that range of biophysical consequences? These two issues push ecology and economics to their analytical limits. The physical questions demand a sophisticated understanding of a complex and interconnected natural system in which changes are difficult to observe and can arise over large scales (both geographical and temporal). The economic questions require the application of non-market valuation tools because the goods and services damaged under an NRD theory are public goods not owned or traded in markets. Not only that, but the two forms of analysis must somehow be integrated, so that biophysical damage measures can be translated into economic costs.

The science of ecosystem service assessment is relevant because it has the same goal as NRD assessment: translate biophysical changes into biophysical outcomes that are economically meaningful and then place an economic value on changes in those outcomes.

3 CURRENT DAMAGE ASSESSMENT PRACTICES

In practice, government trustees have found it difficult – understandably – to measure lost ecosystem goods and services in this manner. As an alternative, the agencies have focused on a more practical route to damages – namely reliance upon resource replacement cost as the measure of damages. In the 1989 case *Ohio v. Department of Interior* the court strongly favoured use of restoration costs as the basis for damages even if restoration is more expensive than available monetary estimates of lost value.[11] In 1996, NOAA followed the 1994 Department of Interior (DOI) rules with rules of its own, to be applied to assessments authorized under OPA.[12] For example, if an oil spill damages sea grass, the objective is to replace the sea grass. What does it cost to replace the sea grass? That 'procurement cost' becomes the measure of damages. Superficially, this strategy avoids the need to measure lost social wealth, because the point is to simply 'replace the wealth' via restoration. And clearly it is much easier to solicit restoration bids and use those monetary costs as a concrete focus in damage negotiations (as opposed to conduct a broad ecological and economic assessment of lost goods and services).

There are drawbacks to the replacement cost approach to damage assessment, however. The most obvious is that replacement costs have nothing to do with the actual social damage caused (the benefits of goods and services foregone). In some cases, replacement costs may vastly exceed the social damages they are meant to

11. The Natural Resource Damage Assessments (NRDA) rules' current focus on the replacement cost of resources, rather than their estimated market value, is a direct outgrowth of these cases. H.R. Conf. Rep. No. 653, 101st Cong., 2d Sess. 108 (1990). Also, see discussion in R. Randle, 'The Oil Pollution Act of 1990: Its Provisions, Intent, and Effects', *ELR News and Analysis* 21 (1991): 10119.
12. The rules are codified at 15 CFR 990 (the NOAA rules for OPA damages) and 43 CFR 11 (the DOI rules for CERCLA damages).

repair. This is particularly true if the replacement activity is taken seriously, monitored over time and guarantees a successful restoration outcome. In other cases, replacement costs may vastly under-represent the social damages caused by the proximate injury.

A damaged sea grass bed or coral reef may be restorable at an estimable cost. But it is possible that the bulk of social costs arise from damages to other resources dependent on the sea grass or coral reef. If these resources are not replaced – as they are generally not[13] – replacement costs may fall significantly short of the real social damage. In either case, replacement cost as the damage measure fails to achieve the main legal and economic principle in play – the desire to have tortfeasors internalize the full costs of their behaviour. Economically, basing penalties on the lost social benefits is the correct and most precise way to 'make the public whole'.

A related drawback is that restoration costs as damages (by design) over-simplify the government trustee's analytical problem. By removing the need to analyse ecological and economic effects more comprehensively, the government is able to under-invest in the analytical capability required to measure real damages. For environmentalists and economists concerned about real measures of natural wealth lost and gained, reliance upon replacement cost has been barrier to scientific progress.

The deterrent and compensatory functions of maritime accident law thus demand continued development of methods to measure real social damages that take into account biophysical causation and subsequent economic impacts. This is the goal of ecosystem services assessment generally. We now turn to a description of the analytical framework, science and data necessary to translate proximate marine injuries into changes in social wealth. To the extent ecosystem services assessment becomes more practical, there may be important implications for the marine commercial and governmental sectors. The OPA does not limit damages to those that can be directly measured in markets or that are based on observable resource uses. As long as the agencies' own damage assessment rules are adhered to, there is a 'rebuttable presumption' of the analyses' correctness and legal validity.[14] This presumption provides agencies with considerable latitude in their choice of damage assessment methods. At least in the United States, where NRDs based on lost economic wealth are well established in theory, there is a firm legal basis to base damage calculations ecosystem services lost and gained as assessment methods improve.

4 ECOSYSTEM SERVICES ASSESSMENT

Colloquially, ecosystem services are the benefits of nature to households, communities and economies. The term is interpreted in a variety of ways, but conveys

---

13. See C.A. Jones & K.A. Pease, 'Restoration-Based Measures in Natural Resources Liability Statutes' (October 1997) *Contemporary Economic Policy* XV; also see M.P Medina, 'Just Do It' (July/August 2001) *The Environmental Forum*.
14. CERCLA § 107(f)(2)(C); OPA § 1006(e)(2).

an important idea: ecosystems are a tangible source of economic wealth. This is intuitively obvious, and consistent with the entire concept of public trust resources. What is less obvious is how that wealth is to be measured.

What in nature do people value and how much do they value it? These questions occupy economists engaged in the valuation of natural resources. Because environmental goods and services are often public goods not traded in markets, economists lack information on the prices paid for those goods and services – because we do not explicitly pay a 'price'. Of course, just because something does not have a price, does not mean it is not valuable. The challenge, then, is to get people to reveal the values they place on goods and services that are un-priced.

Broadly, there are two ways to value un-priced goods and services. First, we can get people to state their preferences by asking them questions designed to elicit value. Second, we can look to people's behaviour and infer natural resource benefits from that behaviour. Houses near beautiful scenery sell for more than houses without scenery, for example. When people spend time and money travelling to enjoy natural resources, they signal the value of those resources. These are called revealed preferences. The definition of environmental commodities is central to both methods.

One of the nice things about markets is that they not only tell us the prices people pay for things, but also about the units (the quantities) people place value on. A grocery store is full of cans, boxes, loaves and bunches. The number of these units bought yields a set of quantity measures to which prices can be attached. But public, non-market environmental goods and services do not come in convenient quantity units. Put another way, what are the physical damages that give rise and can be attached to economic losses?

In conventional markets we take the difference between physical quantities and the price or value of those commodities for granted. At the store, we are presented with rows and rows of quantity units (the goods and services we buy) and we pay their price at checkout. We take this for granted because markets define both the quantities and prices for us. For public, non-market goods, however, the distinction between quantities and their value is not obvious.

Ecosystem services analysis explicitly demands a linkage between ecological outcomes and economic consequences. This elevates the importance of getting units right – or at least clarifying why we use the units we do. The ultimate goal is for natural scientists and economists to describe ecological changes in the same, relevant and meaningful units.

We now turn to the definition of these 'quantity units' and define what we term 'ecological endpoints'. Decomposition of complex natural systems into more discrete commodity units is typically important to valuation. The principles that guide that decomposition are the subject of the next section. These endpoints are a subset of the biophysical consequences of an ecological change (due to restoration or damage). They are the linkage between ecological assessment and economic evaluation.

4.1 Ecological Endpoints

Natural systems are composed of features, things and qualities (e.g., commodities) that interact with other physical features, things and qualities and are thereby changed via physical processes. These natural processes convert inputs – some natural, some made by humans – into outputs (endpoints) such as air and water quality and species populations. Ecologists and economists refer to these processes as biophysical production functions.[15] Ecological endpoints are defined as features and qualities of the natural environment that matter directly to people. Although many things in nature matter to people – and therefore should be counted – endpoints create a distinction between environmental features that are 'directly' and 'indirectly' valuable.

The notion of direct versus indirect goods and services is conventional in the economics of traditional markets. Intermediate goods, for example, are those used to produce final goods. Final goods are what we count in gross domestic product (GDP), and the value of final goods embodies the value of intermediate goods used to produce them. This in no way implies that intermediate goods are less valuable than final goods. But it does mean that we need not count everything in nature, only those final ecosystem goods and services that embody the value of the whole system.

The definition of ecological commodities and the way in which ecological changes can be integrated with economic assessment is described in more detail elsewhere.[16] Three distinct economic principles are central to the definition:

(1) Clearly and consistently distinguish outcomes – in this case ecosystem goods and services – from the value or weight placed on those outcomes. In economic terminology, this means we must clearly distinguish quantity measures from price or value measures.
(2) Choose quantity measures – the ecological endpoints – whose value can be credibly revealed through choices or human preferences. In practice, this means choosing endpoints that have direct, concrete meaning to people.

---

15. U.S. Environmental Protection Agency (EPA). *Valuing the Protection of Ecological Systems and Services: A Report of the EPA Science Advisory Board* (EPA-SAB-09-012), Washington, DC: U.S. EPA, 2008; G. Daily & P. Matson, 'Ecosystem Services: From Theory to Implementation', *Proceedings of the National Academy of Sciences* 105, no. 28 (2008): 9455–9456; J. Boyd, 'Counting Non-Market, Ecological Public Goods: The Elements of a Welfare-Significant Ecological Quantity Index'. Discussion Paper 07-42. Washington, DC: Resources for the Future, 2007.
16. S. Banzhaf & J. Boyd, 'The Architecture and Measurement of an Ecosystem Services Index', Discussion Paper 05-22 (Washington, DC: Resources for the Future, 2005); J. Boyd, 'The Nonmarket Benefits of Nature: What Should Be Counted in Green GDP?', *Ecological Economics* 61 (2006): 716–723; J. Boyd, 'Counting Non-Market, Ecological Public Goods: The Elements of a Welfare-Significant Ecological Quantity Index', Discussion Paper 07-42. Washington, DC: Resources for the Future, 2007; J. Boyd & S. Banzhaf, 'What Are Ecosystem Services?', *Ecological Economics* 63, nos 2–3 (2007): 616–626.

(3) Treat the quantity units as units in an economic accounting system. This implies that all of the units that contribute to welfare should be counted, but counted only once. In practice, it requires a distinction between final and intermediate goods and services.

The distinction between final and intermediate goods and services arises in any economic accounting system. Final goods are not necessarily more important or valuable than intermediate goods. Rather, the distinction arises because of the fundamental accounting identity – count everything, but count it only once. We must distinguish between final and intermediate if we are to avoid double-counting. Nature is a system composed of complex interrelationships. Even in purely biophysical terms, the ecological 'output' of one process will almost always be an 'input' to another ecological process. This means that most ecological features and qualities can be thought of in dual terms – as input and output. This creates a huge challenge for ecological accounting. Is there any way we can address the double counting issue to improve the quality – and validity – of the accounting system?

Nature presents us with an uncountable number of such inputs. Does that mean we must count all of nature's features and qualities to comprehensively depict its contributions to well-being? The answer is no. Much as GDP does not count all of the units exploited and traded in the market economy, an ecological quantity index need not count all of the elements of nature. In both cases, a utilitarian approach to accounting allows us – requires us, actually – to focus on final units of consumption or enjoyment.

End products can be defined economically as the results of complex processes involving multiple inputs, but where the end product is what is perceived, consumed and appreciated by people. From a valuation standpoint (and thus from the standpoint of damage calculation) end products have a wonderful economic property – the social value of end products includes the value of all the inputs and processes needed to create them.

Consider the issue of double-counting in conventional economic accounts. If we counted both cars and the steel used to make them and then weighted carts and steel by their market prices, we will have double-counted the value of the steel. The reason is that the steel's value in car production is embodied in the value of the car. Thus, in the calculation of GDP, final and intermediate goods are distinguished in the following way. If a good or service's value adds to the value of a good or service subsequently sold in the market, it is an intermediate good. Otherwise it is a final good. Cars are final goods. The labour, leather, steel and human capital required to make the car are intermediate goods.

What does 'final goods' mean in an ecological context? Consider an example. What is the value of a natural process (nutrient cycling) and inputs (vegetative cover) that lead to cleaner water? This question can be answered only by placing a value on the clean water itself – the ecological endpoint. If you ask people to state the value of nutrient cycling, they will first respond by asking for an explanation of nutrient cycling? Once that is explained, they will conclude that, 'the value of nutrient cycling is the clean water it makes available to me'. Only the clean water has direct value to people. The value of nutrient cycling, in contrast, must be derived from the value of the clean water provided.

Another example is a particular fish population in a marine environment. The fish population is an ecological endpoint if people harvest the fish commercially, seek it out recreationally or value its existence if threatened. Of course, fish populations depend in turn on many things (inputs to the production of fish), including water temperature, currents, nutrient upwelling and habitat quality and quantity, along with the lifecycle characteristics of the species and its food chain. From an economic accounting perspective, these inputs are valuable, but their value as inputs is captured – embodied – in the value of the fish population, the ecological endpoint.[17]

To illustrate the importance of measuring endpoints, consider the following common measures of oil spill contamination and recovery: molecular biomarkers; benthic indices; contaminant levels in water, sand and soil; and 'conspicuous mortality' of certain species. These measures are all scientifically desirable measures and are practical to apply. However, if the goal is to link damages to economic losses, they are less than ideal. Imagine an economist or lawyer translating biomarker changes into an 'economic loss'. The problem is that these biophysical measures are not economically meaningful endpoints. If physical damages are measured in these terms, there remains a huge gap between them and physical measures that are economically meaningful – such as species population effects, healthy swimmable beaches and safe food consumption.

In the scientific response to the *Prestige* accident, population changes were detected, including reductions in inter-tidal animals and fishery production. These measurable physical damages (because they are meaningful to non-scientists) are more amenable to economic assessment than other measured impacts, such as changes in immune response and the reproductive capacity of mussels. These measures may be leading indicators of collapse in the shellfish population, but economic assessment of 'immune responses' and 'reproductive effects' prompts the question as to how these measures ultimately affect the population.

This is why it is important to focus ecological damage assessment on the measurement of ecological endpoints. First, by design, endpoints are concretely meaningful to society and thus amenable to valuation. Second, endpoints include the value of systems and inputs on which they depend. Third, if we do not distinguish between endpoints and inputs, the damage calculation may double-count damages.

4.2 Causality and Biophysical Production Functions

Once a set of environmental features and qualities is identified as being important to welfare (e.g., the ecological endpoints), natural science is required to tell us how

17. They may be valuable inputs to other endpoints, of course, including water quality.

those endpoints were affected by a marine accident or other resource-damaging incident. Natural science is also important to establishing causality. In the case of marine damages to public natural resources the causal linkage between an event and damages can be described by a set of biophysical production functions. These functions must be demonstrated both to establish a causal linkage between a liability generating action or accident and damages and to estimate the magnitude of those damages.

Production functions describe the manner in which an output is related to the quantity and nature of inputs used to create it.[18] For example, a production function describes the way in which a company can deploy labour and capital investments to manufacture a final product. Ecological analysis, of course, is concerned with the biological, chemical and hydrological relationships that determine biological production.[19] An ecosystem's structure, such as size, vegetation and boundaries, and its functional aspects, such as ability to absorb floodwater or remove contaminants from surface water, are biophysical contributors – as inputs – to the services the habitat generates.

Biophysical production functions describe the effect of a proximate NRD to subsequent damages to other resources that may arise at a distance or over time. In marine cases the biological and physical linkages between an oil spill and the consequences of the oil spill must be described via these production functions. This goes beyond the question of 'fate, transport and deposition' of pollutants that arises in other legal settings. In the recent words of Alaskan trustees working on the aftermath of the *Exxon Valdez* spill:

> Through hundreds of studies conducted over the past twenty years, we have come to understand that the Prince William Sound ecosystem is incredibly complex and the interactions between a changing environment and the injured resources and services are only beginning to be understood.[20]

It is worth reflecting on why this is true. First, many marine accidents result in damages over a wide geographical area (in the *Valdez's* case, 200 miles of shoreline were 'obviously' affected, but measurable biological effects have been found over 1,300 miles of coastline) and over long time periods (twenty years after the spill fewer than half the species affected have 'recovered' to pre-spill levels).[21] The water quality effect of such a spill has a range of 'side-effects' that develop

18. See S.N. Durlauf & L.E. Blume, *The New Palgrave: A Dictionary of Economics* (Basingstoke: Palgrave Macmillan, 1998). 1998, 995: 'The traditional starting point of production theory is a set of physical technological possibilities, often represented by a production or transformation function. The development of the theory . . . leads to input demands (and output supplies) constructed from an explicit consideration of the underlying technology.'
19. There is also a long history of integrated economic and biological production function analysis in agricultural and natural resource economics. Among other things, agricultural studies show how substitution of one farm input for another (e.g., land for fertilizer, tractors for man-hours) affects production levels or how landscape characteristics affect yields. For a general overview, see J. Penson et al. (eds), *Introduction to Agricultural Economics* (New Jersey: Prentice Hall, 1999).
20. <www.evostc.state.ak.us/recovery/status.cfm>.
21. <www.evostc.state.ak.us>.

over a period of years or decades. Oil spills in the short term will deplete herring and other cornerstones of the marine food chain. In turn, this effect on food stocks affects the viability of species dependent on them, such as certain bird species. These biological effects can take years to play out. This means, again in turn, that human uses dependent on these ecological endpoints may be affected for years.

Consider damages to habitat. Habitat can be thought of as a natural asset, where restoration represents investment in the asset and damage represents divestment. Similar to man-made capital assets, habitats are durable, implying that changes today affect physical and economic rates of return in the future. Damages (and restoration) therefore are inter-temporal phenomena in both physical and economic terms. Habitat regeneration takes time. Regeneration difficulties associated with the *Valdez* are an example. Often restoration projects replicate natural system functions only after a period of years. Even if full replication of a more natural system is ensured, there are damages associated with the interim loss in the ecological asset's function.

Also of special importance to the science of ecological damages (both the physical and economic science) is the importance – and difficulty – of geographical measurement. From an ecological perspective, geography matters because natural systems are in constant motion. Air and water circulates, surface waters run downhill, species migrate and seeds and pollen disperse. Moreover, movement of one thing (e.g., water) tends to trigger movement of other things (e.g., birds and fish). This movement creates a significant empirical challenge for damage assessment. With nature in a constant state of flux, it is difficult to measure baseline conditions. It is even more difficult to relate changes in conditions to a particular oil spill or other incident. But precisely because there are so many biophysical linkages across space, a full accounting of damages must somehow take the geography of effects into account.

The challenge for ecosystem scientists and managers is to empirically demonstrate cause and effect when the cause-and-effect relationship is spatial. These spatial production functions tell us how an action (good or bad) in one place affects the production of ecosystem goods and services in another. Of particular importance in the marine environment are notions of resource patchiness and heterogeneity, biophysical linkages and dispersal mechanisms connecting patches.[22] This research has led to a new class of bio-economic models that are composed of a group of subpopulations distributed across a set of patches linked by dispersal processes.[23]

---

22. The spatial nature of ecological relationships is an important topic in ecology generally (not just in marine applications). For example, the quality of a habitat asset can be highly dependent on the quality and spatial configuration of surrounding land uses. The ability of areas to serve as migratory pathways and forage areas typically depends on landscape conditions over an area larger than the habitats relied upon directly by the migratory species. The contiguity of natural land cover patches has been shown for many species to be an indicator of habitat quality and potential species resilience.
23. See, e.g., J.N. Sanchirico & J.E. Wilen, 'Bioeconomics of Spatial Exploitation in a Patchy Environment', *Journal of Environmental Economics and Management* 37 (1996): 129–150;

In the marine context, we need spatial production functions that describe:

- the dependence of fish species on the configuration of habitats – marine, estuarine, and freshwater, depending on the species – needed for their reproduction, forage, and migration;
- the dependence of avian and inland species on marine food stocks (fish and smaller organisms);
- the dependence of coastal groundwater supplies and quality on near-shore water quality and wetlands;
- the fate and transport of pollutants as a function of temperatures, currents, and natural attenuation;
- biological uptake of pollutants, particularly toxins, and consequences for the food chain;
- the biological effects of non-native populations introduced by shipping.

The statistical demonstration of causality – the relationship between a marine incident and economically meaningful endpoint changes that can arise over large areas and long time periods – is the missing piece that inhibits a full, accurate accounting of social damages. These causal relationships can only be tested using rigorous, data-intensive empirical and scientific methods. These methods are difficult and costly to perform. Second, non-uniformity means that even if you establish a causal relationship in one location, that relationship may not hold in other locations.

Particularly in the immediate aftermath of a spill or grounding, the prediction of effects is difficult to make with confidence. NRD assessment, though, allows for long-run monitoring of effects and adaptive management and restoration in response new information. These long-run assessments are fundamental to the science underlying damage calculations. Long-run monitoring geared toward economically meaningful endpoints is the key to economic damage assessment.

4.3 From Physical to Economic Damages

When changes in ecological endpoints are measures (losses or gains), how are those biophysical losses and gains translated into economic damages? Alone, ecological endpoints tell us nothing about their relative importance or value. Somehow, we must pair endpoints with information relating to their economic importance. In order to translate ecological endpoints into social welfare measures, economists require knowledge of both economic production functions and society's underlying preferences. Underlying preferences are not directly observable – hence the use of non-market valuation techniques. If economic

---

and J.N. Sanchirico & J.E. Wilen, 'A Bioeconomic Model of Marine Reserve Creation', *Journal of Environmental Economics and Management* 42 (2001): 257–276.

experts are called on to estimate the social costs of a biophysical damage they will rely on two general methods: revealed and stated preference methods.[24]

Revealed preference studies look at the price people are willing to pay for marketed goods that have an environmental component. From those prices, inferences about the environmental benefits associated with the good can be made. For example, when people purchase a home near an aesthetically pleasing ecosystem, home prices reflect that environmental amenity.[25] Alternatively, when people spend time and money travelling to recreation, they reveal a willingness to pay the time and travel costs to access the recreational services. 'Travel cost' studies are used to make a benefit estimate based on such expenditures.[26] The travel cost method requires data and analysis linking the number of trips to a site with the quality, size or location of a site. Changes in these attributes can be valued if there is a perceptible change in the number, length or cost of trips taken to the site.

Of course, not all environmental benefits are captured in market prices or in easily observable individual choices. One way around this problem is to move away from reliance on preferences revealed in markets. Stated preference studies are one such alternative. Stated preference studies ask people, in a highly structured way, what they would be willing to pay for a set of environmental improvements. Contingent valuation (CV) studies are an example. Stated preference surveys are expensive and controversial and are most reliable when the questions concern specific ecological services provided in specific contexts. The more complex and holistic the improvement or change, the more difficult the methodological challenge. A principal drawback to this approach is the risk that people may misunderstand the precise service being valued when undisciplined by the need to spend their own real money. For the same reason, they may also overstate their willingness to pay.[27]

For this reason, an independent panel was convened in 1993 to assess the validity of the CV methodology to measure non-use values. The so-called 'NOAA Panel' established a set of guidelines for the use of CV methods and concluded that CV can provide a valid economic measure of value associated with resources people do not actually use but whose existence they may nevertheless value.[28] The rules now permit CV for estimating use and non-use values but only

24. For a good overview of these methods see A.M. Freeman, *The Measurement of Environmental and Resource Values: Theory and Methods* (Washington, D.C.: Resources for the Future, 1993).

25. Hedonic analysis is used in this type of study. See, e.g., B. Mahan, S. Polasky & R. Adams, 'Valuing Urban Wetlands: A Property Price Approach,' *Land Economics* 76 (2000): 100.

26. There is a large methodological literature on this subject. See, e.g., K. McConnell, 'On-Site Time in the Demand for Recreation', *American Journal of Agricultural Economics* 74 (1992): 918.

27. See generally R.J. Kopp & V.K. Smith (eds), *Valuing Natural Assets* (Washington, D.C.: Resources for the Future, 1993), presenting a collection of articles relating to the contingent valuation method.

28. The panel concluded that '[contingent valuation] produces estimates reliable enough to be the starting point of a judicial process of damage assessment, including passive-use values (i.e., nonuse values).' Report of the NOAA Panel on Contingent Valuation, 58 FR 4601, 15 Jan. 1993, at 4610.

when 'no use values can be determined'.[29] The method is controversial, however. A common feature of US NRD 'reform efforts' is the restriction or elimination of recovery for lost non-use values[30] and prohibitions on the use of the CV method to value NRDs.[31] Some US states limit the methodologies that can be used by states to calculate the scale of NRDs.[32] It should also be noted that the most controversial valuation methods, such as CV, have in actuality been employed infrequently.[33] Nevertheless, these methods are a distinct improvement relative to evaluation techniques that ignore social preferences.[34]

A more general problem is that econometric, monetary assessment of non-market goods is expensive, technically, difficult and time-consuming. The methods used to place dollar estimates on natural resources are specific to certain kinds of benefits. This means that a comprehensive assessment of benefits (or costs) requires deployment of a wide range of economic techniques. Accordingly, it is nearly impossible to monetize with econometric techniques the full range of public good losses arising from a particular incident.

The second problem is that studies may have limited applicability from one case to another. Even if the physical damage (changed ecological endpoints) is identical in two different locations, the economic damage arising from them may be very different. Consider the *Valdez* and *Prestige* cases. Even if the physical damages were identical (they were not), the *Prestige* spill's location in the populated Bay of Biscay meant that many economic impacts would be greater than in the relatively remote Alaskan Prince George's Sound.

---

29. 43 CFR 11.83(c) (2) (vii).
30. See s. 8, Part 403, 105th Cong., A Bill to Reauthorize and Amend the Comprehensive Environmental Response, Liability, and Compensation Act of 1980, introduced 21 January 1997 (there will be 'no recovery under this Act for any impairment of nonuse values'), and s. 1537, 106th Cong., Superfund Amendments and Reauthorization Act of 1999, introduced in the Senate 5 Aug. 1999.
31. See s. 8, *supra* n. 27, which states that a trustee's claim for recoverable assessment costs 'may not include the costs of conducting any type of study relying on the contingent valuation methodology'.
32. Michigan law establishes liability for natural resource injury (Act 451, Michigan Compiled Laws (MCL) 324.20126a). However, contingent valuation methods cannot be used for damage calculations 'unless a determination is made by the department that such a method satisfies principles of scientific and economic validity and reliability and rules for utilizing a contingent nonuse valuation methods or a similar nonuse valuation methods are subsequently promulgated.' MCL 324.20104(3).

    The state of Texas lists a set of acceptable valuation methods. The list includes contingent valuation but requires that contingent valuation studies be undertaken only in accordance with guidelines established by NOAA. 31 TAC § 20.32(f).
33. See testimony of Douglas Hall, NOAA, Subcommittee on Water Resources and Environment, House of Representatives, 11 Jul. 1995. 'There have only been six contingent valuation studies completed to date, and only one in which the Federal Government was involved in litigation.'
34. See R. Carson, F. Nicholas & M. Norman, 'Contingent Valuation: Controversies and Evidence', *Environmental and Resource Economics* 19 (2001): 173–210, for a review and defence of contingent valuation's role in the evaluation of environmental goods and services.

In the translation of physical damages to economic damages, geography is again a central part of the analytical challenge. A property of ecosystem goods and services is that society cannot easily move these goods and services. Ecosystem goods and services are like real estate. Like houses and factories, a coral reef, fishery or bird population cannot simply be moved to where it is most valuable. These kinds of goods are not 'spatially fungible', in economic parlance. As with homes, economic value is as much a function of neighbourhood as of house's physical structure. As the saying goes in real estate, three things matter to value: location, location, location. The same is true for most ecosystem goods and services. Their economic value depends on their location relative to communities, businesses, recreators and other users.

The consumption of services often occurs over a large scale; examples include recreation and commercial harvests of fish or game, where users may not be confined to the ecosystem good's physical location. This triggers the need for social analysis of users who may travel to or otherwise experience the ecosystem service at a distance. Another reason the location of ecosystem services matters economically is that the benefit of services often depends on where and when the complements to and substitutes for those services arise. As an example, marine areas may have recreational value only if there are complementary assets such as public beaches, trails or docks present.[35] Substitutes have the opposite effect. Many ecosystem services have no substitutes. For example, there is no clear substitute for the existence value of wilderness or an endangered species. Other services do have substitutes, however. If shellfish beds are abundant, the damage to a particular shellfish bed will – all else equal – be less than if shellfish beds are scarce.

In other words, ecosystem services exhibit the basic properties of any economic good. All else equal, greater scarcity or rarity and a relative lack of both natural and human-made substitutes increases the value of an ecosystem service – and economic damages if the good or service is damaged. Because the value of ecosystem services is inversely related to resource scarcity, valuation requires a 'marginal' analysis of ecosystem benefits, which examines the way in which a service's benefits vary with the aggregate level of the service available.

These complications in the economic damage assessment phase create hurdles to econometric damage calculation. There are alternatives to monetary damage estimation worth considering, however.

4.4 Economic Damage Indicators

An alternative to econometric estimation of economic damages – particularly when juries are involved in the determination of damages – is the use of damage indicators. These indicators are social and physical data that help describe the

35. Other types of benefit, such as the existence benefit of a wilderness area do not require and indeed may be reduced by the presence of such features.

economic impact of ecological damages. Because they are indicators, rather than derived monetary estimates, they can be deployed more cheaply and easily to give courts information relevant to economic damages. Location-specific economic damage indicators (EDIs) can help courts understand the larger social and physical landscape to better assess the economic importance of particular losses to ecosystem goods and services.

EDIs are a marriage of easily measured social and physical data to basic principles of economic assessment. For example, all else equal, we can always say the following:

- The scarcer an ecological feature, the greater its value.
- The scarcer are substitutes for an ecological feature, the greater its value.
- The more abundant are complements to an ecological feature, the greater its value.

As noted earlier, a damaged beach implies larger economic damages if the beach is geographically scarce. Geographical scarcity is straightforward to measure with modern geospatial mapping and measurement tools.[36] Substitutes are also easy to measure and convert into indicators, such as the number of non-marine public parks and recreational opportunities in an area. Complements, too, such as access points, can often easily be measured.

Depending on the ecological feature we can often go further than this. For example, the social value of some environmental features is often a direct and increasing function of the number of people with access to them. Similarly, the social value of some environmental features is often a direct and increasing function of the economic value they protect or enhance. Accordingly, we can often – but not always – say the following:

- The larger the population benefiting from an ecological feature, the greater its value.
- The larger the economic value protected or enhanced by the feature, the greater its value.

All else equal, a beach with 10,000 visitors a year is more recreationally valuable than a park with 100 visitors a year. Similarly, a recreational fishery's value will be greater the larger the number of households able to access it. Wetlands that provides flood protection to property worth United States Dollar (USD) 10 million is more valuable than one protecting property worth USD 1 million, all else equal.

The reason to consider EDIs is that they are relatively simple measures to develop and apply to a particular case. Data can be drawn from geo-coded data sets, including satellite imagery; local, regional and national growth, land use or transportation plans; environmental agencies; private conservancies and nonprofits; and

36. J. Boyd & L. Wainger, 'Landscape Indicators of Ecosystem Service Benefits', *American Journal of Agricultural Economics* 84 (2002); J. Boyd & L. Wainger, 'Measuring Ecosystem Service Benefits: The Use of Landscape Analysis to Evaluate Environmental Trades and Compensation', Discussion Paper 02-63 (Washington, DC: Resources for the Future, 2003).

census data. Short of full-blown economic assessment, they provide economic information useful to courts. They are a practical, quantitative middle ground between no economic analysis and full-blown econometric value estimation.

5 CONCLUSION

Maritime law often allows for the measurement and assignment of damages arising from lost public natural resources. Damaged environmental resources, even if those resources are public goods, can trigger subsequent economic damages to individuals, businesses and communities. Given the goal of liability laws to 'make the public whole' the NRDs question boils down to this: how do physical damages propagate through the physical ecosystem and thereby create subsequent biophysical damages that reduce human welfare? Courts must understand biophysical damages that can take years to develop may arise over broad geographical scales. They must also be able to translate those biophysical damages into economic terms.

In the United States, where NRD law is well-established and applicable to marine incidents, linked ecological and economic assessment has not yet been able to fully meet this aspiration for damage assessment. Instead, government trustees have relied on more practical – and thus entirely defensible – alternatives to the calculation of 'real social costs'. The focus on restoration of resources obviously damaged is a natural first step and yields concrete damage estimates based on the costs of restoration. The ecological problem with this approach is that not all biophysical damages are obvious or predictable. This raises the possibility that physical damages are under-estimated given the challenge of demonstrating causally related effects. The economic problem with the approach is that 'costs are not the same as benefits'. A focus on restoration costs as the measure of damages can lead to both over-deterrence and under-deterrence, depending on the relationship of restoration costs to the true social cost of the physical damages.

The measurement of 'ecosystem goods and services' is an increasingly important theme in both academic and government inquiry into environmental assessment. It involves both biophysical analysis of actions and causes and their effects in the biophysical realm and the economic interpretation of those effects. It is therefore focused directly on the analytical problem posed by NRD calculations designed to 'make the public whole' as a result of environmental accidents. The paper has described the broad issues being addressed in the ecosystem goods and services movement.

Ecosystem goods and services analysis is not easy. It is still not practical except where money for large-scale monitoring and statistical assessment is possible. However, the development of such methods is proceeding. When the physical and social science of ecosystem goods and services develops into a more mature phase, the implications for marine liability damages will be direct and material to plaintiffs, trustees and courts.

The insights and principles behind ecosystems services research are also of immediate relevance to trustees who want to be in a position to calculate the most accurate damages possible (in order to serve the deterrent and compensatory goals of liability law). This chapter advocates assessment based around ecological endpoints. It argues that the development of ecological endpoints will lead to more coordinated, comprehensive and cost-effective biophysical and economic analysis of damages.

The chapter also emphasizes the importance of ecological production functions to damage assessment that depict ecological gains and losses across physical geography and over time. The associated economic agenda is to use these ecological endpoints as a standard way to depict ecological outcomes. This is fundamental if we are to improve the statistical underpinnings of damage calculations. Finally, the chapter describes the increasing availability social and physical damage indicators that can be used by courts to depict the economic consequences of physical damages. Together, ecological endpoints, biophysical production functions and economic damage indicators create a quantification and scientific agenda for resource trustees.

References

Banzhaf, S. & J. Boyd. 'The Architecture and Measurement of an Ecosystem Services Index'. Discussion Paper 05-22. Washington, DC: Resources for the Future, 2005.

Barbier, E.B., et al. 'Coastal Ecosystem-Based Management with Nonlinear Ecological Functions and Values'. *Science* 319 (2008): 321–323.

Boyd, J. 'The Nonmarket Benefits of Nature: What Should Be Counted in Green GDP?'. *Ecological Economics* 61 (2006): 716–723.

Boyd, J. 'Counting Non-Market, Ecological Public Goods: The Elements of a Welfare-Significant Ecological Quantity Index'. Discussion Paper 07-42. Washington, DC: Resources for the Future, 2007.

Boyd, J. & S. Banzhaf. 'What Are Ecosystem Services?'. *Ecological Economics* 63, nos 2–3 (2007): 616–626.

Boyd, J. & L. Wainger. 'Landscape Indicators of Ecosystem Service Benefits'. *American Journal of Agricultural Economics* 84 (2002).

Boyd, J. & L. Wainger. 'Measuring Ecosystem Service Benefits: The Use of Landscape Analysis to Evaluate Environmental Trades and Compensation'. Discussion Paper 02-63. Washington, DC: Resources for the Future, 2003.

Carpenter, S.R., et al. 'Millennium Ecosystem Assessment: Research Needs'. *Science* 314 (2006): 257–258.

Carson, R., F. Nicholas & M. Norman, 'Contingent Valuation: Controversies and Evidence'. *Environmental and Resource Economics* 19 (2001): 173–210.

Daily, G. *Nature's Services: Societal Dependence on Natural Ecosystems.* Washington, DC: Island Press, 1997.

Daily, G. & P. Matson. 'Ecosystem Services: From Theory to Implementation'. *Proceedings of the National Academy of Sciences* 105, no. 28 (2008): 9455–9456.

Durlauf, S.N. & L.E. Blume. *The New Palgrave: A Dictionary of Economics.* Basingstoke, UK: Palgrave Macmillan, 1998.

Freeman, A.M. *The Measurement of Environmental and Resource Values: Theory and Methods.* Washington, D.C.: Resources for the Future, 1993.

Jones, C.A. & K.A. Pease. 'Restoration-Based Measures in Natural Resources Liability Statutes'. *Contemporary Economic Policy* 15 (1997): 111–112.

Kopp, R.J. & V.K. Smith (eds). *Valuing Natural Assets.* Washington, D.C.: Resources for the Future, 1993.

Mahan, B., S. Polasky & R. Adams, 'Valuing Urban Wetlands: A Property Price Approach'. *Land Economics* 76 (2000): 100–113.

McConnell, K. 'On-Site Time in the Demand for Recreation'. *American Journal of Agricultural Economics* 74 (1992): 918–925.

Medina, M.P. 'Just Do It'. *The Environmental Forum* (July/August 2001).

Penson, J., et al. (eds). *Introduction to Agricultural Economics.* Upper Saddle River, NJ: Prentice Hall, 1999.

Polasky, S., et al. 'Conserving Species in a Working Landscape: Land Use with Biological and Economic Objectives'. *Ecological Applications* 15, no. 4 (2005): 1387–1401.

Randle, R. 'The Oil Pollution Act of 1990: Its Provisions, Intent, and Effects'. *ELR News and Analysis* 21 (1991): 10119.

Ricketts, Taylor H., et al. 'Economic Value of Tropical Forest to Coffee Production'. *Proceedings of the National Academy of Sciences* 101 (2004): 12579–12582.

Sanchirico, J.N. & J.E. Wilen. 'Bioeconomics of Spatial Exploitation in a Patchy Environment'. *Journal of Environmental Economics and Management* 37 (1996): 129–150.

Sanchirico, J.N. & J.E. Wilen. 'A Bioeconomic Model of Marine Reserve Creation'. *Journal of Environmental Economics and Management* 42 (2001): 257–276.

Spyridon, G. & S. LeBanc. 'The Overriding Public Interest in Privately Owned Natural Resources: Fashioning a Cause of Action'. *Tulane Environmental Law Journal* 6 (1993): 287.

U.S. Environmental Protection Agency (EPA). *Valuing the Protection of Ecological Systems and Services: A Report of the EPA Science Advisory Board* (EPA-SAB-09-012). Washington, DC: U.S. EPA, 2008.

Wainger, L.A., et al. 'Wetland Value Indicators for Scoring Mitigation Trades'. *Stanford Environmental Law Journal* 20, no. 2 (2001): 413–478.

Chapter 5

Liability and Compensation for Environmental Damage Caused by Ship-Source Oil Pollution: Actionability of Claims

*Proshanto K. Mukherjee**

1 INTRODUCTION

The notion of environmental damage in international law relating to liability and compensation arising from ship-source marine pollution is fraught with uncertainties. The convention regimes contained in the 1992 versions of the Civil Liability Convention[1] (CLC) and Fund Convention,[2] and the Hazardous and Noxious Substances (HNS) Convention of 1996[3] addressing the issue of compensability of claims for environmental damage are inchoate at best. The principal deficiency in the convention regimes is the absence of any express provision establishing the

* World Maritime University, Sweden and Colombo International Nautical and Engineering College, Sri Lanka.

1. Protocol of 1992 to amend the International Convention on Civil Liability for Oil Pollution Damage, 1969.
2. Protocol of 1992 to the International Convention on the Establishment of an International Fund for Compensation for Oil Pollution Damage, 1971.
3. International Convention on Liability and Compensation for Damage in Connection with the Carriage of Hazardous and Noxious Substances by Sea (HNS), 1996.

Michael G. Faure, Han Lixin & Shan Hongjun, *Maritime Pollution Liability and Policy: China, Europe and the US*, 75–95.

polluter's liability for environmental damage and the corresponding legal basis on which a claim for such damage may rest and subsist. Yet, in the innocuous definition of 'pollution damage' that was arrived at after extensive and arduous deliberations within the Legal Committee of the International Maritime Organization (IMO) and the diplomatic conference leading up to the 1992 Protocols to the CLC and Fund Convention, a remedy is provided that essentially specifies the quantification methodology that is to be used in the event a claim is admitted by the relevant International Fund or a tribunal bound by the applicable convention law. The inclusion of a remedial provision without any express consideration of the liability for environmental damage and the actionability of a claim relating thereto creates an anomaly in the convention regimes and points to the assumption that the precluded matters are left to be addressed by domestic law. The reason for this preclusion may be attributed to the lack of international uniformity in domestic jurisdictions regarding the validity of a cause of action in respect of claims pertaining to environmental damage.

This paper attempts to probe this perceived lacuna in the law by exploring various co-related issues and legal implications associated with the establishment of liability for environmental damage and cognizability of claims for such damage. The discussion is restricted to consideration of damage to the marine environment resulting from ship-source oil pollution, and therefore only the CLC and Fund Convention are contextually relevant.

It is submitted that actionability in respect of environmental damage is a function of the *locus standi* or standing of the claimant and is the determinant factor in the adjudication of the claim. Environmental damage springs from recognition of the fact that the environment is susceptible and vulnerable to harm or damage. It is further submitted that there are two dimensions to the environment; one is tangible, the other is intangible. The tangible aspect is characterized by things readily perceptible even if invisible and can be conveniently captioned as natural resources, living or non-living. The intangible aspect is circumscribed by the aesthetic attribute of the environment, which translates into the human enjoyment factor, or where damage is alleged, deprivation of it. When damage occurs, the tangibles can be quantified more easily than the intangibles.

The premise being so established, the chapter proceeds to focus on the issue of the *locus standi* of a claimant purporting to claim compensation for environmental damage. At the outset, the point needs to be made that *locus standi* in respect of a claim alleging deprivation of enjoyment of the marine environment is a thornier issue to grapple with than a claim relating to the tangible attributes of that environment. There is no doubt that in virtually all jurisdictions, a proprietary right or interest vested in the claimant will afford him standing with respect to the prosecution of a claim for environmental damage. In many jurisdictions, this is the sole basis for a claimant's *locus standi*. In attempting to examine the legal positions of both public and private entities, the principal thrust of this chapter is to explore what other legal avenues are available in the absence of a proprietary interest.

The discussion primarily revolves around the application in such cases of the public trust doctrine and extends to a contextual examination of the doctrine of *parens patriae*, which in essence is a variation of the public trust doctrine and subsumed in it. As the discussion unfolds, it will become apparent that for the most part, *locus standi* of public bodies for claims in respect of environmental damage are mainly derived from statute, and in this respect, the United States is in the forefront in terms of domestic legislation and related case law jurisprudence, although much of it pertains to natural resources generally rather than other specific elements of the marine environment. Ironically, the United States does not subscribe to the relevant international conventions. Even so, it is instructive to refer to that domestic regime, especially from the perspective of state parties to the international conventions who are disadvantaged by the lacuna in the conventions, resulting in resort to domestic law, and whose national legal systems are relatively less developed in this field.

A summary of the state of the law is provided in the conclusion, together with suggestions for improvement and clarification of the subject through modification of the current regime.

2 ENVIRONMENTAL DAMAGE UNDER THE CONVENTION REGIME

Claims relating to damage to the marine environment will typically include, among others, economic losses arising out of deprivation of legitimate sea use, clean up costs, and perhaps above all, damage to the natural resources of the marine environment, including the flora and fauna. The effects of pollution frequently linger over lengthy periods of time following the completion of clean up operations, and pollution victims remain uncompensated.[4] This is often due to impediments in the law and policy relating to compensability of claims for environmental damage.

The issue of compensability for such claims, albeit in an incomplete way, is addressed in the definition of 'pollution damage'. It would appear that only a loss or damage including environmental damage, that squarely fits the definition of pollution damage and is consistent with all its elements, would be compensable under the CLC and Fund Convention. Notably, in the original CLC of 1969, 'pollution damage' was defined as loss or damage caused outside the ship carrying oil by contamination resulting from the escape or discharge of oil from the ship, wherever such escape or discharge may occur and includes the costs of preventive measures and further loss or damage caused by preventive measures.[5]

By contrast, the definition of 'pollution damage' in the current 1992 version of the CLC reads as follows:

(a) loss or damage caused outside the ship by contamination resulting from the escape or discharge of oil from the ship, wherever such escape or discharge may occur, provided that compensation for impairment of the

4. C.M. De La Rue & Ch.B. Anderson, *Shipping and the Environment: Law and Practice* (London: LLP, 1998), 503.
5. Article I, para. 6, International Convention on Civil Liability for Oil Pollution Damage, 1969.

environment other than loss of profit from such impairment shall be limited to costs of reasonable measures of reinstatement actually undertaken or to be undertaken;

(b) the costs of preventive measures and further loss or damage caused by preventive measures.[6]

A number of observations are relevant with regard to these two definitions and are discussed in detail below.

First, there is no clear indication in the wording of the definition of pollution damage in the 1969 version of CLC that recoverability for environmental damage was contemplated, although it is arguable that the generality of the words 'loss or damage . . . may occur' could conceivably accommodate a claim for environmental damage.

Second, the words 'loss or damage caused outside the ship' in both definitions may be construed as loss or damage caused to the marine environment itself or to a loss or damage arising out of or consequential to the damage caused to the environment. The latter construction raises implications in respect of compensability for economic loss as do the words 'other than loss of profit from such impairment' in the 1992 version of the definition. This issue warrants thorough and comprehensive analytical treatment that is beyond the scope of this chapter except in a peripheral way. In the context of the words 'loss of profit', one commentator has expressed the following view:

> The new definition, moreover, could on its wording be taken to allow in principle for recovery of purely economic loss, because, by excluding loss of profits from the limitation regarding recovery for damages to the marine environment, it can be implied that loss of profits (which are an example of purely economic loss) are, with respect to damages to the marine environment and by extension to all other damages, fully recoverable under the general words 'loss or damage'.[7]

It is further noted by this commentator that the summary records of the diplomatic conference leading up to the 1984 Protocols to the CLC and Fund Convention (which are now defunct and replaced by the 1992 versions of the Conventions) indicate support for the above interpretation regarding recoverability for economic losses.[8]

The third point is that in the current 1992 definition, the words 'compensation for impairment of the environment . . . shall be limited to costs of reasonable measures of reinstatement actually undertaken or to be undertaken' exemplify a leapfrog into the remedy or, to be precise, the quantification methodology, without any express consideration of the liability, as already pointed out in the introduction

6. Article 2, para. 6, CLC. In the Fund Convention pollution damage is defined by cross reference to the definition in the CLC.
7. D. Silverstone, 'Ship Source Oil Pollution Damage: A Canadian Perspective on Recoverability of Economic Losses and Damage to the Marine Environment', *Marine Policy* (April 1985): 118.
8. *Ibid.*

of this chapter. It is submitted that the question of liability is tied to the validity of the claim, which in turn hinges on the *locus standi* of the claimant. As mentioned earlier, the silence of the convention provisions regarding this matter creates the anomaly, which is the principal drawback of the current convention regime in terms of the treatment of claims for environmental damage. It is notable in this context as well that the Guidelines issued by the Comité Maritime International (CMI), the object of which is presumably to provide assistance in the interpretation of the definition, particularly with respect to environmental damage, do not elaborate on the questions of liability and *locus standi.*

In the view of the present author, this convoluted provision represents a product of compromise emanating from intense debate typical of negotiations at diplomatic conferences. The obscurity of the definition of pollution damage is exemplified by the following observations.

The definition as presently articulated would be meaningful in respect of environmental damage only if the following conditions exist:

- the claimant has *locus standi* in respect of a claim for 'impairment of the environment' as the holder of a proprietary interest in the natural resources of the environment or is a public trustee by virtue of common or customary law or statute and is entitled to be compensated;
- the claimant derives a profit from those resources and is deprived of it by reason of the impairment;
- the compensation envisaged is in respect of the impairment and not any consequential damage such as an economic loss suffered by a person who has no proprietary interest or is not a public trustee, such as a fisherman who has suffered a loss of profit.

This begs the question whether loss of profit as particularized in the definition and economic loss in general is at all compensable under the convention regime contrary to the view expressed by the commentator cited earlier. The obtuse reference to 'loss of profit' compounds the confusion. Although the provision is silent on the question of compensability of economic losses, under International Oil Pollution Compensation (IOPC) Fund policy, in which four criteria are set out and followed in recent court decisions, certain kinds of economic losses are compensable.[9] Economic losses that are too remote or are considered secondary or relational are not compensable. Be that as it may, in the opinion of this author it is a deficiency of the convention regime that this matter is not expressly provided for and left to the vagaries of national courts and the IOPC Fund. At any rate, it is

9. The four criteria are the geographic proximity between the claimant's activity and the contamination, the degree to which a claimant is economically dependent on an affected resource, the extent to which a claimant has alternative sources of supply or business opportunities and the extent to which a claimant's business forms an integral part of the economic activity within the area affected by the spill. Account is also taken of the extent to which the claimant can mitigate his loss. See M. Jacobsson, 'The International Regime on Liability and Compensation for Tanker Oil Spills', World Maritime University, 4 Jul. 2008, 8. On recent court decisions, see G. Gauci, 'The *Sea Empress* and the Special Status of Fishermen' *JBL*, November Issue (2003): 609–615.

submitted that, the exclusion of loss of profit from the quantification methodology formula is superfluous because one has little to do with the other.

Fourth, it is to be noted that the 1992 definition makes no mention of damage to the environment or environmental damage[10] but rather refers to 'impairment of the environment'. This at once raises the question of whether 'damage' and 'impairment' are synonymous or are meant to be synonymous in the given context. It is submitted that impairment may from a scientific viewpoint have a different meaning to damage, which in turn may or may not constitute damage in a legal sense. At any rate, a micro-surgical treatment of this phenomenon from both of those perspectives is beyond the scope of this discussion. Suffice it to say that damage for the purposes of determining the quantum of compensation is more susceptible to quantifiability, for example, in the case of damage to fish stock, than impairment in terms of, for example, water quality.[11]

Finally, purely from the viewpoint of legislative draftsmanship, it is submitted that the second segment of the definition of pollution damage in the 1992 version, that is, the so-called proviso, does not constitute a definition at all but is rather a statement of law that should be cast separately for the sake of clarity and drafting precision. The words in question are:

'Compensation for impairment of the environment other than loss of profit from such impairment shall be limited to costs of reasonable measures of reinstatement actually undertaken or to be undertaken.'

With a view to addressing all of the points discussed earlier, the following modifications to the 1992 CLC are suggested. First, the definition of pollution damage may be re-crafted as follows:

> 'Pollution damage' means
>
> Loss or damage caused outside the ship by contamination resulting from the escape or discharge of oil from the ship, wherever such escape or discharge may occur including environmental damage;
>
> The costs of preventive measures and further loss or damage caused by preventive measures.

Second, flowing from the above and for the sake of completeness, a new definition should be included as follows:

> 'Environmental damage' includes damage to natural resources, loss of amenities and deprivation of quiet enjoyment in relation to the environment.

Third, the following substantive provision should be inserted in an appropriate place to address the question of compensability for environmental damage:

(1) The *locus standi* of a claimant in respect of a claim for environmental damage may be based on a proprietary interest vested in the

10. See definition of 'Damage to the environment', Art. 1, para. (d), International Convention on Salvage, 1989.
11. E.H.P. Brans, *Liability for Damage to Public Natural Resources: Standing, Damage, and Damage Assessment* (London: Kluwer Law International, 2001), 17–20.

claimant, or the doctrine of public trust or *parens patriae* as may be appropriate.

(2) Compensation for environmental damage shall be limited to costs of reasonable measures of reinstatement actually undertaken or to be undertaken, and may include loss of profit suffered by the claimant as a consequence of the environmental damage.

3 *LOCUS STANDI*

The concept of *locus standi* is viewed by some as being peculiar to the common law. In essence, any legal proceedings, judicial, arbitral or otherwise, involve two elements. The tribunal must have jurisdiction over the subject matter under dispute and the parties appearing before it and the litigants must have *locus standi* or standing to appear in that tribunal to plead their respective cases. The two elements must meet each other in confluence for the action to proceed. It may well be that the tribunal has jurisdiction over the subject matter of the dispute or action but not over one or the other party or both. It may also be the case that the court has jurisdiction but the litigant plaintiff or defendant has no standing in that tribunal. A tribunal may decline jurisdiction over a matter or person appearing before it on grounds provided by the law, including discretion, or a litigant (the defendant) may challenge the jurisdiction of the tribunal to hear the case. A defendant may challenge the *locus standi* of the plaintiff, or the tribunal may rule that the plaintiff or defendant or both do not have standing in that tribunal. The concept of jurisdiction is, by and large, appreciated by all regardless of the legal system to which they subscribe, but *locus standi* is a phenomenon that does not always enjoy widespread appreciation and familiarity.

With respect to a claim for damage to the marine environment, the central question is – who has the legal right to be compensated; in other words, who has standing? This is of particular significance where the ownership of the damaged element of the environment is in doubt or where it is established that the thing in question belongs to no one, that is, it is *res nullius*. In the absence of a proprietary interest in the subject matter, the claim might be denied.[12]

In this Chapter, *locus standi* is addressed primarily from the perspective of litigants purporting to claim for environmental damage who are public bodies. Although the issue is relevant in respect of private litigants as well, the problems appear to be less severe for reasons that appear in the following discussion. With respect to public bodies, including governments both at sovereign as well as non-sovereign levels, the problem of *locus standi* for environmental damage claims is largely due to a demonstrable lack of proprietary interest in the environment or at least in certain elements of it.[13]

---

12. G. Gauci, *Oil Pollution at Sea: Civil Liability and Compensation for Damage* (Chichester: Wiley, 1997), 253.
13. De La Rue & Anderson, *supra* n. 4, 503–504.

3.1 PROPERTY RIGHTS OR PROPRIETARY INTEREST

It should be obvious from the last statement in the previous section that property rights or a proprietary interest in the environment is the bed-rock of *locus standi* with regard to claims for environmental damage. In terms of property law generally, ownership is the ultimate form of right in property. The term 'title' signifying right of or entitlement to ownership is often used interchangeably with the term 'ownership' with respect to property of all kinds. The law usually requires ownership to be evidenced by documentation such as a title deed or deed of conveyance in the case of realty or by a bill of sale in the case of a ship. A proprietary interest is a legal interest in property that may reside at a lower threshold than outright ownership, although it may include ownership. Examples of proprietary interests short of ownership are leases, mortgages and liens and in the case of regimes governing waters, riparian rights, fishing rights, rights to extract underwater minerals and the like.

Whether the marine environment can be owned as property is a valid query at this juncture of the discussion. As stated by authors Colin de la Rue and Charles B. Anderson in reference to claims for environmental damage:

> In most countries the absence of any proprietary interest in the marine environment would be an insuperable obstacle to a claim of this nature.[14]

As mentioned above, the tangible dimension of the environment consists of elements that are readily perceptible; that is, their presence can be felt or their existence can be confirmed by objective means. The natural resources in the marine environment consisting of the waters, the seabed, the flora and fauna, fish and other living resources and minerals and other non-living resources are tangible and can be the subject of legal ownership. Where such ownership can be demonstrated, a claimant in respect of environmental damage inflicted on these tangible elements of the environment can successfully plead *locus standi*. It is arguable that by virtue of ownership, a person can claim compensation for deprivation of enjoyment derived from the aesthetic attributes of the environment caused by environmental damage. This is akin to compensability of non-pecuniary losses such as pain and suffering well known to the law of torts.[15] However, quantifiability of the damage associated with the intangible dimension of the environment is difficult to say the least. The above-noted authors also point out that – the very notion of 'damage' – routinely used and understood in relation to property of static physical and chemical composition – can seriously mislead when applied to biological communities, which are in a state of constant flux.[16]

In jurisdictions in which property rights in a body of water and its constituents, including the bed and subsoil beneath it, are vested in private entities through ownership or lease, the owner or lessee, as a claimant for environmental damage

14. *Ibid.*, 504.
15. *Ibid.*, 506.
16. *Ibid.*, 504.

will, all other things being equal, enjoy *locus standi* in a tribunal conferred with appropriate jurisdiction to entertain proceedings. In many countries, private ownership of islands together with rights over their associated appurtenances consisting of proprietary interests is not uncommon. In such countries, there will likely be no impediment to a private entity's *locus standi* in respect of a claim for environmental damage as long as ownership of the damaged property can be established. Indeed in such countries it will probably be easier for a private claimant to found jurisdiction and establish actionability than it would be for a public entity as a claimant in the absence of express property rights vested in that entity. In some jurisdictions, however, private ownership of property may not include a proprietary interest in resources such as water and fish adequate to support a claim for damage caused to the resource itself.[17] In common law jurisdictions this is a reflection of the rule relating to animals in the wild, that is, *ferae naturae* as natural resources that they are not property until reduced to possession by a captor.[18] The same rule applies to fish in non-private waters except shell fish and varieties of ground fish that cling to the bottom.[19] Fish as a natural resource becomes property only after being caught by the fisherman.[20]

A public entity making the same claim will enjoy the same benefit of *locus standi* if it can demonstrate ownership or a proprietary interest in the environment in question. Unfortunately, except in authoritarian socialist regimes in which ownership of land and water is categorically vested in the state,[21] generally in most other countries the law does not provide for public ownership of property not specifically owned by anyone, including the government and of the unexploited resources therein. This point is driven home in the following passage in reference to state parties to the CLC of 1969 and Fund Convention of 1971:

> In most contracting states it is likely that the absence of any proprietary interest in the marine environment, coupled with difficulties in valuing the environment or damage to it by methods which are not unduly abstract or arbitrary, would constitute insuperable obstacles to such a claim.[22]

Property rights over elements of the environment, specifically the natural resources, can be vested in public bodies through statute. In such instances the body in question will no doubt have standing in respect of a claim for environmental damage. However, in many cases the statutory rights are not property rights per se but rather custodial or fiduciary rights. Such custodial and fiduciary rights conferred through statute flow from the doctrine of public trust, of which the notion of *parens patriae* is a part. Where public entities are unable to establish a property

17. *Ibid.*, 513.
18. *Pierson v. Post* 3 Cai. R. 175, 2 Am. Dec. 264[1] (N.Y. 1805); *Missouri v. Holland* (1919), 252 US 416, 424. It is contended that the rule has its roots in the Roman law doctrine of *res nullius*. See T.C. Sandars, *The Institutes of Justinian* (1865), 160. See also Gauci, *supra* n. 12, 253.
19. *McKee v. Gratz* (1922), 260 US 127.
20. De La Rue & Anderson, *supra* n. 4, 513.
21. Gauci, *supra* n. 12, 253.
22. De La Rue & Anderson, *supra* n. 4, 507.

right or other kind of proprietary interest by virtue of statute or otherwise, they may resort to the previously mentioned doctrines to assert *locus standi* in respect of a claim for environmental damage. The actionability of such a claim will depend on whether these legal principles are recognized in the jurisdiction in which the claim is brought or by the IOPC Fund where a claim is submitted for receiving compensation.

3.2 Public Trust Doctrine

3.2.1 Principles of Trust Law

The temporal evolution of the public trust doctrine points to its development along the lines of the principles of classical trust law. Conceptually, the trust that is a product of the English law of equity is a trilateral legal mechanism designed to protect assets. The settlor as owner of assets settles a trust in favour of a designated beneficiary by appointing a trustee through deed or contract, who is then charged with the fiduciary and custodial duty of protecting the assets and dealing with them accordingly.[23] In a trust involving public resources, the public comprising the citizenry at large has the status of settlor and may also be the beneficiary; or the beneficiary may be a designated subset of the community, whereas the state or a public body so identified by statute or otherwise is the trustee.[24] In such an arrangement the fundamental principles of trust law will apply, exemplified by the requirement of substantive care to be exercised in the course of protecting and preserving the assets and ensuring due observance of correct procedures coupled with corresponding prohibitions against unfair dealing and dissipation of the assets.[25] These principles elaborated further as the trustee's duty to protect and preserve the corpus of the trust against invasion and to prevent its diminution and act with diligence, fairness and faithfulness are also referred to as presumptions in the law of trusts. In the event of damage to the environment, the effect of these presumptions is to shift the onus of proof to the person responsible for causing the damage to the corpus of the trust if 'necessity' is pleaded by that person.[26]

3.2.2 Ancient Origins of the Doctrine

The antiquated roots of the public trust doctrine in the maritime context are to be found in the ancient law of Rome. Under the law, the waters of harbours and the bottoms beneath them were open to public use devoid of any express manifestation of state ownership. Roman law placed seashores under the guardianship of the

---

23. See generally, P. Todd, *Textbook on Trusts*, 4th edn (London: Blackstone Press Limited, 1999).
24. G. Osherenko, 'New Discourses on Ocean Governance: Understanding Property Rights and the Public Trust', *Journal of Environmental Law and Litigation* 21 (2006): 366.
25. De La Rue & Anderson, *supra* n. 12, 254.
26. V. Yannacone, B. Cohen & S. Davison, *Environmental Rights and Remedies* (1972), 14.

Roman people. The notion of guardianship was manifested through the Roman State's assertion of jurisdiction over the seashores and their supervision. Thus the State exercised jurisdictional and supervisory rights over these maritime elements as distinguished from ownership rights.[27] In essence, these maritime elements, namely, the seas, their beds and subsoil and the seashores were, in accordance with the *jus naturale*, things 'common to all men' under the Roman law and therefore constituted *res communis*, or things belonging to the community of peoples.[28] It appears that the public trust doctrine in the way it originated in Roman law went into disuse in subsequent times when bodies of water including tidal areas that were previously considered public came under the ownership of feudal lords and local aristocracies.[29]

3.2.3 Roots of the Doctrine in English Law

Although the origin of the public trust doctrine is rooted in Roman law, as noted above, its evolution in subsequent times has also been entrenched in the common law of England. Before the advent of the Magna Carta, the monarch as sovereign had title to the bed of navigable waters, which was characterized as 'private and alienable',[30] meaning that the Crown had power to dispose of it as private property through grant or otherwise. In 1215, when the Magna Carta was signed, the concept of guardianship of the people was imported from Roman law with respect to tidelands and navigable waters, effectively making them subject to a public trust. The citizenry were thus granted fishing and navigation rights in waters previously held as private monarchical property.[31] Needless to say, the historical pre-eminence of the Magna Carta established the notion of public trusteeship of navigable waters, tidelands and their associated natural resources as a doctrine of constitutional stature insofar as the law of England is concerned.

As early as in 1741, in the case of *Ward v. Cresswell*, [32] the court held that all subjects of the Crown had the right to fish in the high seas and made void any exclusive rights based on prescription.[33] In a landmark decision rendered by the House of Lords in 1865, over 120 years later, it was held that:

> The bed of all navigable rivers where the tide flows and reflows, and of all estuaries or arms of the sea, is by law vested in the Crown. But this ownership

27. V.P. Nanda & W.K. Ris Jr, 'The Public Trust Doctrine: A Viable Approach to International Environmental Protection', *Ecology Law Quarterly* 5 (1976): 297. In fns 38–40 at that page, the original sources for this information cited by the authors are T.C. Sanders, *The Institutes of Justinian* (1865), 159 and W.A. Hunter, *Roman Law* (1897): 311.
28. De La Rue & Anderson, *supra* n. 12, 253.
29. V.P. Nanda & W.K. Ris Jr, *supra* n. 27, 297. For the original source, see Note, 'The Public Trust in Tidal Areas: A Sometime Submerged Traditional Doctrine', *Yale Law Journal* 79 (1970): 762.
30. B.S. Cohen, 'The Constitution, the Public Trust Doctrine and the Environment', *Utah Law Review* 388 (1970): 389.
31. Nanda & Ris, *supra* n. 27, 297–298.
32. 125 Eng. Rep. 1165 (C.P. 1741).
33. Nanda & Ris, *supra* n. 27, 298.

of the Crown is for the benefit of the subject, and cannot be used in any manner so as to derogate from, or interfere with the right of navigation, which belongs by law to the subjects of the realm.[34]

Although the connection with Roman law is acknowledged, as an English common law concept, the doctrine of public trust had more to do with 'the sovereign rights of the crown than on the public interest'.[35]

3.2.4 Application of Public Trust Doctrine in Various States

Flowing from its historical evolution presented in the previous section, in the modern context, it is evident that the public trust doctrine prevails in several jurisdictions around the world. As a general rule, the doctrine recognizes that certain varieties of natural resources are held in trust by a state authority or public body where the beneficiary of the trust corpus is the citizenry. By virtue of the trust, a right of action or *locus standi* is conferred on the state authority or public body as trustee in the event of damage inflicted on the protected resources. It is particularly significant in the cases of claims arising from marine pollution damage because of the historical application of the doctrine to navigable waters, submerged lands and tidelands, as well as sea and aquatic uses for commercial and recreational purposes such as navigation, fishing and water sports.[36]

The flexibility of application of the doctrine, which is in contrast with the rigidity of the tenets of classical trust law, even though it has inherited several of its attributes, makes it a suitable device for meeting the needs of states in the international marine environmental arena, particularly because its roots can be traced back to both the civil and common law traditions. Indeed, with respect to tidelands and navigable waters, the public trust doctrine is a 'part of the heritage of the civil law'.[37] It is instructive to note that the fundamental principles entrenched in the doctrine, albeit with variations of it in terms of public use and custodial and fiduciary rights, responsibilities and competences, are to be found in *inter alia* Mexico in Central America, France, Spain, Portugal, Greece, Italy, Germany and Austria, as well as other countries in the European continent, all of which are civil law states.[38]

---

34. *Gann. v. Free Fishers of Whitestable* (1865), 11 Eng. Rep. 1305 (H.L.).
35. B.C. Mank, 'Standing and Future generations: Does Massachusetts *v.* EPA open Standing for Generations to Come?', *Colum. J. Envtl. L* 34 (2009): 1, 36. See also R.J. Lazarus, 'Changing Conceptions of Property and Sovereignty in Natural Resources: Questioning the Public Trust Doctrine', *Iowa Law Review* 71 (1986): 631, 635.
36. De La Rue & Anderson, *supra* n. 4, 517. See also K.M. Ward & J.W. Duffield, *Natural Resource Damages: Law and Economics* (New York: John Wiley and Sons, 1992), § 2.20.
37. Nanda & Ris, *supra* n. 27, 296, 304.
38. See *ibid.*, 304–306 for more detail.

3.2.5 Application of the Public Trust Doctrine in the United States

It is in the United States that the use and application of the public trust doctrine is relatively advanced. In that country, the courts have over the years through case law jurisprudence, developed the doctrine of public trust on the basis of the English common law referred to earlier.[39] As a general rule, in the absence of statute law, the judiciary sets American common law standards for the protection of the public interest in maritime matters such as navigation, fisheries and the marine environment through the operation of the public trust doctrine.[40] In *McCready v. Virginia*[41] for example, it was held:

> The State owns the tide-waters themselves, and the fish in them so far as they are capable of ownership while running. For this purpose the State represents its people and the ownership is that of the people in their united sovereignty.

It is notable in the context of this decision that whereas reference is made to the state as representing the people and that ownership resides in the people in their united sovereignty, the dictum is an expression of the traditional legal theory of 'direct ownership' of the state as the manifestation of the state's interest in the living resources of the waters situated within its geographical perimeter, and therefore the state has standing to sue as proprietor.[42]

The doctrine is also reflected in the constitutions of several of the American states and in federal and state legislation.[43] As noted by de la Rue and Anderson, there are some sixteen states in the United States where legislation on oil spills contain express provisions on recovery for damage to natural resources.[44] The principal federal statutes conferring *locus standi* on governments and other public bodies to recover damages in respect of damage to maritime natural resources are the Clean Water Act, 1972 (CWA), the Comprehensive Environmental Response, Compensation and Liability Act, 1980 (CERCLA) otherwise known as the 'Superfund' legislation[45] and the Oil Pollution Act, 1990 (OPA 90).[46] The public trust doctrine provides the basis for recovery by federal and state governments in respect of damage to natural resources under OPA 90 and CERCLA.[47] The expression 'natural resources' is defined in both CERCLA and OPA 90 to include such marine elements as fish, water and groundwater, where, in both cases, the resources belong to, are managed or held in trust by, or are otherwise

39. Osherenko, *supra* n. 24, 365–366.
40. See Gauci, *supra* n. 12, 254 and R.W. Johnson, 'Oil and the Public Trust Doctrine in Washington', *University of Puget Sound Law Review* 14 (1991): 671–708.
41. (1876), 94 US 391 at 394.
42. De La Rue & Anderson, *supra* n. 4, 516–517.
43. Osherenko, *supra* n. 24, 365–366.
44. De La Rue & Anderson, *supra* n. 4, 518, fn. 77 at that page.
45. 42 USC.
46. 33 USC.
47. De La Rue & Anderson, *supra* n. 4, 517.

under the control of the United States government, any state or local government or Indian tribe or even a foreign government.[48] The polluter is expressly made liable to the entities mentioned above.[49]

It is apparent from the earlier-noted provisions that in the event of damage to the named natural resources by ship-source marine pollution, the named entities will enjoy *locus standi* by virtue of statute. Damage from oil spills is not covered by CERCLA because the definition of 'hazardous substance' in that legislation explicitly precludes petroleum[50] but recovery can be had under CWA[51] and OPA 90. Under the National Contingency Plan[52] the Secretary of Commerce is designated as the trustee for marine resources and their habitat and the Secretary for the Interior is designated as trustee for *inter alia* certain species of fish, marine mammals and fresh water resources.[53] The requirement by the President for the designation of federal officials as trustees of the public trust in respect of natural resources is contained in section 1006(b) of OPA 90. Furthermore, pursuant to this legislation, regardless of ownership or management of natural resources being vested in public trustees, private persons enjoy *locus standi* to sue and recover for loss of use, lost profits or loss of earning capacity resulting from damage to those resources.

3.3 *Parens Patriae*

The concept of *parens patriae* is a branch or variation of the doctrine of public trust. The literal meaning of the expression is 'parent of the country'[54] The term is also construed as 'father of the people'.[55] Its origin lies in English law, pursuant to which, under the authority of the King, minors and mentally disabled persons and individuals otherwise incapacitated were entitled to protection.[56] The state, in its manifestation as sovereign, acted in *loco parentis*, that is, in the capacity of a guardian vis-à-vis such persons. Thus, the doctrine of *parens patriae* is historically rooted in English constitutional law.[57] In the United States, where the doctrine is most prevalent, over the last 100 years or so, since the beginning of the last century, its use of the doctrine by courts has expanded to include protection of quasi-sovereign interests of states in matters of health and welfare of the public and their natural resources.[58]

---

48. § 9601(16) and § 2701(20), respectively.
49. § 9607(f)(1) and § 2702(b)(2), respectively.
50. § 9601(14).
51. De La Rue & Anderson, *supra* n. 4, 520–521.
52. 40 CFR § 300.600 (1997)
53. 40 CFR § 300.600(b)(1) and (2).
54. De La Rue & Anderson, *supra* n. 4, 517; see fn. 71 at that page.
55. See *Fields v. Kirton*, 961 So. 2d 1127 (Fla. 4th DCA 2007).
56. Mank, *supra* n. 35, 34.
57. De La Rue & Anderson, *supra* n. 4, 517. See in particular fn. 71 at that page.
58. Mank, *supra* n. 35, 34.

The doctrine provides *locus standi* to the state to sue for damage to resources on behalf of its people.[59] In that context, *parens patriae* as a legal principle is clearly a close affiliate of the public trust doctrine and there are conspicuous overlaps even though they developed separately where the public trust doctrine rooted in Roman law readily embraced protection of navigational and fishing rights in the course of its evolution as observed earlier. In present times, at least in the United States, the two doctrines can be invoked simultaneously, and sometimes are, in court actions by states seeking to protect natural resources. However, there are some distinctions between the two. The public trust doctrine imposes certain express fiduciary obligations on the state circumscribed by some of the basic tenets of trust law, as mentioned earlier; by contrast, *parens patriae* is more a matter of judicial discretion.[60]

In line with judicial thinking in cases exemplified by the decision in *Missouri v. Illinois*,[61] which involved sewage pollution of the Mississippi River emanating from Illinois, it was thought that the doctrine of *parens patriae* could provide only injunctive relief, and was thus viewed as a limitation discouraging litigants from invoking it.[62] However, as exemplified in *Maine v. M/V Tamano*,[63] *parens patriae* actions are not limited to injunctive relief pleas per se; they may be invoked in pollution cases.[64] As pointed out by de la Rue and Anderson, *parens patriae* is often invoked as a procedural device[65], and, whereas it can operate in the absence of any proprietary interest of the state in the damaged resource, two conditions need to be met for it to be successfully invoked. First, it must be based on a state interest separate from damage suffered by individuals of the public such as when damage from pollution is suffered by state-owned property or public welfare is detrimentally affected as in the application of the common law tort of public nuisance. Second, a substantial part of the citizenry must have suffered from the incident.[66] In the view of one commentator, limitations arising out of the cases render the doctrine 'an incomplete authority for recovering damages for harm done to public resources'.[67]

3.4 Statutes and Case Law on Public Trust and *Parens Patriae*

In the United States, there are a host of statutes besides the CWA, CERCLA and OPA 90, in both federal and state domains, providing for recovery in respect of

59. 'Parens patriae' <http://en.wikipedia.org/wiki/Parens_patriae>, 31 Mar. 2008.
60. Mank, *supra* n. 35, 36.
61. (1901), 180 US 208.
62. Th. Newlon, 'Defining the Appropriate Scope of Superfund Natural Resource Damage Claims: How Great an Expansion of Liability?', *Va. J. Nat. Resources L.* 5 (1985): 197, 201–202.
63. [1993] A.M.C. at 1136–1137.
64. De La Rue & Anderson, *supra* n. 4, 518.
65. De La Rue & Anderson, *supra* n. 4, 517, fn. 71 at that page.
66. De La Rue & Anderson, *supra* n. 4, 517–518.
67. F.B. Cross, 'Natural Resource Damage Valuation', *Vanderbilt Law Review* 42 (1989): 278.

damage to natural resources caused by pollution emanating from ships and other sources. The relevant provisions are based on public trustee status or direct ownership conferred on governments or other public bodies by statute. There is also an abundance of US case law on the application of the public trust and *parens patriae* doctrines in reference to statutes in various jurisdictions within the United States.

In *Maine v. M/V Tamano*,[68] a state's claim to ownership of land beneath the waters of the marginal seas was held to be based on the Submerged Lands Act, 1953,[69] which grants title to the states in respect of all lands lying under navigable waters within three nautical miles off the coast. In this Act, the definition of 'natural resources' in section 1301(e) includes varieties of living resources, together with sponges, kelp and other forms of marine life. By virtue of the statute, the right and power to manage and develop the natural resources in the waters covered by that legislation are vested in the states.[70]

In *Montauk Oil Transport Corporation v. Steamship Mutual Underwriting Association (Bermuda) Ltd*[71] a waterway had to be closed because of an oil spill. The federal government claimed compensation for loss of use of the waterway by invoking the *parens patriae* doctrine. Arguments were made on behalf of the ship's interests contending that commercial vessel traffic was not a public use in the same sense as recreational use such as fishing or swimming. They further submitted that private suit by the owners who were deprived of the use of the waterway was the proper mode for recovery of damages. Contrary to the submissions of the vessel interests, it was held that private recovery was inadmissible on two grounds; namely, that compensation for loss of use of the waterway was for the benefit of the government and not of private parties, even though the latter may have suffered loss of use of the natural resources, which they enjoyed free of cost. The government had standing to sue as owner of the resources. Second, a *parens partiae* action by government would provide compensation for an actual loss, thus compelling private party defendants to bear the full costs of their actions given that private tort actions could not be maintained because of a number of impediments. These included problems with proof of individualized damages, class action problems and the common law rule not allowing *locus standi* to a party without any proprietary interest in the damaged property.[72]

The case that is most well known internationally in the maritime field is the landmark decision of the US Court of Appeals on appeal from the District Court of Puerto Rico in the *Zoe Colocotroni* case.[73] The tramp oil tanker *Zoe Colocotroni*, owned by a Panamanian company and controlled and managed by their agents,

68. [1993] A.M.C. 1131.
69. 43 USC § 1301.
70. [1993] A.M.C., *supra* n. 63.
71. 79 F.3d 295 (2nd Cir. 1996).
72. Authority for the last proposition cited by the court was *Robins Dry Dock & Repair Co. v. Flint* (1927), 275 US 303, 308–309.
73. *Commonwealth of Puerto Rico et al., Plaintiffs, Appellees, v. The SS Zoe Colocotroni, Her Engines, Appurtenances, etc., et al., Defendants, Appellants,* US Court of Appeals, 628 F. 2d 652 (1st Cir. 1980).

Colocotroni Brothers, ran aground, causing extensive pollution damage to the coast of Puerto Rico. The finding of the trial court was that the vessel was unseaworthy. The Commonwealth of Puerto Rico and the Environmental Quality Board (EQB) of Puerto Rico as plaintiffs, brought actions under Puerto Rican statutes, the Solid Waste Disposal Act, the Water Pollution Control Act and the Public Policy Environmental Act to recover for damage to natural resources and pollution clean-up costs. The District Court awarded damages in the amount of United States Dollar (USD) 6,164,192.09 with costs. Limitation of liability was denied on a finding of privity on the part of the defendants.[74]

The District Court dealt with the *locus standi* of the plaintiffs in the following terms:

> As a sovereign it [Puerto Rico] represents the collective interests of the People of this jurisdiction. The Commonwealth holds title in trust to the public property and domain, and it is charged with the protection of the People's interest in the same. . . . In the execution of these endeavours the Commonwealth may bring legal actions in court to protect its property and recover damages to the same. . . . The Commonwealth has title to all beaches and to the maritime terrestrial zone abutting the navigable waters. . . . The Commonwealth is thus the owner of both the living and nonliving resources located in the navigable waters of the Commonwealth and those on the bottom and its subsoil, as well as those located within the referred to maritime-terrestrial zone. . . . The Commonwealth therefore has standing to sue to recover for oil pollution to Bahia Sucia and related resources because it has proprietary interest in the same . . . and because it is the trustee of the public trust in these resources. . . . Additionally, in its capacity as *parens patriae*, the Commonwealth has a sovereign interest in the general welfare of its citizens which transcends any injury which may be caused to its proprietary interests or to the property of its individual citizens. Particularly when a nuisance of disastrous proportions occurs such as in the case of a maritime oil spill, the special status of the body politic vis-à-vis its citizens gives rise to a right to seek redress on behalf of the collective community which is not limited to the abatement of the nuisance, but which can allow for recovery of damages by the body politic.[75]

The defendants challenged the *locus standi* of the plaintiffs in the Court of Appeals. The First Circuit Court of Appeals circumvented the issue of whether the Commonwealth had *locus standi* under common law and held that it had undisputed authority pursuant to the pertinent statutes of Puerto Rico, to recover damages relating to damage suffered by the environment and its natural resources where the damage resulted from a violation of statutory prohibition against the discharge of any pollutant into its territorial seas.[76]

74. See E.D. Brown, 'Making the Polluter Pay for Oil Pollution Damage to the Environment: A Note on the Zoe Colocotroni Case', *LMCLQ* (1981): 328.
75. 456 F. Supp. 1327 (1978) at 1336–1337.
76. De La Rue & Anderson, *supra* n. 4, 523.

Outside of the United States, the Patmos was a case involving the Italian courts and the IOPC Fund. The oil spill resulting from a collision in the Straits of Messina involving the Greek tanker *Patmos* caused damage to the coast of Sicily, which subsequently came to litigation in the Court of First Instance in Messina. In the action the Italian government alleged violation of its sovereignty over its territorial seas but the court held that the right so alleged was not a right of ownership and dismissed the claim after referring to the 1971 Fund Assembly Resolution of 1980.[77]

This decision is a reflection of judicial attitude in jurisdictions in which proprietary interest is the only basis on which government claims for environmental damage can successfully rest. In this context it is also interesting to note that as a matter of Fund policy, recovery for damage to unexploited natural resources is not admissible. In the words of the former Director of the IOPC Fund:

> the 1992 Fund has taken the position that it would be inappropriate to admit claims for compensating damage to unexploited natural resources which have no owner.[78]

In the United Kingdom title to the foreshore[79] and the seabed underlying the waters of the territorial seas is vested in the Crown. However, it appears that the Crown does not engage in the direct exploitation of the seabed and foreshore but grants leases and licenses through Crown Estate Commissioners. If the Crown suffers a loss of revenue from damage to resources caused by oil pollution, it is likely that it will be actionable against the polluter.[80]

3.5 *Locus Standi* of Private Entities

Whereas it is apparent from the earlier discussion that a government or other public body can enjoy *locus standi* in respect of environmental damage even in the absence of proprietary rights, it is at best unclear whether a private entity, as an individual or an interest group possesses a similar right of action. It is evident that natural resources such as wild animals, fish or subterranean minerals may not be subject to a private property owner's ambit of proprietary interests but rather may be vested in the state or a government.[81]

It is interesting to note that the Law Reform Commissions of the Provinces of Ontario and British Columbia in Canada have made recommendations toward the

77. *Ibid.*, 508.
78. M. Jacobsson, 'The International Oil Pollution Compensation Funds and the International Regime of Compensation for Oil Pollution Damage', World Maritime University, 27 Jun. 2003, 12. It is notable that this statement does not appear in successive papers of Mr Jacobsson delivered at the World Maritime University.
79. The foreshore is the land lying between the mean high and low water marks on the coastline which is exposed and visible at low water.
80. Brown, *supra* n. 74, 326.
81. Gauci, *supra* n. 12, 257.

creation of civil statutory remedies comprising damages payable to compensate the public for environmental damage independent of compensation payable to individuals or corporations. Through this kind of remedy a person could enjoy *locus standi* not for his own personal benefit but rather for the benefit of the public at large, which has an interest in the protection of the environment. Such reform would point in the direction of *locus standi* in favour of interest groups devoid of any proprietary or pecuniary interests.[82]

4 SUMMARY AND CONCLUSION

The private law of ship-source marine pollution essentially consists of two components, namely liability, a qualitative concept defined by conduct repugnant to the law, and remedy in the form of damages or compensation, which is, by contrast, a quantitative concept. The compensability of a claim under the CLC/Fund convention regime depends on whether the claim meets the definition of 'pollution damage'. In this chapter, the regime of liability and compensation in respect of environmental damage has been critically addressed. It is submitted that there are two dimensions to the compensability equation in respect of environmental damage, one being the actionability of the cause, that is, the *locus standi* or standing of the claimant, and the other, the quantifiability of damages once standing has been positively established. This chapter focuses analytically on the first dimension, with particular reference to the position of governments and other public bodies.

The ground rule prevailing in virtually all jurisdictions is that the claimant, whether it is a public or private entity, must have a proprietary interest in the environment in order to establish *locus standi*. With respect to the standing of private individuals, one noted author made the following insightful and perceptive comments:

> the point of departure should be that one should not *a priori* exclude the right of an individual to protect the natural environment and it is suggested that future environmental legislation should specify that *locus standi* in these matters should not be dependent on proprietorial rights. . . . Such rights of suit would be over and above the rights of an individual to claim compensation for damage to private property and would be a recognition of the importance of the possible contribution by an individual to the environment.[83]

Where, in the case of the state or a public body, proprietary interest is absent, the claimant may resort to certain non-proprietary rights vested in it. In some jurisdictions, the public trust doctrine or *parens patriae*, a variation of it, may be invoked

82. Report on Environmental Harm of the Ontario Law Reform Commission, 1990 as cited in *supra* n. 12 at 258. It is to be noted that in 1992 the Federal Court Act of Canada was amended and the issue of standing in public interest litigation for environmental damage has been addressed.
83. Gauci, *supra* n. 12, 257–258.

by virtue of the common law or by statute supported by case law jurisprudence. In the United States, where the law in this area is most advanced, the state or a public body is conferred with explicit statutory rights in that regard as a trustee of the public trust.

In the view of the present author, these non-proprietary legal rights may be explored at the international level and incorporated into convention law to fill the lacuna that presently exists and bring about uniformity. The public trust doctrine is not only a flexible device devoid of many of the rigours of traditional trust law but is one that may be acceptable to most jurisdictions to meet international needs because of its roots in both Roman law and English common law tradition. Indeed the basic concept is found to reside in a variety of national jurisdictions.[84] Embracing it through the vehicle of convention law would give the public trust and *parens patriae* doctrines, the universal and unified recognition that they deserve. It is recognized, however, that these doctrinal matters may impinge on the constitutional law of a nation state and it may be difficult if not impossible to entrench them into convention law, even if it is proposed in the interest of international uniformity. Nevertheless, all feasible attempts should be made to address this inchoate element in the international regime for liability and compensation in relation to claims for environmental damage caused by ship-generated pollution.

References

Brans, E.H.P. *Liability for Damage to Public Natural Resources: Standing, Damage, and Damage Assessment*. London: Kluwer Law International, 2001.

Brown, E.D. 'Making the Polluter Pay for Oil Pollution Damage to the Environment: A Note on the Zoe Colocotroni Case'. *LMCLQ* Part 3 (1981): 328.

Cohen, B.S. 'The Constitution, the Public Trust Doctrine and the Environment'. *Utah Law Review* 3 (1970): 388–389.

Cross, F.B. 'Natural Resource Damage Valuation'. *Vanderbilt Law Review* 42 (1989): 278.

De La Rue, C.M. & Ch.B. Anderson. *Shipping and the Environment: Law and Practice*. London: LLP, 1998.

Gauci, G. 'The *Sea Empress* and the Special Status of Fishermen'. *JBL* (November 2003): 609–615.

Gauci, G. *Oil Pollution at Sea: Civil Liability and Compensation for Damage*. Chichester: Wiley, 1997.

Jacobsson, M. 'The International Regime on Liability and Compensation for Tanker Oil Spills'. World Maritime University, 4 July 2008, 8.

Jacobsson, M. 'The International Oil Pollution Compensation Funds and the International Regime of Compensation for Oil Pollution Damage'. World Maritime University, 27 June 2003, 12.

84. Nanda & Ris, *supra* n. 27, 292, 306.

Johnson, R.W. 'Oil and the Public Trust Doctrine in Washington'. *University of Puget Sound Law Review* 14 (1001): 671–708.

Lazarus, R.J. 'Changing Conceptions of Property and Sovereignty in Natural Resources: Questioning the Public Trust Doctrine'. *Iowa Law Review* 71 (1986): 631, 635.

Mank, B.C. 'Standing and Future generations: Does Massachusetts *v.* EPA Open Standing for Generations to Come?'. *Columbia Journal of Environmental Law* 34 (2009): 1, 36.

Nanda, V.P. & W.K. Ris Jr. 'The Public Trust Doctrine: A Viable Approach to International Environmental Protection'. *Ecology Law Quarterly* 5 (1976): 297.

Newlon, Th. 'Defining the Appropriate Scope of Superfund Natural Resource Damage Claims: How Great an Expansion of Liability?'. *Virginia Journal of Natural Resources Law* 5 (1985): 197, 201–202.

Osherenko, G. 'New Discourses on Ocean Governance: Understanding Property Rights and the Public Trust'. *Journal of Environmental Law and Litigation* 21 (2006): 366–367.

Todd, P. *Textbook on Trusts*. 4th edn. London: Blackstone Press Limited, 1999.

Sandars, T.C. *The Institutes of Justinian*. 3rd edn. London: Longmans, Green and Co., 1865.

Silverstone, D. 'Ship Source Oil Pollution Damage: A Canadian Perspective on Recoverability of Economic Losses and Damage to the Marine Environment'. *Marine Policy* (April 1985): 118.

Ward, K.M. & J.W. Duffield. *Natural Resource Damages: Law and Economics.* New York: John Wiley and Sons, 1992.

Yannacone, V., B. Cohen & S. Davidson. *Environmental Rights and Remedies.* Rochester, NY: Lawyers Co-operative Publishing Company, 1972.

Chapter 6

Funds' Admissibility Criteria for Pure Economic Loss Claims versus British Judicial Stand: Complementary or Contradictory?

*Baris Soyer**

1 INTRODUCTION

With an intention of providing compensation for the victims of oil spills from ships that carry oil in bulk as cargo, and also spills of bunker oil from unladen tankers that have residues of a persistent oil cargo aboard, the international community put in place a legal regime at the end of 1960s. The fundamental principles of this international regime were initially laid down by the International Convention on Civil Liability for Oil Pollution Damage 1969 (Civil Liability Convention 1969) and the International Convention on the Establishment of an International Fund for Compensation for Oil Pollution Damage 1971 (Fund Convention 1971).[1] These Conventions were amended by two protocols adopted in 1992. The revised texts

* Professor of Commercial and Maritime Law, member of the Institute of International Shipping and Trade Law, Swansea University, United Kingdom.

1. It should be noted that oil-spill damage from bunkers and lubrication oils from ships not employed for the carriage of oil do not come under the scope of these Conventions. Compensation for damage caused by such spills is now provided by the Bunker Oil Pollution Convention (BOPC) 2001, which entered into force on 21 Nov. 2008. As of 31 Aug. 2009, 44 states have

Michael G. Faure, Han Lixin & Shan Hongjun, *Maritime Pollution Liability and Policy: China, Europe and the US*, 97–109.

are usually referred to as the Civil Liability Convention 1992 and the Fund Convention 1992.[2] The Civil Liability Conventions 1969 and 1992 set out a strict liability regime for the shipowners. A shipowner can be exempted from liability only if he proves that the damage: (1) resulted from an act of war, hostilities, civil war, insurrection or a natural phenomenon of an exceptional, inevitable and irresistible character; or (2) was wholly caused by an act or omission done with intent to cause damage by a third party; or (3) was wholly caused by the negligence or other wrongful act by public authorities for the maintenance of lights or other navigational aids.[3] Shipowners under the regime are required to carry liability insurance, and the possibility of bringing a direct action against the liability insurer is open to the victims. However, the liability of the shipowners under these Conventions is limited. The liability of shipowners ranges from Special Drawing Right (SDR) 4.51 million for tankers up to 5000 gross tonnes to SDR 89.77 million for large tankers. The Fund Conventions 1971 and 1992 establish a system providing additional compensation to victims when compensation under the respective Liability Convention is inadequate.[4] The compensation payable for each single incident under the Fund Convention 1992 is limited to an aggregate amount of SDR 203 million, including the sum actually paid by the shipowner or his liability insurer under the Civil Liability Convention 1992. A third tier of compensation in the form of a Supplementary Fund was established on 3 March 2005 by means of the Supplementary Fund Protocol adopted in 2003. The Supplementary Fund is available only in those states that are party to it,[5] and its rationale is to provide higher levels of compensation for the victims of an oil spill. The amount payable under the Supplementary Fund is SDR 547 million, in addition to the amount available under the Civil Liability and Fund Conventions 1992. An intergovernmental organization is established under each of these Fund Conventions, the 1971 Fund, the 1992 Fund and the Supplementary Fund, collectively referred to as the International Oil Pollution Compensation (IOPC) Funds.

---

ratified the convention, representing around 79% of the world's tonnage (*Source*: <www.imo.org>).

2. These Conventions have received worldwide support. As of 31 Aug. 2009, 104 states have ratified both the Civil Liability Convention 1992 and the Fund Convention 1992, representing over 94% of the world tonnage. It is worth mentioning that the United States is not party to this international regime, opting to regulate this area of law with national legislation. The main Federal Statutes the deal with the matter are: Clean Water Act (the Federal Water Pollution Control Act) 33 USC § 1251, the Oil Pollution Act of 1990 33 USC § 2701 (hereinafter referred to as the OPA 1990) and the Comprehensive Environmental Response, Compensation and Liability Act 42 USC § 9601.
3. Article 1(6)(a)–(b) of the Civil Liability Convention 1992.
4. The Fund Convention 1971 ceased to be in force on 24 May 2002 and does not apply to incidents occurring after that day. However, the 1971 Fund has not yet been wound up, because some claims are still pending that resulted from incidents occurring before 24 May 2002.
5. Any state that is party to the Fund Convention 1992 may become party to the Supplementary Fund Protocol. As of 31 Aug. 2009, twenty-four states have become parties to the Supplementary Fund Protocol (*Source*: <www.imo.org>).

For the purposes of these Conventions, the concept of pollution damage includes not only property damage, cleanup costs and economic damage but also costs of reasonable measures taken to prevent or minimize pollution damage.[6] The international regime also seems to allow recovery for pure economic losses.[7] However, no indication appears in these conventions as to which criteria should be applied in order to arrive at a reasonable delimitation of the right of compensation in the context of pure economic loss claims. The matter is left entirely to the discretion of national courts and thus creates a potential for discrepancy in states implementing the international regime into their legal systems. This is, in fact, the position in the United Kingdom. The Merchant Shipping Act (MSA) 1995, the legislation implementing the Civil Liability and Fund Conventions 1992 into British law,[8] does not provide any guideline as to which type of pure economic loss claims is recoverable. The Funds, on the other hand, have developed certain principles on the admissibility of pure economic loss claims. The central focus of this chapter is pure economic loss claims. First, the admissibility criteria developed by the Funds for pure economic loss claims will be considered. Then, the stand taken by British courts on pure economic loss claims falling under the scope of these conventions will be analysed. At the end of the chapter, answers are suggested to the following questions: (1) whether there is any difference between the approach adopted by the Funds and British courts in the assessment of pure economic loss claims; (2) whether any such difference is desirable and (3) which legal stand British courts are likely to take in future on the point of pure economic loss.

2 FUNDS' APPROACH TO PURE ECONOMIC LOSS

Under the Funds' admissibility criteria, pure economic loss claims suffered by the fisheries, mariculture and fish processing sectors qualify for compensation if there

6. The term 'pollution damage' has been defined in Article 1(6) of the Civil Liability Convention 1992 as: '(a) loss or damage caused outside the ship by contamination resulting from the escape or discharge of oil from the ship, wherever such escape or discharge may occur, provided that compensation for impairment of the environment other than loss of profit from such impairment shall be limited to costs of reasonable measures of reinstatement actually undertaken or to be undertaken; (b) the costs of preventive measures and further loss or damage caused by preventive measures.' The same definition has been adopted by the Fund Convention 1992 (Art. 1(2)) and also the Supplementary Fund (Art. 1(6)).
7. Pure economic loss relates to consequential economic loss suffered by people whose property has not been contaminated by the oil spill. A classic example would be the loss of earnings of a fisherman who has been prevented from fishing for a period of time because of the contamination, although his fishing equipment suffered no physical loss or damage.
8. Part III, ss 152–172 of the MSA 1995, as amended by the Merchant Shipping (Oil Pollution) (Bunker Convention) Regulations 2006 (SI 2006/1244), and Part IV, ss 173–181 of the MSA 1995.

is a sufficiently close link of causation between the contamination and the loss of damage. When considering whether such a close link exists, the following factors are taken into account:

- The geographic proximity of the claimant's business activity to the contaminated area (e.g., whether a fisherman operates predominantly in the affected area or whether a fish farm or processing facility is located on or very close to the affected coast).
- The degree to which a claimant's business is economically dependent on an affected resource, such as a polluted fishing ground (e.g., whether a fisherman also exploits a nearby, unaffected fishing ground, or is able to exploit an alternative fishing ground to the one affected without being economically disadvantaged).
- The extent to which a claimant had alternative sources of supply or business opportunities (e.g., whether a fish processor was able to find alternative sources of fish).
- The extent to which a claimant's business forms an integral part of the economic activity within the area affected by the spill (e.g., whether a claimant's business is located or has assets in the affected area, or provides employment for people living there).[9]

The Funds would allow any pure economic loss claim brought by fisheries and fish processors that satisfies a majority of the factors expressed in the *Claims Manual*. For instance, following the *Sea Empress* disaster in Wales, a shellfish processor based in Newquay, some 80 km by road from the north of the area covered by the fishing ban, was allowed recovery for the loss of business by the Funds essentially on the grounds that (1) the factory was located close to the area covered by the fishing ban; (2) the claimant was highly dependent on the supplies from the affected area and (3) the claimant had only limited possibilities to obtain supply from other sources.[10] Furthermore, the Funds have been accommodating to pure economic loss claims brought by businesses that depend on the fisheries and mariculture activities. Accordingly, loss of business claims brought by suppliers of fuel and ice, fish porters, fish wholesalers and retailers will be paid by the Funds as long as the claimants manage to establish sufficiently a link of causation between their activities and the contamination resulting from the oil spill.

In similar vein, pure economic loss claims brought by the businesses in the tourism sector would qualify for compensation under the Funds' admissibility criteria as long as a sufficiently close link of causation between the contamination

9. 3.3.4 of the *Claims Manual*, International Oil Pollution Compensation Fund 1992, December 2008 Edition.
10. 71FUND/EXC.49/9Add.1, Executive Committee of the International Oil Pollution Compensation Fund, 49th session.

and the loss or damage exists. When considering whether such a close links exists, account is taken of the following factors:

- The geographic proximity of the claimant's business activity to the contaminated area (e.g., whether a tourist hotel, campsite, restaurant or bar is located on or close to the affected coast).
- The degree to which the claimant's business is economically dependent on an affected coastline (e.g., whether a hotel or restaurant located close to an affected coast caters solely or predominantly to leisure visitors or to the business community).
- The extent to which a claimant had alternative sources of supply or business opportunities (e.g., whether a reduction in income from tourists was offset by income from those involved in the response to an oil pollution incident, such as cleanup personnel and representatives from the media).
- The extent to which the claimant's business forms an integral part of the economic activity within the area affected by the spill (e.g., whether the business is allocated or has assets in the area, or employs people living there).[11]

The Funds would, therefore, be accommodating to claims brought by claimants who sell goods or services directly to tourists, such as the owners of hotels, campsites, bars and restaurants and tourist shops. For example, a sculptor in Morhiban was allowed recovery by the 1992 Fund for economic losses suffered as a result of a decline in the sales of his art work due to the reduction of the tourism in the area caused by the Erika disaster. Again, in the aftermath of the *Erika* disaster, the 1992 Fund compensated businesses letting sailing boats to tourists in the area affected by contamination. The 1992 Fund was also receptive of the claims put forward by estate agents in Vendée for losses suffered in their activity of seasonal lettings of furnished apartments and villas, as a result of the reduction in the number of tourists in the affected area.[12] However, the Funds would not normally pay pure economic loss claims advanced by businesses who provide goods and services to other businesses in the tourist industry but not directly to tourists, on the ground that there is not a sufficiently close link of causation between the contamination and any losses suffered by such claimants. Those who would fall under this category would be wholesalers, manufacturers of souvenirs and postcards and hotel launderers.

Perhaps surprisingly, the Funds might allow claims for costs of measures taken to prevent or minimize pure economic loss. Such measures, such as marketing campaigns, would normally be aimed at counteracting the negative impact of

---

11. 3.4.1 of the *Claims Manual*, International Oil Pollution Compensation Fund 1992, December 2008 Edition.
12. 92FUND/EXC.30/6/Add.2, Executive Committee of the International Oil Pollution Compensation Fund 1992, 30th Session.

the pollution on the fishery and tourism sectors. In order to qualify for compensation the measures should fulfil the following requirements:

(a) The cost of the measures should be reasonable.
(b) The cost of the measures should not be disproportionate to the further damage or loss that they are intended to mitigate:
    (i) The measures should be appropriate and offer a reasonable prospect of being successful (e.g., measures to restore confidence in seafood products should normally only be undertaken once fishing grounds are cleared of contamination and there is little or no risk of further contamination).
    (ii) In the case of marketing campaigns, the measures should relate to actual targeted markets (e.g., measures to counteract the negative effects on tourism in a particular area should normally be focused on the normal visitor client base of that area).[13]

3 BRITISH COURTS' APPROACH TO PURE ECONOMIC LOSS

Even though over the years there have been several incidents causing oil pollution damage in states that are party to the international regime, one is hard pressed to find court decisions on the interpretation of the Conventions and particularly their definition of 'pollution damage'.[14] This is due partly to the efficiency of the Funds in handling claims and the tendency of the protection and indemnity clubs (liability insurers) to settle cases out of court. However, British courts in the space of four years were called in on several occasions to determine the standard for recovery in pure economic loss claims.

The leading case on the subject is the Scottish Second Division Inner House Court of Session decision in *Landcatch Ltd v. International Oil Pollution Compensation Fund.*[15] The facts of the case are straightforward. On 5 January 1993, the *Braer*, laden with crude oil, grounded off Shetland. 84,700 tonnes of crude oil and 1600 tonnes of bulk oil escaped or were discharged from the tanker. Three days later, the Secretary of State for Scotland issued the Food Protection (Emergency Prohibitions) (Oil and Chemical Pollution of Fish) Order 1993, designating an area that could be affected by oil or other chemical substances that were likely to create a hazard to human health if fish or shellfish drawn from within the area were consumed. Another order issued 17 January 2004 replaced this order, enlarging the designated area of exclusion to about 400 square miles. Landcatch Limited,

13. 3.4.1 of the *Claims Manual*, International Oil Pollution Compensation Fund 1992, December 2008 Edition.
14. The notable exception to this is France, where courts have been so far involved in more than a hundred cases stemming from the *Erika* disaster, which caused massive oil pollution along the Atlantic coastline of the country in 1999.
15. [1999] 2 Lloyd's Rep 316 affirming the decision of the Outer House of the Court of Session delivered by Lord Gill.

a company whose business was to rear salmon from smolt in freshwater conditions and then to sell them for raising to maturity in seawater conditions, contended that as a result of oil pollution and the exclusion zone designation they suffered losses, namely: (1) smolt culled because of the lack of sales, (2) reduced selling price of smolt sold in 1993 and in 1994 when prices had not resumed their normal level and (3) additional rearing costs incurred in making special arrangements for raising 266,000 smolt. The principal place of business of Landcatch was Ormsary, Argyll, about 500 km from Shetland.

Landcatch's claim was rejected by the Court, which placed emphasis on the need for the pure economic loss claim to be linked directly and immediately to the contamination resulting from the discharge or escape of oil from the ship. Viewed from this angle, Landcatch's claim was regarded as 'indirect relational pure economic loss which is too remote from the causal factor which makes the causer liable'.

In the same year, *Skerries Salmon Ltd v. The Braer Corporation*[16] presented another opportunity to the judiciary to evaluate the test laid down in the *Landcatch* case. Skerries Salmon Co. was a company based in Shetland whose principal activity was salmon farming. Even though their fish farm was outside the exclusion zone imposed by the Government following the incident, and their equipment and stocks suffered no contamination, they claimed for loss of profit as a result of a fall in demand and consequent lower prices obtained for Shetland-produced fish and fish products following the incident. The pursuers attempted to distinguish their case from Landcatch on four grounds: (1) the proximity of the pursuers' fish farm to the exclusion zone; (2) the fact that the pursuer's business is aquaculture; (3) the fact that the pursuers shared the same market with fish farmers within the exclusion zone and (4) the fact that Shetland salmon is a recognized product with a special market identity. The essence of their argument was that these factors provide the necessary proximity between the contamination and their loss, thus satisfying the 'directness' requirement as set out in Landcatch. Lord Gill was not convinced that these factors provide any material ground of distinction. Their claim fell to be judged as no more than a claim based on relational economic loss, because the cause of the financial loss the pursuers had suffered from was damage to other parties' property.[17] In effect, the pursuers were judged to be claiming economic loss because of the world-wide publicity the pollution damage attracted.

The same reasoning was adopted in *P&O Scottish Ferries Ltd v. The Braer Corporation* and another and International Oil Pollution Compensation Fund[18] to turn down the pure economic loss claim brought by a company (P&O Scottish Ferries) whose activities included operation of passenger and freight ferry services between Shetland and the United Kingdom mainland. Their claim for compensation was based on the contention that the pollution damage affected their business in that they suffered a decline in passenger and freight revenue. Lord Gill was

16. [1999] SLT 1196.
17. *Ibid.*, at 1201–1202.
18. [1999] 2 Lloyd's Rep. 535.

adamant that these losses were not a direct consequence of the escape of oil. They were simply an indirect consequence of adverse publicity affecting the image of Shetland as a source of fish and fish products and as a holiday destination. In other words, the pursuer's claim for compensation was based on the fact that parties who would otherwise have entered into contracts with them failed to do so, resulting in the operation of their business becoming less profitable than it otherwise would have been. His Lordship did not find this sufficient to create the necessary relationship of proximity between the loss and contamination as specified by the Landcatch case.[19]

A few years later, it was this time the English courts' turn to tackle the matter. In *Algrete Shipping Co Inc. and another v. International Oil Pollution Compensation Fund and others* (*The Sea Empress*)[20] a claim was brought by Tilbury against the owners of the *Sea Empress*, which ran aground off Milford Heaven in Wales in February 1996. Tilbury, based in Exmouth, Devon, was a company engaged in the business of fish processing. Roughly half of Tilbury's sales were accounted for by processed whelks, obtained mostly from Wales. Tilbury had supply contracts with Welsh fishermen who fished the whelks in the area where the incident occurred. Following the oil spill, a fishing ban was enforced from February to September 1996. As a result of the ban, Tilbury claimed that it lost profit of about GBP 644,000 which it would otherwise have made by selling the processed whelks to the lucrative Korean market. At the first instance, David Steel, J., adopting the reasoning in *Landcatch*, rejected the claim on the basis that lack of proximity rendered the claim too remote. Tilbury's claim, in the words of David Steel, J., 'was secondary, derivative, relational and/or indirect'.[21]

Before the Court of Appeal, the counsel for Tilbury made an attempt to distinguish the present case from *Landcatch* by virtue of the close physical association between the contamination and Tilbury's loss. The crux of the argument was that the contamination of the relevant sea area affected the whelks and prevented Tilbury from receiving whelks from the contaminated waters of the very sea area from which it was to receive such whelks. In contrast, in *Landcatch* the contamination never physically affected the pursuers' smolt; it merely prevented them from being ordered from the pursuers and supplied to the Shetland fish farmers in the contaminated area under the usual annual contracts. Put another way, the suppliers in *Landcatch* did not have any physical involvement with anything contaminated, whereas this was not so in the present case. To boost their argument, the counsel further attempted to demonstrate that Tilbury's claim was different to potential loss of earning claims that could be brought by London wholesalers or restaurateurs, because the latter had no interest in the contaminated waters or in any contaminated whelks; their only interest would be in whelks as processed by Tilbury in Devon and sold on to them. Obviously, the purpose behind the supplementary submission was to demonstrate that allowing recovery to Tilbury would

19. *Ibid.*, at 539–540.
20. [2002] High Court of England and Wales (EWHC) 1095 (Admlty); [2003] 1 Lloyd's Rep. 123.
21. *Ibid.*, at [34].

not open the floodgates so that claims could be brought by potential claimants below Tilbury in the supply chain.

Although the Court of Appeal acknowledged that there were minor differences between the facts of the present case and *Landcatch*, it considered such differences insufficient to lead the court to a different result in this case. Mance, L.J. (as he then was), emphasized the point that the relevant statute in defining the term 'damage' is focused on physical contamination and its consequences. Tilbury's interest was not in the whelks in their natural habitat but instead in landed whelks. The loss Tilbury suffered was, therefore, a result of its inability to carry out processing and packaging and deliveries of processed and packed whelks at a point far away from the contaminated areas. This type of loss was deemed to be secondary economic loss not falling under the relevant provision of the statute.[22] It is apparent that the geographical distance between the place of contamination and the place of business of Tilbury played an important role in the decision of the court.[23]

4 RELATIONSHIP BETWEEN THE FUNDS' ADMISSIBILITY CRITERIA AND THE STAND TAKEN BY BRITISH COURTS

At the outset, it is appropriate to identify the legal status of the funds' admissibility criteria on pure economic loss claims in jurisdictions that are party to the international regime set up by the Civil Liability and Fund Conventions. It is submitted that the admissibility criteria has no binding effect on courts in those jurisdictions. The admissibility criteria should simply be viewed as a formulation developed by the funds over the years in order to assist in the consideration of the claims submitted to this body. There is no doubt that the criteria have implications on the funds, but one should not lose sight of the fact that the decisions of the funds are themselves subject to the judicial control of the courts in member jurisdictions. It was, in fact, the view of the members of the Court of Appeal in the *Sea Empress* case that the admissibility criteria of the funds would not provide any guidance in the consideration of pure economic loss claims advanced before a British court. In cases stemming from the *Erika* disaster, French courts have taken a similar line. In some of these cases, the courts made the point that these criteria were not

22. [2003] East-West Center Association (EWCA) Civ 65; [2003] 1 Lloyd's Rep. 327, at [33].
23. L.J. Mance, *ibid.*, at [29], referring to the judgment of Lord McCluskey in *Landcatch*, made the following observation:

> In all these passages Lord McCluskey was emphasising a fisherman's close physical involvement with the contaminated waters. It is difficult to see how the present appellant [Tilbury] in Exmouth, Devon can claim any matchingly close involvement with the contaminated waters off Dyfed. It is true that it has a strong economic interest in such waters in continuing to be available for fishing by its local Welsh fishermen suppliers, and so in continuing to be able to process and profit out of the Korean contracts. But that is a much less direct and physical link with the contaminated waters or with any whelks which may have been contaminated.

binding on the courts but provided a useful reference. Conversely, some courts chose to ignore the criteria altogether.

British courts seem to make use of the notion of 'proximity' in determining which types of economic loss claims are recoverable. The requirement of proximity involves the notion of nearness or closeness and embraces not only the physical proximity between the person or property of the claimant and the physical contamination but possibly also causal proximity in the sense of closeness or directness of the causal connection between the physical contamination and the loss sustained. A potential claimant group, for example, is commercial fishermen whose livelihood might be dependent on the contaminated area. Commercial fishermen operating in the area would naturally be in close physical involvement with the contaminated waters, and it is likely that a causal proximity between the contamination and their economic loss can be easily established if they are prevented from fishing in the area due to contamination. The same is likely to be the case if commercial fishermen incur additional expenses in search of an alternative fishing ground if they are prevented from fishing in an area where they usually fish, as a result of contamination. The reasoning in *Landcatch* and *The Sea Empress* leaves no doubt that the statutory regime would be accommodating for pure economic loss claims in the shape of loss of earning put forward by commercial fishermen.

In relation to the position of fishermen, the test laid down by the British courts seems to be in harmony with the Funds admissibility criteria. However, when it comes to fish processors and other businesses that rely on the fishing sector, it cannot be confidently suggested that pure economic loss claims put forward by them will be treated favourably by British courts. Pure economic loss claims put forward by such businesses are indirect, in the sense that it is the inability of the fishermen to supply the products and not the contamination itself that causes the loss in such cases. Therefore, policy concerns about indeterminate liability might surface in such instances that might tilt the balance against recovery.

The reasoning in the British cases discussed leaves little doubt that the traders who regularly supply fishermen with diesel and fishing nets, and also the wholesalers who supply diesel in bulk to the trader, or the net makers who sell the traders nets for onward sale to fishermen, will not be able to recover their economic loss even though they are bound to be economically prejudiced by the disruption of their markets. The same would hold true for those who provide goods and services to the hotels (assuming that hotels are directly influenced from contamination), such as meat producers and launderers. When it comes to such businesses, it is evident that British courts would not regard contamination as the 'direct and immediate' cause of the loss. Instead, in such cases the contamination, borrowing the words of Lord McCluskey,[24] sets in train a chain of events that eventually results in the claimant's suffering loss or damage. Undoubtedly, the driving force here is the desire to avoid indeterminate liability. Allowing recovery for pure economic claims put forward by second-line victims would have an adverse

24. *Landcatch Ltd v. International Oil Pollution Compensation Fund* [1999] 2 Lloyd's Rep. 316, at 332.

financial impact on those who were directly affected by the contamination, because it is inevitable that the amounts received by the first-line victims will be attenuated given the financial limitation of the funds available.

The analysis gets more blurred when it comes to pure economic loss claims coming from the tourism sector. Although tourist shops are allowed recovery under the Funds' criteria, the likelihood of success for such shops under the 'proximity requirement' set by British courts is very remote. There is no doubt that those shops might well enjoy a geographical proximity to the waters contaminated and be negatively affected economically as a result of a reduction in the number of tourists visiting the region; however, what makes their case rather problematic is the difficulty in differentiating their position from the position of other business entities operating in the region, such as markets, supermarkets and newsagents providing service to the tourism sector. In other words, the issue of indeterminate liability shows its ugly head again in this context. Conversely, the position of hotels, campsites and even restaurants located in close geographical proximity to the contaminated waters might be rather different. It is indisputable that businesses of this kind would have a strong economic interest in the contaminated waters, but obviously their interest is not as direct as that of commercial fishermen. In the case of a contamination affecting their business, should they be entitled to make a claim for their economic loss under the statute? Naturally, the answer to this question will depend on what factors courts will take into consideration in applying the 'proximity' requirement to any given set of circumstances. However, it should be noted that the 'proximity requirement' can be attributed a wide meaning by the courts to afford recovery in some cases as long as policy considerations regarding indeterminate liability can be addressed. In other areas of tort law, pure economic loss claims have been allowed in cases in which it is evident that an ascertainable class suffers as a result of a tortious act. This point has been made explicit in the judgment of Gibbs, J., in *Caltex Oil (Australia) Pty Ltd v. The Dredge (Willemstad)*:[25]

> In my opinion it is still right to say that as a general rule damages are not recoverable for economic loss which is not consequential upon injury to the plaintiff's personal property. The fact that the loss was foreseeable is not enough to make it recoverable. However, there are exceptional cases in which the defendant has knowledge or means of knowledge that the plaintiff individually, and not merely as a member of an unascertained class, will be likely to suffer economic loss as a consequence of his negligence, and owes the plaintiff a duty to take care not to cause him such damage by his negligent act.[26]

25. [1976] 136 CLR 529, at [36].
26. A similar development took place in Canada in which a railway company was allowed to recover economic loss for damage on a bridge used largely, though not exclusively, by the railway company although the bridge was not owned by them; see, *Canadian National Railway Co. v. Norsk Pacific Steamship Co.* [1992] 1 SCR 1021; (1992) 91 DLR (4th) 289.

Although the *Caltex* case was concerned with an economic loss claim associated with negligence, it is submitted that its reasoning equally applies in the present context. It is an inevitable consequence of any major oil pollution incident that the economic life in the region will suffer a significant blow and an indeterminate number of businesses will suffer economic losses. The 'proximity' requirement will, certainly, restrict the number of businesses that can bring an action for loss of profit. The decision of the High Court of Australia in the Caltex case demonstrates that the members of this normatively justified class (those satisfying the proximity requirement) should be allowed to recover for their economic loss as long as they are uniquely foreseeable as likely to be harmed as a result of contamination (or at least foreseeable as one of a fairly small group). Put another way, businesses that have strong geographical and economic links with contaminated waters should be able to recover as long as they are reasonably ascertainable. On that basis, it is submitted that hotels, campsites and even restaurants located close to a contaminated area might be able to recover loss of earnings caused by contamination as they form part of an ascertainable class.

A final question relates to the recoverability of preventive measures taken by authorities to prevent or minimize pure economic loss. Applying the proximity requirement developed by the British courts one could argue that the relationship between the contamination and such expenses is too remote. Put another way, such expenses seem to be a rather indirect consequence of the contamination. Although such expenses could be recovered from the Funds in certain cases, it would come as no surprise if the British courts' approach to them is rather adverse under the 'proximity requirement'.

5 CONCLUSION

Given the international context of the matter, it would have been desirable that the claims for pure economic loss are dealt with in a similar manner in all jurisdictions that are party to the Civil Liability and Fund Conventions. However, the draftsmen's conscious decision not to specify the precise criteria in the relevant convention regime means that courts in different jurisdictions have been given an open cheque in determining which types of pure economic loss claims would be recoverable in their jurisdiction. The law in states that are party to the international regime could have been harmonized if the courts decided to follow the admissibility criteria laid down by the Funds for pure economic loss claims. However, it is evident that the Funds' criteria will not be followed in legal systems that are party to the international regime, essentially on the premise that this is a private arrangement that is developed by the Funds in dealing with the claims submitted to them and cannot be treated as imposing binding obligations on courts in various legal systems. One should also not lose sight of the fact that the Funds' decisions themselves are subject to legal scrutiny in each legal jurisdiction that has subscribed to the international regime.

The analysis in this chapter leaves no doubt that there are some similarities between the Funds' admissibility criteria for pure economic loss claims and the British courts' stand on the matter. It is remarkable that a degree of flexibility is inherent in both types of approaches, which undoubtedly allows the Funds and the British courts to adopt a pragmatic approach when dealing with pre-economic loss claims. However, it is obvious that the 'proximity requirement' adopted by the British courts is much narrower and is aimed at dealing primarily with the 'indeterminacy problem' that can often arise in the context of pure economic loss claims. That is not to say that the Funds are not concerned with the indeterminacy problem. However, it is evident the driving force for the Funds in the development of the admissibility criteria is pragmatism so as to enable them to achieve the most appropriate outcome depending on the geographical location of the oil spill, the economic climate in that region and also political factors.

In summary, potential claimants that are affected from an oil spill should be aware that if their claim for pure economic loss is rejected by the Funds, which consider such claims from a more generous point of view, it is very likely that British courts will take a similar approach on the matter. This outcome, in fact, did occur in the cases that faced the British courts in recent years. In all of these cases, the Funds' decision not to pay for the pure economic loss claims was endorsed by the courts.

Chapter 7

Is the Polluter Pays Principle Really Fundamental? An Economic Explanation of the Relative Unimportance of Environmental Liability and Taxes in US Environmental Law

*Jason Scott Johnston**

1 THE POLLUTER PAYS PRINCIPLE AND THE PARADOXICAL INSIGNIFICANCE OF ENVIRONMENTAL TAXES AND LIABILITY

Originating in a 1972 Organisation for Economic Co-operation and Development (OECD) document,[1] the 'polluter pays' principle is now found in the

* Jason Scott Johnston, Henry L. and Grace Doherty Charitable Foundation Professor of Law, Nicholas E. Chimicles Research Professor in Business Law and Regulation, University of Virginia School of Law, 580 Massie Road, Charlottesville, VA 22903, (434) 243-8552, E-mail: jsj8q@virginia.edu.

1. The principle apparently first appeared in 1972 in the *OECD Guiding Principles Concerning International Economic Aspects of Environmental Policies*, which said that the principle meant that polluters 'should bear the expenses of carrying out . . . measures decided by public authorities to ensure that the environment is in an acceptable state. In other words, the cost of these

Michael G. Faure, Han Lixin & Shan Hongjun, *Maritime Pollution Liability and Policy: China, Europe and the US*, 111–119.

environmental laws of many countries[2] and also in various international environmental treaties.[3] While there are various statements of the principle, they all rest upon the idea that environmental regulatory instruments should force polluters to pay for the environmental harm that they cause, thus internalizing what economists would call the social cost of pollution. In its strict, or strong form, the polluter pays principle would dictate environmental laws that would hold polluters legally responsible not only for the cost of pollution abatement or prevention measures, but also for all of the environmental damages caused by their pollution, including the cost of repairing or restoring the natural environment if possible.[4] The most straightforward regulatory instruments for implementing this strong form of the polluter pays principle are taxes and liability rules. However, although many countries formally include the polluter pays principle as a guiding principle for environmental law, most national environmental laws do not rely primarily on taxes or environmental liability to implement the polluter pays principle. Instead, in many countries, the predominant type of environmental regulation is a technology-based standard that tells polluters in a particular industry that they must reduce their emissions to a certain level, where that level can typically be attained by installing certain known types of pollution abatement equipment at their plants or facilities.

The polluter pays principle is supported by very fundamental economic reasoning: When property rights in environmental assets are not well defined, or the transaction costs of bargaining over the use of those environmental assets are high, the market alone will not cause polluters to internalize the environmental cost of their pollution. Under such circumstances, a seemingly direct way to accomplish cost internalization – thus creating incentives for polluters to reduce pollution to lower, socially optimal levels – is through a system of taxes or *ex post* liability for environmental harm. When the emission tax or liability payment is set equal to the actual environmental harm, then the polluter's benefit from pollution reduction – lower taxes or lessened liability – is precisely the same as society's benefit – lower environmental harm.

Given this powerful economic logic, the fact that environmental liability is not the rule but the exception in the laws of many developed countries stands as a paradox requiring explanation.[5] In this brief essay, I present an economic

---

measures should be reflected in the costs of goods and services which cause pollution in production and/or consumption.'

2. For example, the laws of all of the OECD countries include the polluter pays principle as a 'guiding principle' of environmental policy. See F. Biermann et al., *The Polluter Pays Principle under World Trade Organization (WTO) Law: The Case of National Energy Policy Instruments*, Environmental Research Report 201 19 107, German Federal Ministry of the Environment, Nature Conservation and Nuclear Safety 5 (2003).
3. The polluter pays principle now appears as Art. 174 of the EU Treaty (1997) and is acknowledged as a 'general principle of international environmental law by the International Convention on Oil Pollution Preparedness, Response and Co-operation. In 1992, the United Nations Convention on Environment and Development included the polluter pays principle as its Principle 16.
4. C.S. Pearson, 'Testing the System: General Agreement on Trade and Tariffs (GATT) + PPP =?', *Cornell International Law Journal* 27 (1994): 553, 556–559.
5. It should be noted that in some developed countries, such as the Netherlands, effluent fees have played a more significant role. See, e.g., J.T.A. Bressers & K.R.D. Lulofs, 'Industrial Water

explanation for the apparently paradoxical insignificance of environmental liability as an instrument of environmental regulation. This key to my explanation is the great divide between the theoretical ideal of environmental liability and the reality of environmental liability. While this reality varies greatly across legal systems – so that a full explanation of the relative insignificance of environmental liability as a regulatory tool really requires more sustained comparative legal analysis – for brevity, I use the federal US environmental laws as an example to illustrate the main points.

2 ENVIRONMENTAL TAXES AND LIABILITY IN THE UNITED STATES

The United States is one of the few developed countries not to have codified the polluter pays principle. In the United States, neither environmental taxes nor environmental liability play a significant role in federal environmental regulation. Emission taxes in particular are very rare. Perhaps the most important emission tax in the United States is also one of the least known. This effluent tax consists of the fees that industrial dischargers of polluted wastewater must pay to municipal water treatment systems into whose system they discharge. While a part of these fees are fixed, another part is paid based on the strength and volume of waste being discharged, and recent work[6] has shown that at least in some municipalities in the United States, these fees (so-called waste surcharge fees) are high enough so that they approximate an effluent tax.

Environmental liability is by no means a dominant instrument in US federal environmental regulation, but there are at least two statutory schemes where it is very important. Perhaps most well-known is Superfund or, more technically, the Comprehensive Environmental Response, Compensation and Liability Act of 1980 (CERCLA). This statute assigns strict, joint and several liability for the costs of cleaning up sites contaminated by hazardous waste to a broad class of potentially responsible parties including current and past owners and operators of facilities where waste was disposed, landowners of polluted sites, generators of hazardous waste, and those who have either arranged for the disposal of or disposed of the hazardous waste. Liability under CERCLA is retroactive, in that private parties have CERCLA liability for acts that occurred before the law was passed in 1980. Finally, and importantly, CERCLA also includes a provision empowering the Environmental Protection Agency (EPA) to issue unilateral administrative orders compelling private parties to undertake cleanups or else face the risk of potential punitive damage liability.

Pollution in the Netherlands: A Fee-Based Approach, in *Choosing Environmental Policy: Comparing Instruments and Outcomes in the United States and Europe*, ed. W. Harrington et al. (Washington, D.C.: Resources for the Future, 2004).

6. See W. Harrington & P. Nelson, *Public Treatment of Private Waste: Industrial Use of Municipal Wastewater Treatment 31–37* (Washington, D.C.: Resources for the Future, 2006).

The other US federal environmental regulatory regime that relies significantly on liability as an instrument is the Oil Pollution Act (OPA) of 1990.[7] The OPA imposes liability for removal costs and damages resulting from an incident in which oil is discharged into navigable waters or adjoining shorelines or the exclusive economic zone. Under the OPA, the vessel owner is required to initiate the response to a spill and to provide funding. The law limits liability at USD 1500 per gross tonne (about USD 240 million for the largest tankers) but removes all limits if the incident was caused by gross negligence, wilful misconduct or violations of Federal safety, construction or operating regulations.

Importantly, under both CERCLA and the OPA, polluters face potential liability not just for the cost of cleanup or remediation, but also for harm to natural resources.[8] These statutes contain very similar language authorizing claims for 'damages for injury to, destruction of, or loss of natural resources, including the reasonable costs of assessing such injury, destruction, or loss' resulting from the release of hazardous substances. Federal, state and tribal governments may, in certain circumstances, act on behalf of the public as 'trustees' for injured natural resources and seek to recover the cost of restoring, replacing or acquiring the equivalent of the damaged resource. Under regulations promulgated by the National Oceanic and Atmospheric Administration of the Department of the Interior, natural resource damages can include not only any decline in the market value of natural resources and the lost use value of those resources, but also a decline in the 'existence value' of natural resources.[9] Existence value, moreover, can be measured by using the contingent valuation technique, a methodology that involves surveys which ask people how much they have been willing to pay to avoid the harm to the resource.[10]

Natural resource damages are increasingly sought by American states in CERCLA-type hazardous waste site cleanup lawsuits. In 2006, for example, Occidental Petroleum agreed to pay New York state USD 12 million in natural resource damages for the lost benefits due to restrictions on recreational fish consumption due to high levels of contaminants in fish. The New Jersey Department of Environmental Protection (NJDEP) is seeking to collect up to USD 950 million in natural resource damages from sixty-six corporations, and filed over 100 natural resource damage lawsuits during the summer of 2007. Since 2002, the state of New Jersey has recovered about USD 29 million from the settlement of natural resource damage claims.

---

7. 33 USC §§ 2701–2761 (2005).
8. There is also potential liability for natural resource damages under the Clean Water Act. CERCLA limits natural resource damages to USD 50 million 'for each release of a hazardous substance or incident involving release of a hazardous substance.' 42 USC § 9607(c)(1)(D).
9. The regulations appear at 43 CFR 11.
10. The contingent valuation method and the inclusion of existence value were approved by an expert panel convened by National Oceanic and Atmospheric Administration (NOAA), whose report is available at US Dept of Commerce, National Oceanic and Atmospheric Admin., 'Natural Resource Damage Assessments under the Oil Pollution Act of 1990,' 58 Fed. Reg. 4601–4614 (1993).

3 LIMITS TO IMPLEMENTING THE POLLUTER PAYS PRINCIPLE WITH EFFLUENT TAXES AND *EX POST* LIABILITY: ECONOMIC LESSONS FROM THE AMERICAN EXPERIENCE

3.1 Environmental Taxes

From an economic point of view, the infrequency of true environmental taxes in the United States is not surprising. When polluters comply with technology-based command and control regulatory standards, they incur the cost of pollution abatement, but do not have to pay for the remaining harm from pollution that still occurs. When polluters face an effluent or pollution tax, even if the tax incentivizes them to choose precisely the same level of pollution abatement as they would under the command and control regime, they face an additional cost, as they must pay the tax even if they lower their pollution optimally. Hence if a pollution tax is set at a high enough level to replicate the pollution reduction that command and control regulations would achieve, then polluters always have higher costs under the pollution tax than under a command and control regime of technology-based pollution standards.

On this reasoning, we would expect that polluters would always prefer command and control environmental standards to an environmental tax of equivalent stringency. Polluters would go along with environmental taxes only if they are very low (if low enough, environmental taxes actually generate lower costs for polluters than would a relatively stringent command and control regime). One might expect that environmentalists would prefer taxes, because taxes – as the polluter pays principle dictates – not only encourage polluters to abate their pollution but also force polluters to pay for their pollution. However, some environmentalists seem to believe that environmental taxes imply that as long as polluters pay a price for their pollution – the tax – then they have a right (entitlement) to pollute. Inasmuch as environmentalists want a legal regime that eliminates any notion of an entitlement to pollute, and that insists upon eliminating pollution if possible, they might well prefer a regulatory regime with feasibility-based pollution standards rather than a regime of environmental taxes.

Hence, somewhat surprisingly, both polluters and environmentalists might prefer a regime of command and control regulatory standards to a regime of environmental taxes. Such a theoretical possibility is certainly consistent with the American environmental regulatory system, where environmental taxes are quite rare. The few environmental taxes in the American system – such as (arguably) the waste surcharge fees paid by industries that discharge their wastewaters into municipal treatment systems – are consistent with this analysis. Such surcharges are generally quite low, and the existing evidence seems to indicate that there would be significant opposition on the part of industrial dischargers to increasing the level of such taxes and making them more widely used.

3.2 *Ex Post* Environmental Liability

Economics teaches that the effectiveness of *ex post* liability as a tool of environmental regulation depends upon primarily upon: (1) the ability to identify and trace causation and thus hold polluters liability for the harm that they have caused; and, (2) the ability to accurately measure and then assess natural resource damages. Defects in either establishing causation or assessing damages can cause an environmental liability regime to either undeter or overdeter. If causation cannot be determined, so that polluters often escape liability, and/or if damages are severely underestimated, then *ex post* liability will cause polluters to internalize only a small portion of the actual environmental harm that they cause, so that liability leads to too little pollution reduction. If, on the other hand, polluters are held liable even for harm that they did not cause, or held liable for damages that are far greater than the harm they caused, then an environmental liability regime will cause polluters to spend too much to reduce pollution. Even worse, such as an overly expansive liability regime will cause rational actors to simply stay away from activities or markets that might trigger potential environmental liability.

Certainly problems in establishing causation and measuring damages are one of the reasons that environmentalists have traditionally strongly preferred federal regulation over common law liability as a way to reduce pollution. Especially when pollution is widespread – as it had become in the late 1960's in the US – it is difficult to say that any particular polluter is actually causing harm, since reducing pollution at one source has little impact when there are many polluting sources. The difficulty of establishing causation may be an inherent limitation of environmental liability as a way to cleanup a heavily polluted environment. Under such conditions, a regulatory regime that tells polluters to do the best that is technologically possible (and/or economically feasible) may be the only way to succeed in large scale pollution reduction. Thus given that the goal of modern American environmental legislation was to get significant reductions in a pollution in an environment that had become heavily polluted by many polluters as a consequence of post World War II industrial expansion, it is not surprising that environmental liability was not viewed as a very important instrument.

The two American environmental statutes where liability is an important instrument – CERCLA and the OPA – do not have pollution reduction as their direct goal. Rather, the direct goal of these states is to legally compel private parties to cleanup the harm caused by their pollution. Indirectly, of course, these statutes do create incentives for pollution reduction: since polluters face potential liability for an oil spill or hazardous waste site contamination, they have an incentive to spend money to reduce the chance of an oil spill or hazardous waste contamination. While rigorous evidence is difficult to find, it is generally believed that CERCLA has created an incentive to manage hazardous wastes more carefully.[11]Similarly, the dramatic decrease in oil tanker spills since the

11. See K.N. Probst & P.R. Portney, *Assigning Liability for Superfund Cleanups: An Analysis of Policy Options 25–27* (Washington, D.C.: Resources for the Future, 1992).

passage of the OPA is often attributed to the deterrent effect of potential liability under that law.[12]

However, unlike common law liability, which is often too weak in deterring pollution, the liability regime implemented by CERCLA in particular illustrates the risk of overdeterrence from environmental liability. By eliminating the causation requirement, and imposing retroactive, strict and joint and several liability, CERCLA essentially makes a polluter's ability to pay – rather than its actual responsibility in causing environmental harm – the primary determinant of liability. There is evidence that the magnitude of private liability under CERCLA is determined not primarily by the actual environmental harm caused, but rather by a highly politicized regulatory determination of how much a particular party is able to spend on site cleanup.[13] Unsurprisingly, the average cost of a CERCLA cleanup is huge, recently estimated at between USD 20 and USD 40 million per site.[14] The amount being spent on Superfund cleanup sites far exceeds what would be recommended by a reasonable balancing of the costs and benefits of site cleanup: in a relatively recent study, Kip Viscusi and Jay Hamilton[15] have found that EPA cleanups of Superfund sites cost an average of almost USD 12 billion for every cancer case prevented, with 99.5% of the cancer cases prevented by the first 5% of agency cleanup expenditure.

The effect of the risk of massive liability under CERCLA for hazardous waste site cleanup has had a profound impact on land markets, for a time virtually completely eliminating the market in old, industrial properties with a risk of hazardous waste contamination.[16] This is precisely as economics would predict, as the risk of highly uncertain and potentially unlimited liability for having anything to do with a contaminated property caused rational firms to simply stay away from such properties. Indeed, economic incentives for the purchase and reuse of old, potentially contaminated industrial sites only returned in earnest when many industrial

12. For such a view, see the Oil Companies International Marine Forum, the Oil Pollution Act of 1990, Why has it Been So Successful at Reducing Spills? (19 Feb. 2003) available at <www.ocimf.com/view_document.cfm?id=383>, which notes that since the OPA became law, 'the volume of oil spilled from tankers into US waters has fallen from an average of 70,000 barrels per year to an average of 4,000 barrels per year – a decrease of 95%.'
13. The vast majority – 7 out of 10 – cleanups are conducted by private parties, not the EPA. There is evidence that EPA selects cleanup options based in part on the identity and bargaining power of the private responsible parties. H. Sigman, 'Liability Funding and Superfund Clean-Up Remedies', *J. Envl. Econ. & Manage.* 35 (1998): 205.
14. See, e.g., International Organization for Standardization (ISO), Superfund and the Insurance Issues Surrounding Abandoned Hazardous Waste Sites, available at www.iso.com/index.php?option=com_content&task=view&id=1016&Itemid=211.
15. W.K. Viscusi & J.T. Hamilton, 'Are Risk Regulators Rational? Evidence from Hazardous Waste Cleanup Decisions', *American Economic Review* 89, no. 4 (1999): 1010–1027.
16. P.J. Patchin, 'Contaminated Properties and the Sales Comparison Approach', *Appraisal J.* 62 (1994): 402, observed that as of 1994, contaminated properties had only recently become marketable at all, after years when there simply was no market for such properties.

states passed legislation providing guarantees of limited liability for those who purchased and redeveloped such sites.[17]

The addition of natural resource damage liability has exacerbated the risk of overdeterrence under CERLCA even further: private parties now face potential natural resource damage liability for groundwater contamination at sites where they have already spent huge amounts of money on site cleanup and remediation. While the same risk of essentially double liability does not exist in quite the same way under the OPA, the measurement and assessment of natural resource damages under both statutes is not only uncertain, but also very likely to exceed actual environmental harm. This is because federal regulations now allow natural resource damages to include the estimated harm to the 'existence value' of the natural resource, and they allow these estimates to be based on contingent valuation surveys. As a number of economists have recognized, contingent valuation surveys are very unlikely to actually give the true economic price for harm to natural resources.[18] Rather, because contingent valuation surveys elicit what people say they would hypothetically pay, rather than what people actually have to pay, they are very likely to generate inflated estimates for the harm to existence values.

4 CONCLUSION

The purest way to implement the polluter pays principle is through either environmental taxes or environmental liability. While some developed countries do have a fair degree of reliance on environmental liability (e.g., the Netherlands), environmental taxes and liability are generally of secondary importance as environmental regulatory instruments. Economic theory suggests that environmental taxes will face big political and economic obstacles. Economics also explains why traditional, common law forms of *ex post* liability for environmental harm were ineffective instruments for effecting dramatic reductions in pollution. Two notable US federal environmental regulatory statutes – the Oil Pollution Act, and Superfund, or CERCLA – adopt liability for environmental cleanup and for natural resource damages. These statutes attempt to overcome the weakness of common law liability. But natural resource damages are difficult to calculate and under current methodologies, harm to so-called 'existence value' is very likely to be overestimated. In the case of CERCLA (Superfund), by dispensing with the traditional common law causation requirement and imposing strict, retroactive, joint and several liability for massive potential damages that bear no relationship to the

17. See K. Wernstedt et al., 'Incentives for Private Residential Brownfields Development in US Urban Areas' *J. Envl. Plan. & Manage.* 49 (2006): 101, who reports that developers who were surveyed said that protection from liability was a crucial factor in deciding to redevelop Brownfields sites.
18. P.A. Diamond & J.A. Hausman, 'Contingent Valuation: Is Some Number better than No Number?', *J. Econ. Perspec.* 8 (1994): 45.

actual costs and benefits of site cleanup, the statute has deterred the purchase and cleanup of contaminated sites (Brownfields), precisely the opposite result desired. Brownfields sites are now purchased, redeveloped and cleaned up in the United States, but only because CERCLA cleanup costs have become more predictable (indeed, more like regulation than liability), and because states have passed Brownfields laws that limit the liability of private parties who purchase and redevelop such sites. Thus, the lesson from US experience with environmental liability under CERCLA is that while liability may indeed generate some strong incentives to avoid liability, on balance *ex post* environmental liability may be a very expensive and often counterproductive instrument for environmental protection. Thus the polluter pays principle is likely to continue to be of quite limited practical significance.

References

Biermann, F., et al. *The Polluter Pays Principle under World Trade Organization (WTO) Law: The Case of National Energy Policy Instruments*. Environmental Research Report 201 19 107, German Federal Ministry of the Environment, Nature Conservation and Nuclear Safety 5 (2003).

Bressers, J.T.A. & K.R.D. Lulofs. 'Industrial Water Pollution in the Netherlands: A Fee-Based Approach. In *Choosing Environmental Policy: Comparing Instruments and Outcomes in the United States and Europe*, edited by W. Harrington, et al. Washington, D.C.: Resources for the Future, 2004.

Diamond, P.A. & J.A. Hausman. 'Contingent Valuation: Is Some Number Better Than No Number?'. *J. Econ. Perspec.* 8 (1994): 45–64.

Harrington, W. & P. Nelson. *Public Treatment of Private Waste: Industrial Use of Municipal Wastewater Treatment 31–37.* Washington, D.C.: Resources for the Future, 2006.

Patchin, P.J. 'Contaminated Properties and the Sales Comparison Approach'. *Appraisal J.* 62 (1994): 402.

Pearson, C.S. 'Testing the System: General Agreement on Trade and Tariffs (GATT) + PPP = ?', *Cornell International Law Journal* 27 (1994): 553, 556–559.

Probst, K.N. & P.R. Portney. *Assigning Liability for Superfund Cleanups: An Analysis of Policy Options 25–27.* Washington, D.C.: Resources for the Future, 1992.

Sigman, H. 'Liability Funding and Superfund Clean-Up Remedies'. *Journal of Envl. Econ. & Manage.* 35 (1998): 205–224.

Viscusi, W.K. & J.T. Hamilton. 'Are Risk Regulators Rational? Evidence from Hazardous Waste Cleanup Decisions'. *American Economic Review* 89, no. 4 (1999): 1010–1027.

Wernstedt, K., et al. 'Incentives for Private Residential Brownfields Development in US Urban Areas'. *J. Envl. Plan. & Manage.* 49 (2006): 101–119.

Part II

The Bunker Convention

Chapter 8

The Era after the Bunker Convention: Is the Gap in China's Regime for Compensating Victims of Vessel-Source Oil Spills Filled?

*Shan Hongjun**

1 PREFACE

The International Convention on Civil Liability for Bunker Oil Pollution Damage, 2001 ('Bunker Convention') came into force as of 21 November 2008. From 9 March 2009, the Bunker Convention became effective in China.[1]

The Bunker Convention aims to fill the one remaining significant gap in the international regime for compensating victims of vessel-source oil spills. Specifically, because the existing conventions covering oil spills do not include bunker oil spills from vessels other than tankers, the Bunker Convention has established a liability and compensation regime for spills of oil carried as fuel in ships' bunkers.

As noted, the Bunker Convention came into force in China as of 9 March 2009. However, was the gap in China's regime for compensating victims of vessel-source oil spills filled after China ratified the Bunker Convention? In order to clearly and

* Professor of Dalian Maritime University Law School, Arbitrator of China Maritime Arbitration Commission, Maritime Lawyer of Liaoning Sunshine Law Firm, Council Member of China Maritime Law Association. E-mail: hjshan@gmail.com.

1. Note that for the purpose of this chapter, 'China' refers to mainland China only.

Michael G. Faure, Han Lixin & Shan Hongjun, *Maritime Pollution Liability and Policy: China, Europe and the US*, 123–143.

correctly answer this question, this chapter will first focus on studying the general situation of the Chinese regime for compensating victims of vessel-source oil spills before the Bunker Convention to consider the inherent gap. Next the chapter will briefly examine the contents of the Bunker Convention. The chapter will then comment on whether the gap in China's regime for compensating victims of vessel-source oil spills is filled or not. Finally, we will consider some suggestions on ways to improve the Chinese regime for compensating victims of vessel-source oil spills.

2 GENERAL SITUATION OF CHINA'S REGIME BEFORE THE BUNKER CONVENTION

Briefly, the Chinese regime for compensating victims of vessel-source oil spills before the Bunker Convention consists kind of a two-fold path.

China became a party the International Convention on Civil Liability for Oil Pollution Damage, 1969 (hereinafter referred to as 'CLC 69') on 30 January 1980. CLC 69 became effective in China as of 29 April 1980. In order to grant more compensation to the victims of a pollution incident, the International Maritime Organization (IMO) passed the 1992 Protocols of CLC 69 on 27 November 1992 in London. The 1992 protocols came into force on 30 May 1995. After having been modified by the 1992 Protocols, CLC 69 was titled as the International Convention on Civil Liability for Oil Pollution Damage, 1992 (hereinafter referred to as 'CLC 92'). China joined the CLC 92 on 5 January 1999. CLC 92 entered into force in China as of 5 January 2000, and on the same day CLC 69 ceased to be in force in China. Notably, China did not join the International Convention on the Establishment of an International Fund for Compensation for Oil Pollution Damage, 1992 (hereinafter referred to as 'Fund Convention 92')[2] and the 2003 Protocols.[3]

In China, in broad terms, the issue of the application of law in relation to foreigners in civil cases is determined by the provisions of the General Principles of Civil Law (GPOCL), specifically *inter alia* Article 142. Article 142 of the GPOCL provides:

If any international treaty concluded or acceded to by the People's Republic of China contains provisions differing from those in the civil laws of the People's Republic of China, the provisions of the international treaty shall apply, unless the provisions are ones on which the People's Republic of China has announced reservations. International practice may be applied to matters for which neither the law of the People's Republic of China nor any international treaty concluded or acceded to by the People's Republic of China has any provisions.

2. Hong Kong Special Administrative Region joined the Fund-Convention-92 in January 1999. The Convention became effective in Hong Kong as of January 2000.
3. With the entry into force of the 2003 Protocols on 3 Mar. 2005, the Supplementary Fund introduced by the 2003 Protocols came into existence.

Accordingly, where a victim claims against an international polluter for compensating pollution damage caused by the spill of crude oil carried by a tanker within Chinese waters, CLC 92 shall apply to the case.

CLC 92 does not draw a distinction between vessels for domestic service and vessels for international service. The Convention should be applied so long as the vessel and oil involved are that of the 'vessel' and 'oil' defined by the Convention, respectively. As far as the oil pollution damage without a foreign-related factor is concerned, because it does not deal with the performance of an international obligation, whether the Convention is applicable or not shall be decided by the domestic laws of a contracting party. Given that Chinese law has no relevant rules in this respect, whether CLC 92 applies to coastal tankers is in dispute both in practical and research circles.[4]

What is the situation of the application of law with respect to the purely domestic vessel-source oil pollution cases in China? In other words, does CLC 92 apply to a purely domestic vessel-source oil pollution case, other than a CLC 92 pollution case? Historically, there were two conflicting positions. One position advocated that because China was a Member State of CLC 92, CLC 92 was a component part of Chinese domestic law. Therefore, CLC 92 should accordingly apply to purely domestic oil pollution cases.[5] Another position maintained that CLC 92 was only for the purpose of addressing liability and compensation in respect to foreign-related vessel-source oil pollution cases. CLC 92 should therefore not necessarily apply to purely domestic oil pollution cases. The prevailing position was that CLC 92 did not apply to purely domestic oil pollution cases.[6] It was not until December 2005, when the Summary of the Second National Work Conference on Foreign-related Commercial and Maritime Trial (hereinafter referred to as 'the Summary') was announced by the People's Supreme Court, that the disputes ceased. Article 141 of the Summary provides:

The [CLC 92] joined by China shall apply to the foreign-related oil pollution caused by a ship owned by a Contracting State of the Convention, including the oil pollution within the sea waters of China caused by a Chinese-flag ship engaging in the international service. The Convention shall not apply to the oil pollution within the sea waters of China caused by a Chinese-flag ship engaging in the domestic service.

Therefore, in accordance with the provisions of Article 141 of the Summary, CLC 92 shall not apply to the purely domestic oil pollution cases.

What was the governing law applicable to oil-spill damage within Chinese waters from the bunker of a foreign ship other than a CLC 92 ship before the Bunker Convention? In broad terms, China's domestic law governed the oil-spill damage within China waters.

---

4. See T/V 'Yanjiuyou 2' case, the Civil Judgment of the People's Supreme Court (2002), MinSiTiZi No.3.
5. In the China Ship Banker Supply (FuJian) Company case, the trial court applied the CLC in determining the amount of liability limits of the polluter.
6. Prof. Y. Si, *Application of Law in Oil Pollution Damages Compensation concerning Coastwise Vessels*, LAW COMMENTARY (Dalian: Dalian Maritime University Press, 2002).

In view of the above, the vessel-source oil pollution before the Bunker Convention was categorized into two different types, namely, CLC 92 oil pollution and non-CLC 92 vessel-source oil pollution. Accordingly, the applicable laws differ.

For the purpose of this chapter, unless otherwise mentioned, CLC 92 oil pollution means the oil pollution within the territory of China, including the Exclusive Economic Zone, caused by a foreign-flag tanker or a Chinese tanker engaging in international service, the primary governing law of which is CLC 92. Non-CLC 92 vessel-source oil pollution means the oil pollution within the territory of China, including the Exclusive Economic Zone, caused by a foreign-flag non-CLC 92 ship or a Chinese coastwise ship, the governing laws of which is China's domestic law and other applicable conventions different than CLC 92.

2.1 Pollution Damages Governed by CLC 92

Generally, under CLC 92, the owner of a ship is liable for any pollution damage caused by the ship as a result of the incident.[7] For the purpose of identifying the admissible damages, CLC 92 further lays down six tests as follows:

> The pollution damage means loss or damage caused outside the ship by contamination resulting from the escape or discharge of oil from the ship, wherever such escape or discharge may occur, provided that compensation for impairment of the environment other than loss of profit from such impairment shall be limited to costs of reasonable measures of reinstatement actually undertaken or to be taken and the costs of preventive measures and further loss or damage caused by preventive measures.[8]

The incident means any occurrence, or series of occurrences having the same origin, that causes pollution damage or creates a grave and imminent threat of causing such damage.[9]

The preventive measures mean any reasonable measures taken by any person after an incident has occurred to prevent or minimize pollution damage.[10]

The pollution damage should be caused in the territory, including the territorial sea, of China, and in the EEZ of China. The preventive measures can be taken wherever with the purpose of preventing or minimizing the pollution damage.[11]

The ship refers to any sea-going vessel and seaborne craft of any type whatsoever constructed or adapted for the carriage of oil in bulk as cargo, provided that a ship capable of carrying oil and other cargoes shall be regarded as a ship only when it is actually carrying oil in bulk as cargo and during any voyage following

7. Article 3(1) of CLC 92.
8. Article 1(6), *ibid.*
9. Article 1(8), *ibid.*
10. Article 1(7), *ibid.*
11. Article 2, *ibid.*

such carriage unless it is proved that it has no residues of such carriage of oil in bulk aboard.[12]

The oil means any persistent hydrocarbon mineral oil such as crude oil, fuel oil, heavy diesel oil and lubricating oil, whether carried on board a ship as cargo or in the bunkers of such a ship.[13]

Summarily, only when all of the foregoing six tests are satisfied at the same time, shall the damages suffered by an oil pollution victim within the territory of China fall within the type of admissible damages of CLC 92.

In view of the above, given that there are six tests for triggering the application of CLC 92, not all kinds of foreign-related oil pollution fall within the governing scope of CLC 92.

2.2 Pollution Damages Governed by Domestic Laws

The Marine Environment Protection Law of 1982 (hereinafter referred to as 'MEPL 82')[14] is the first special and comprehensive law for China to address the pollution issues in respect of the marine environment. The primary purposes of MEPL 82 are to protect the marine environment and resources, prevent pollution and promote marine scientific research. Briefly, MEPL 82 applies within the China internal waters, the territorial sea and all other sea areas under the jurisdiction of China.

In order to effectively implement MEPL 82, several supplementary regulations were later promulgated by the State Council to govern the various sources of marine pollution. Specifically, they are:

- Regulations Concerning Environmental Protection in Offshore Oil Exploration and Exploitation;[15]
- Regulations of China Concerning the Prevention of Pollution of Sea Areas by Vessels of 1983;[16]
- Regulations Concerning the Dumping of Wastes at Sea;[17]
- Regulations Concerning Prevention of Pollution Damage to the Marine Environment by Coastal Construction Projects;[18]
- Regulations Concerning Prevention of Pollution Damage to the Marine Environment by Land-Based Pollutants;[19]
- Regulations Concerning Prevention of Environmental Pollution by Ship-Breaking.[20]

12. Article 1(1), *ibid.*
13. Article 1(5), *ibid.*
14. Promulgated on 23 Aug. 1982, amended on 25 Dec. 1999 by National People's Congress.
15. Effective as of 29 Dec. 1983.
16. Effective as of 29 Dec. 1983.
17. Effective as of 6 Mar. 1985.
18. Effective as of 1 Aug. 1990.
19. Effective as of 1 Aug. 1990.
20. Effective as of 18 May 1998.

Of particular importance is the Regulations of China Concerning the Prevention of Sea Areas by Vessels of 1983 ('Regulations 83'). Regulation 83 equally applies to China-flag and foreign-flag vessels once the ships as well as shipowners and other relevant individuals are within the sea areas or seaports under the jurisdiction of China. Regulation 83 mandates the compliance with MEPL 82 and other regulations relating to the discharge of oils, oily mixtures, wastes and other poisonous or harmful substances.

With the adoption of the Law on the Exclusive Economic Zone and the Continental Shelf in 1998, China clarified its jurisdiction over sea areas in the form of law. In view of Article 10 of this law, the competent authorities of China have the right to take the necessary measures for the purpose of prevention, reduction and control of marine pollution and of protection and preservation of the marine environment in the EEZ and the continental shelf.[21]

Along with the fast-growing economy of China, especially the booming of the economic activities in sea and coastal areas, the marine environment faces great pressure. Accordingly, MEPL 82 became out-dated and does not accommodate the development of new economic activities. For the purpose of updating the marine environment protection law, the new revision of MEPL 82 was enacted on 25 December 1999 (MEPL 99) and entered into effect on 1 April 2000.

MEPL 99 comprises 10 chapters with 98 articles. Chapter 1 contains general provisions. Chapter 2 deals with marine environmental supervision and management. Chapter 3 covers marine ecological protection. Chapter 4 provides for prevention of pollution damage to the marine environment by land-based pollutions. Chapter 5 covers prevention of pollution damage to the marine environment by coastal construction projects. Chapter 6 sets rules to the prevention of pollution damage to the marine environment by marine construction projects. Chapter 7 governs the prevention of pollution damage to the marine environment by dumping of wastes. Chapter 8 deals with the prevention of pollution damage to the marine environment by vessels and relevant operating activities. Chapter 9 provides for the legal liabilities of wrong-doers. Chapter 10 contains supplementary provisions. Among these 10 chapters, Chapter 8 focuses on dealing with vessel-source pollution.

Compared with MEPL 82, MEPL 99 imposes higher standards and thus strengthens the protection of the marine environment from vessel-source pollution by applying to all kinds of ships and requiring the ships to be equipped with anti-pollution equipment and capacities. Also, MEPL 99 lists the ship-operating activities that cannot be carried out without express approval by the competent authorities. More important, MEPL 99 mandates to promulgate the detailed policies and rules concerning oil pollution compulsory insurance and the oil pollution liability fund by taking reference to international customs, and so forth, and delegates rule-making power to the competent authorities.

21. See Art. 10 of the Law on the Exclusive Economic Zone and the Continental Shelf.

China declared by MEPL 99 that it should establish a civil liability and compensation system for oil pollution damages from ships, and also empowered the State Council to work out specific rules for oil pollution damage compulsory insurance and an oil pollution compensation fund system.[22] Based on this declaration, the competent authorities have finished the draft of the Regulations on Levy and Usage Management of Vessel-Source Oil Pollution Compensation Fund ('China Fund Regulations') and Regulations of China on the Prevention and Control of Vessel-Induced Sea Pollution ('Vessel Pollution Prevention Regulations').[23] On 9 September 2009, the State Council declared its approval of the draft Vessel Pollution Prevention Regulations into law by its Order No. 561. Vessel Pollution Prevention Regulations will enter into force as of 1 March 2010. In accordance with China Fund Regulations, China will establish a China Oil Pollution Compensation Fund ('COPC Fund') by levying a tax on the import of oil. Compared with Regulations 83, Vessel Pollution Prevention Regulations directly sets out the rule of liability and compensation for the vessel-source oil pollution damage. Given that the China Fund Regulations are now pending for approval by the State Council and the Vessel Pollution Prevention Regulations are not in force, to date, China has no fund scheme available to indemnify a victim of non-CLC 92 oil pollution.

The MEPL 99 offers only a very general liability and compensation rule for the purpose of an oil pollution damage claim, which is:

> Whoever causes pollution damage to the marine environment shall remove the pollution and compensate the losses; in case of pollution damage to the marine environment resulting entirely from the intentional act or fault of a third party, that third party shall remove the pollution and be liable for the compensation. For damages to marine ecosystems, marine fishery resources and marine protected areas which cause heavy losses to the State, the department invested with power by the provisions of this law to conduct marine environment supervision and administration shall, on behalf of the State, put forward compensation demand to those held responsible for the damages.[24]

Furthermore, Article 150 of the Summary provides:

> The admissible damages for oil pollution compensation include:
>
> – Property damages of any citizen, legal person and other organization caused by vessel-source oil pollution.
> – Removal costs for preventing or mitigating pollution. The calculation of the removal costs shall reasonably base on the scope and severity of pollution,

22. Article 66 of MEPL provides: 'The Nation will make perfect and carry out a civil compensation regime for oil pollution damages from ships. The oil pollution liability insurance and compensation fund system shall be established and shall be based on the ship-and-cargo-in-common-risk rule. The specific methods of carrying out the oil pollution damage insurance and the oil pollution compensation fund regime shall be laid down by the State Council.'
23. Author's translation from the original Chinese text.
24. Article 90 of MEPL 99.

the amount of the oil spilled, charges of staff and equipment and other relevant evidences.
- Fishery resource damage and marine resource damages caused by vessel-source oil pollution, which is limited to costs of reasonable measures of reinstatement actually undertaken or to be undertaken.

Additionally, as far as the removal costs are concerned, in accordance with Article 151 of the Summary,[25] the claims for removal costs and the claims for other pollution damages shall be compensated proportionately in accordance with the amount fixed by the Court.

Notably, for years the Maritime Courts of China have used various domestic laws or regulations in adjudicating relatively similar oil pollution cases.[26]

Moreover, Article 2 of the Law of China on Prevention and Control of Water Pollution (Water Pollution Law) as amended on 28 February 2008[27] holds:

> This law applies to the prevention and control of pollution of rivers, lakes, canals, irrigation channels, reservoirs and other surface waters and underground waters within the territory of the People's Republic of China. The prevention and control of marine pollution shall be governed by the Law of the People's Republic of China on Marine Environment Protection.

It is the Water Pollution Law and hence not the MEPL that governs inland-water pollution.

The Water Pollution Law only offers the very general liability and compensation rule for the purpose of an oil pollution damage claim, which is:

> The victim who suffered from water pollution shall have the right to require the polluter to eliminate the infringement and claim for compensation. Unless the laws have other provision, if the water pollution damage is caused by force majeure, the polluter may be exempted from liability. If the water pollution damage is caused intentionally by the victim, the polluter may be exempted from liability. If the water pollution damage is caused by the gross negligence of the victim, the liability of the polluter may correspondingly be reduced. If the water pollution damage is caused by a third party, the polluter has, after having paid the compensation, a right to recourse against the third party.[28]

---

25. The condition triggering the application of this Article is that the total amount of the claims is more than the amount of the limitation of liability of the vessel involved, namely, the claims are subject to limitation of liability and consequently all the claims can not be fully compensated.
26. The binding law of the two oil pollution cases named T/V 'Jia Cui' and T/V 'Ma Ya' was the Marine Environmental Protection law, The binding rule of T/V 'Yan Jiu You No. 2' and T/V 'Yadian Horizon' was the General Principles of Civil Law, The binding rule of T/V 'Eastern Ambassador' oil pollution damage compensation case was the Maritime Code; The binding rule of the other four oil pollution damage compensation cases named T/V 'Nan Yang', T/V 'Bi Ya Wang', T/V 'Na Yang Feng Shou' and T/V 'Asia Flying Goose' was the International Practice.
27. Effective as of 1 Jun. 2008.
28. Article 85 of WPPCL.

2.3 The Gap in the Chinese Regime before the Bunker Convention

With the above discussion in mind, it is clear that the gap in China regime for compensating victims of vessel-source oil spills before the Bunker Convention is shaped by the following aspects:

2.3.1 CLC 92 Oil Pollution Situation

When China joined CLC 92, it did not join the Fund Convention 92. Similarly, to date, China has not joined the 2003 Protocols. Therefore, in the situation in which a CLC 92 oil pollution damage occurs in China, neither the International Oil Pollution Compensation (IOPC) Fund nor the Supplementary Fund are available to compensate for the losses or damages suffered by the victims.

2.3.2 Non-CLC 92 Oil Pollution Situation

2.3.2.1 No Domestic Compensation Fund

In the situation in which a non-CLC 92 oil pollution damage occurs, there is no domestic fund available to compensate the victims because the Chinese Fund Regulations have not been laid down into law and the COPC Fund is not established.

2.3.2.2 No Compulsory Insurance Rule

Since the Vessel Pollution Prevention Regulations are not in force, the non-CLC 92 oil pollution compensation regime does not have a compulsory insurance rule. Therefore, where the liable person is bankrupt or out of business in a non-CLC 92 oil pollution case, there is no compulsory insurance available to guarantee the compensation for the victims.

2.3.2.3 No Freestanding Liability Limitation Rule

Article 142 of the Summary provides:

> China's Maritime Code, China's Marine Environmental Protection Law and the relevant regulations shall, for the purpose of determining the liability of an involved person, apply to the vessel-source oil pollution compensation disputes not governed by the International Convention on Civil Liability for Oil Pollution Damage, 1992. The responsible party of the oil pollution has the right to enjoy the limitation of liability for maritime claims in accordance with Chapter 11 of China's Maritime Code.

If an oil pollution incident is caused by an incident involving a coastwise ship or coastwise ships, such as ship collision, stranding and so forth, the shipowner or the demise charterer of the ship involved,[29] if being identified as the liable person, may have the right to invoke the liability limitation in accordance with Articles 204, 207 and 208[30] of the China Maritime Code (CMC).[31] Obviously, the non-CLC 92 oil pollution compensation regime does not have the freestanding liability limitation rule.

2.3.2.4 Unlimited Liability for Oil Pollution Caused by Seagoing Ships of Less Than 20 Tonnes Gross Tonnage

Except for the ships governed by CLC 92, given the fact that the ships as referred to in CMC means seagoing ships and other mobile units, but does not include ships or craft to be used for military or public service purpose, nor small ships of less than 20 tonnes gross tonnage,[32] and the fact that there is no liability limitation rule available to the ships of less than 20 tonnes gross tonnage, once an oil pollution incident has been caused by seagoing ships of less than 20 tonnes gross tonnage, the liable person shall assume unlimited liability.

For a better understanding of the limits under both the CMC and the Regulation of Limitation of Liability for Maritime Claims relating to the ships with a gross tonnage not exceeding 300 gross tonnes and those engaging in transport services between the ports of China as well as those for other coastal operation (hereinafter referred to 'Liability Limitation Regulation'),[33] please see Table 8.1.

---

29. In accordance with Art. 4 of the Regulations of the People's Supreme Court on Several Issues of the Case Trial of Collision of Ship, which entered into force as of 23 May 2008, the liability resulted from the collision of ships shall be borne by the shipowner(s). Where the involved ship is in demise charter and the demise charter has been registered in accordance with law, the compensation liability resulted from the collision of ships shall be borne by the demise charterer(s).
30. Article 204 of CMC holds 'Ship owners and salvors may limit their liability in accordance with the provisions of this chapter [Limitation of Liability for Maritime Claims] for claims set out in Article 207 of this Code.'

    Art. 207 of the CMC holds 'Except as provided otherwise in Arts 208 and 209 of this Code, with respect to the following maritime claims, the person liable may limit his liability in accordance with the provisions of this chapter, whatever the basis of liability may be (1) claims in respect of loss of life or personal injury or loss of or damage to property including damage to harbor works, in direct connection with the operation of the ship or with salvage operations, as well as consequential damages resulting therefrom.'

    Art. 209 of the CMC holds 'The provisions of this Chapter shall not be applicable to the following claims: . . . (2) Claims for oil pollution damage under the International Convention on Civil Liability for Oil Pollution Damage to the People's republic of China is a party.'
31. Because CLC 92 does not apply to non-persistent oil such as light diesel oil or petroleum, where an oil pollution incident with foreign factors was caused by light diesel oil discharge from a vessel, the rule of liability limitation of CMC will instead apply to this particular case.
32. See Art. 3 of CMC.
33. Effective as of 1 Jan. 1994.

Table 8.1 Limitation of Liability under Chinese Maritime Law

Types of Ships	*Scope of Gross Tonnages (GT)*	*Calculation of Liability of limitation in Respect to Claims Other Than Loss of Life or Personal Injury (Units of Account or Special Drawing Right)*
CMC ships	300–500	167,000
	501–30,000	167,000 + 167 × (GT of a ship – 500)
	30,001–70,000	167,000 + 167 × (30,000 – 500) + 125 × (GT of a ship – 30,000)
	≥ 70,001	167,000 + 167 × (30,000 – 500) + 125 × (70,000 – 30,000) + (GT of a ship – 70,000)
Liability limitation regulations Ships not engaging in transport services between the ports of China as well as those for other coastal operations	20–21	27,500
	21–300	27,500 + (GT of a ship – 21) × 500
Ships engaging in transport services between the ports of China as well as those for other coastal operations	20–21	27,500 × 50%
	21–300	[27,500 + (GT of a ship – 21) × 500] × 50%
	300–500	167,000 × 50%
	501–30,000	[167,000 + 167 × (GT of a ship – 500)] × 50%

Table 8.1 (cont'd)

Types of Ships	*Scope of Gross Tonnages (GT)*	*Calculation of Liability of limitation in Respect to Claims Other Than Loss of Life or Personal Injury (Units of Account or Special Drawing Right)*
	30,001–70,000	[167,000 + 167 × (30,000 – 500) + 125 × (GT of a ship – 30,000)] × 50%
	≥ 70,001	[167,000 + 167 × (30,000 – 500) + 125 × (70,000–30,000) + (GT of a ship – 70,000)] × 50%

2.3.2.5 *Unlimited Liability for Oil Pollution Caused by Inland-Water Ships*

As noted earlier, in accordance with Article 85 of the Water Pollution Law, the victim of water pollution caused by the spill of oil from an inland-water ship shall have the right to claim compensation from the polluter. Given the fact that inland-water ships do not fall within the categories of CMC ships, in other words, the inland-water ships are not seagoing ships, the liability limitation rules formulated by both the CMC and the Liability Limitation Regulation do not apply to the inland-water ships.

3 THE GAPS IN THE INTERNATIONAL REGIME FILLED BY BUNKER CONVENTION

As mentioned at the beginning of this chapter, the Bunker Convention aims to fill the one remaining significant gap in the international regime for compensating victims of vessel-source oil spills. What is the significant gap in the international regime? Briefly, the gap in the international regime is that there are no international freestanding rules for compensating victims of bunker oil spills.

Compared with the cargo oil pollution incidents, the number of bunker pollution incidents is higher, and the damages thereof are greater.[34] Actually, the issue

34. See <www.moc.gov.cn/zhuzhan/jiaotongxinwen/xinwenredian/200903xinwen/200903/t20090309_564100.html>.

of the bunker pollution was placed on the agenda as early as the formulating CLC 69. On 23 March 2001, the IMO passed the Bunker Convention. Consequently, the gap of non – tanker-source and non – hazardous and noxious substance (HNS) pollution not regulated by CLC 92 or the HNS Convention has been filled.

The Bunker Convention has been formulated with reference to the structure and liability regime of CLC 92 and the HNS Convention, but it has its own specific features.

First, as far as the liable person is concerned, the liable persons under the Bunker Convention include not only the registered ship owner but also the demise charterer, the ship operator and the ship manager. Meanwhile, the victims are entitled to bring up actions directly against the insurer of the shipowner's oil pollution liability or other persons providing financial guarantee.

Second, as far as the limitation of liability is concerned, the Bunker Convention does not formulate an independent liability limitation rule but provides that:

> Nothing in this Convention shall affect the right of the ship owner and the person or persons providing insurance or other financial security to limit liability under any applicable national or international regime, such as the Convention on Limitation of Liability for Maritime Claims, 1976, as amended.[35]

Third, as far as the compulsory insurance and the related certificate are concerned, in accordance with the Convention, the registered owner of a ship having a gross tonnage greater than 1,000 tonnes registered in a State Party shall be required to maintain insurance or other financial security to cover the liability of the registered owner for pollution damage in an amount equal to the limits of liability under the applicable national or international limitation regime, but in all cases, not exceeding an amount calculated in accordance with the Convention on Limitation of Liability for Maritime Claims, 1976, as amended.

With the above special provisions demonstrated by the Bunker Convention, the Convention filled the gap in the international regime for compensating victims of the spills of oil carried by non-CLC 92 ships' bunker.

4 SITUATION OF THE CHINESE REGIME AFTER THE BUNKER CONVENTION

The Bunker Convention came into force in China as of 9 March 2009. Upon becoming a party to the Convention, China declared a reservation on the application of Article 7 to the inland water ships.[36] Accordingly, the Bunker Convention does not apply to ships in China's inland water. So, what is the situation of the Chinese vessel-source oil pollution compensation regime after China became a

35. Article 6 of the Bunker Convention.
36. The reservation is applicable to Macau.

party to the Bunker Convention? In other words, does the gap in China's regime for compensating victims of all kinds of vessel-source oil spills still exist?

4.1 Spills of CLC 92 Oil from CLC 92 Ships

In accordance with Article 4(1) of the Bunker Convention, the Convention shall not apply to pollution damage as defined in CLC, regardless of whether compensation is payable in respect to it under that Convention. Given that China is a party to CLC 92 and the Bunker Convention, and by reference to the provision of Article 4(1) of the Bunker Convention, the Chinese regime for compensating victims of CLC 92 oil spills has not been affected by the Bunker Convention.

4.2 Spills of Bunker Oil from Non-CLC 92 Ships

In accordance with the Circular on Implementing the International Convention on Civil Liability for Bunker Oil Pollution Damage, 2001 issued by Maritime Safety Agency (MSA) of the Ministry of Transport on 23 December 2008, both the foreign-flag ships and the Chinese-flag ships engaging in international navigation whose tonnes gross tonnage are in excess of 1,000 tonnes shall maintain insurance or other financial security as of 9 March 2009. Otherwise, the ships shall not be allowed to come in or out the Chinese ports. These Chinese-flag ships engaged in the coastwise transportation whose gross tonnage is in excess of 1,000 tonnes shall maintain insurance or other financial security as of 1 July 2009. Accordingly, as of 1 July 2009, the Bunker Convention will fill the gap in the Chinese regime for compensating victims of the bunker oil spills caused by non-CLC 92 ships, namely, foreign-flag ships and Chinese-flag ships (not including the China inland-water ships).

Notably, given that a ship as referred to in the Bunker Convention as any seagoing vessel and seaborne craft of any type whatsoever,[37] the Bunker Convention has no restrictions on the tonnage of the qualified ships.

4.3 Spills of Persistent Oil as Cargo from Non-CLC 92 Ships (Coastwise Ships)

Given that the Bunker Convention shall not apply to loss or damage caused outside the ship by contamination resulting from the escape or discharge of oil carried as cargo by a ship, wherever such escape or discharge may occur, provided that compensation for impairment of the environment other than loss of profit from such impairment shall be limited to costs of reasonable measures of reinstatement actually undertaken or to be undertaken, and the costs of preventive measures and further loss or damage caused by preventive measures,

37. Article 1(1) of the Bunker Convention.

the Bunker Convention does not improve the situation of the Chinese regime for compensating the victims of the spill of persistent oil as cargo from non-CLC 92 ships.[38]

4.4 Spills of Non-CLC 92 Oil from CLC 92 Ships

As mentioned earlier, CLC 92 does not apply to spills of non-CLC 92 oil from CLC 92 ships.

In accordance with Article 1(5) of CLC 92, CLC 92 oil means any persistent hydrocarbon mineral oil, such as crude oil, fuel oil, heavy diesel oil and lubricating oil, whether carried on board a ship as cargo or in the bunkers of such a ship. Accordingly, the spill of non-CLC 92 oil from a CLC 92 ship may be the spill of any non-persistent hydrocarbon mineral oil as cargo carried by a CLC 92 ship or the spill of any non-persistent hydrocarbon mineral oil carried by a CLC 92 ship's bunker.

Given that the Bunker Convention applies only to any hydrocarbon mineral oil, including lubricating oil, used for the operation or propulsion of the ship, and any residues of such oil, the Bunker Convention has filled the gap in China's regime for compensating the victims of the spill of any non-persistent hydrocarbon mineral oil carried by a CLC 92 ship's bunker.

4.5 Oil Spills from Ships of Less than 20 Tonnes Gross Tonnage

The CMC does not apply to the ships of less than 20 tonnes gross tonnage. It means that if an oil incident was caused by a ship of less than 20 tonnes gross tonnage, the liable person cannot limit his liability in accordance with the CMC but has to compensate for the pollution damage without limitation.

Given that the ships governed by the Bunker Convention are all seagoing vessels and seaborne crafts, of any type whatsoever, if the bunker oil pollution was caused by a seagoing or seaborne ship of less than 20 tonnes gross tonnage, the Bunker Convention shall apply to the case. However, given the fact that nothing in the Bunker Convention shall affect the right of the ship owner and the person or persons providing insurance or other financial security to limit liability under any applicable national or international regime, such as the Convention on Limitation of Liability for Maritime Claims, 1976, as amended,[39] and given the fact that the applicable national regime of China is the CMC, the Bunker Convention does not

38. For the purpose of this chapter, non-CLC 92 ships means the ships referred to in China Maritime Code that are seagoing ships and other mobile units, but does not include ships or craft to be used for military or public service purposes, nor small ships of less than 20 tonnes gross tonnage.
39. See Art. 6 of the Bunker Convention.

fill the gap in China regime for compensating the victims of the spill of bunker oil from the ships of less than 20 tonnes gross tonnage.

4.6 Oil Spills from Inland-Water Ships (Cargo/Bunker)

Given that China declared reservations on the application of the Bunker Convention to inland-water ships when joining, the Bunker Convention does not fill the gap in China's regime for compensating the victims of the spill of oil from inland-water ships.

For a better understanding of the earlier discussion, please refer to Table 8.2.

Table 8.2 Application of the Bunker Convention in China

Ships	*Pollution Damages*	*Governing Law*	*Independent Liability Limitation Rule*
CLC 92 ships	Pollution damages of CLC 92 oil carried as cargo	CLC 92	Yes. The rule of CLC 92.
	Pollution damages of CLC 92 oil carried as bunker	CLC 92	Yes. The rule of CLC 92.
	Pollution damages of non-CLC 92 oil carried as cargo	China domestic laws	Not independent. The rule of CMC applies simultaneously to both pollution damage and other property damages.
	Pollution damages of non-CLC 92 oil (non-persistent hydrocarbon mineral oil) carried as bunker	The Bunker Convention	Yes. The rule of CMC.

Table 8.2 (cont'd)

Ships	*Pollution Damages*	*Governing Law*	*Independent Liability Limitation Rule*
Non-CLC 92 ships other that of less than 20 tonnes gross tonnage and inland water ships	Pollution damages of oil carried as cargo	China domestic laws	Not independent. The rule of CMC applies simultaneously to both pollution damage and other property damages.
	Pollution damages of any hydrocarbon mineral oil carried as bunker	The Bunker Convention	Yes. The rule of CMC.
Ships of less than 20 tonnes gross tonnage	Pollution damages of oil carried as cargo	China domestic laws	No available rule.
	Pollution damages of any hydrocarbon mineral oil carried as bunker	The Bunker Convention	No available rule.
Inland-water ships	Pollution damages of oil carried as cargo	China domestic laws	No available rule.
	Pollution damages of any hydrocarbon mineral oil carried as bunker	China domestic laws	No available rule.

5 THE GAP TO BE FILLED BY THE VESSEL POLLUTION PREVENTION REGULATIONS AND THE CHINA FUND REGULATIONS

5.1 Liability Limitation Rule for Non-CLC 92 Pollution Damages and Non-Bunker Pollution Damages Will Be Established

Article 52 of the Vessel Pollution Prevention Regulations provides:

> The limitation of liability for vessel-source oil pollution claims shall be in accordance with the Limitation of Liability for Maritime Claims regulated by China Maritime Code. If, however, the oil pollution within the sea waters of China was caused by any vessel-source persistent oil the amount of the liability limitation shall be calculated in accordance with the relevant International Conventions ratified or joined by China. The term 'Persistent Oil' referred to in the preceding paragraph means any persistent hydrocarbon mineral oil.

Evidently, in accordance with Article 52 of the Vessel Pollution Prevention Regulations, even if the ship is a coastwise tanker, as long as the oil spilled is persistent hydrocarbon mineral oil, the amount of the liability limitation shall be calculated in accordance with the relevant International Conventions ratified or joined by China. Accordingly, for pollution caused by any persistent hydrocarbon mineral oil spilled from coastwise ships, calculation of the amount of liability limitation shall be in accordance with CLC 92. For damage caused by any non-persistent oil spilled from coastwise ships, the calculation of the amount of liability limitation shall be in accordance with the CMC.

5.2 Compulsory Insurance Rule for Non-CLC 92 Vessels Except Those of Less Than 1,000 Tonnes Gross Tonnage and Carrying Non-Oil Cargo Will Be Established

The Vessel Pollution Prevention Regulations to Chapter 7 have laid down specific rules in respect to oil pollution liability insurance and other relevant matters. Specifically, in accordance with Article 53 of the Vessel Pollution Prevention Regulations, except for ships of less than 1000 tonnes gross tonnage and carrying non-oil cargo, any ship navigating within the sea waters of China shall establish and maintain oil pollution liability insurance or evidence of other financial security. The amount of oil pollution liability insurance or other financial security shall not be less than the amount fixed in accordance with the CMC or the International Convention ratified or joined by China, namely, CLC 92.

For a better understanding of the provisions of the Vessel Pollution Prevention Regulations, see Table 8-3.

Table 8.3 Effects of the RCPCVSP Amendments

Ships	*Oil*	*Governing Law*	*Compulsory Insurance Rule*	*Liability Limitation Rule*
Non-CLC 92 ships	Non-persistent oil carried as cargo	RCPCVSP Amendments	Yes	Yes. The rule of CMC.
	Non-persistent oil (any hydrocarbon mineral oil) carried as bunker	The Bunker Convention	Yes	Yes. The rule of CMC.
	Persistent oil carried as cargo	RCPCVSP Amendments	Yes	Yes. The rule of CLC 92
	Persistent oil (persistent hydrocarbon mineral oil) carried as bunker	The Bunker Convention	Yes	Yes. The rule of CMC.
CLC 92 ships	Non-CLC 92 oil carried as cargo	RCPCVSP Amendments	Yes	Yes. The rule of CMC.
	Non-CLC 92 oil (any non-persistent hydrocarbon mineral oil) carried as bunker	The Bunker Convention	Yes	Yes. The rule of CMC.

5.3 A COPC Fund Will Be Established

In accordance with the China Fund Regulations, the China Oil Pollution Compensation Fund (COPC Fund) will receive its revenues by levying a tax on the consignee or his agent if he receives, in Chinese waters, oil goods and materials consistently, including crude oil, fuel oil, heavy diesel oil, lubricating oil, or other persistent hydrocarbon mineral oil, transported by sea.[40] The tax rate is China Yuan Renminbi (RMB) 0.40 per tonne.[41] The levy will not apply to non-persistent hydrocarbon mineral oil and the persistent hydrocarbon mineral oil in transit. More importantly, the COPC Fund can only charge one time from the same consignee if the persistent hydrocarbon mineral oil, after he has received in waters of China, should be transported by sea to another place in China.[42]

Notably, in accordance with Article 14 of the China Fund Regulations, the COPC Fund can only be used to compensate or indemnify oil pollution damages exceeding the limits of liability of a shipowner as the responsible party.[43] Specifically, the admissible damage for compensation or indemnity from the COPC Fund shall include:

> damages arising from the oil pollution incident that exceed the limits of liability of a ship owner; damages that the ship owner is exempt from liability; damages that the ship owner is financially incapable of meeting his obligations; and damage caused by an unidentifiable ship.[44]

6 CONCLUSION

In view of the this discussion, it is reasonable to say that the Chinese regime for compensating victims of vessel-source oil spills after the Bunker Convention becomes a three-fold path. Specifically, it consists of the Chinese regime for compensating victims of CLC 92 damages, the Chinese regime for compensating victims of the Bunker Convention damages and the Chinese regime for compensating victims of vessel-source oil spills not governed by CLC 92 and the Bunker Convention.

The implementation of the Bunker Convention in China does improve the current situation of the Chinese regime for compensating victims of CLC 92 damages. To date, both the IOPC Fund and the Supplementary Fund are still not available to compensate for a victim of CLC 92 oil pollution within the territory of China, including the territorial sea and the Exclusive Economic Zone. In order to improve and perfect the China regime for compensating victims of CLC 92

40. See Art. 6 of China Fund-Regulations.
41. See Art. 7, *ibid.*
42. See Art. 10, *ibid.*
43. See Art. 14, *ibid.*
44. See Art. 15, *ibid.*

damages, China should join the Fund Convention 92 and the 2003 Protocols as soon as possible.

The implementation of the Bunker Convention in China has filled the gap in the Chinese regime for compensating victims of the spills of oil from ships' bunkers.

The implementation of the Bunker Convention in China does not fill the gap in the Chinese regime for compensating victims of vessel-source oil spills governed by neither CLC 92 nor the Bunker Convention. Once the China Fund Regulations have been approved and the Vessel Pollution Prevention Regulations have entered into force, given the facts that the freestanding liability rule, the compulsory insurance rule and the Chinese compensation fund will be established accordingly, the situation of the Chinese regime for compensating victims of vessel-source oil spills not governed by CLC 92 and the Bunker Convention will become relatively perfect.

Given that the Chinese regime for compensating victims of oil spills caused by ships of less than 20 tonnes gross tonnage and the inland-water ships is comparatively clumsy, China should formulate special rules for these kinds of oil pollution damages.

Presumably, once the noted gaps have been filled completely, the Chinese regime for compensating victims of vessel-source oil spills will then become a sound regime.

Reference

Si, Y. *Application of Law in Oil Pollution Damages Compensation Concerning Coastwise Vessels*, Law Commentary (Dalian: Dalian Maritime University Press, 2002).

Chapter 9

Discussion on Limitation of Liability and Compulsory Insurance of Compensation for Bunker Oil Pollution Damage from Ships in China

Han Lixin & Wang Dapeng***

1 INTRODUCTION

The International Convention on Civil Liability for Bunker Oil Pollution Damage, 2001 ('Bunker Convention 2001') came into force on 21 November 2008. Until December 2008, there were twenty-five Contracting States to the Convention. The Chinese government approved the accession to the Bunker Convention on 17 November 2008, and from 9 March 2009, the Bunker Convention became effective in China. The Maritime Safety Administration of the Ministry of Transport of the People's Republic of China (PRC) promulgated the Notice on Implementing International Convention on Civil Liability for Bunker Oil Pollution Damage, 2001 (hereinafter referred to the Notification) on 23 December 2008.

* Professor of Law School of Dalian Maritime University, Ph.D. in Law, Ph.D. Supervisor, the Director of Maritime Law Department, and the Director of Institute of International Maritime Law, Arbitrator of CMAC. Tel.: +86 411 8472 8505; e-mail: hlxwhy@gmail.com.

** Ph.D. Candidate in Maritime Law, Law School of Dalian Maritime University, Dalian, 116026, China.

The authors believe that other Contracting States of CLC and Bunker Convention may be facing the same problems discussed here and hope this chapter may have some reference to them.

Michael G. Faure, Han Lixin & Shan Hongjun, *Maritime Pollution Liability and Policy: China, Europe and the US*, 145–158.

Clause 1 of the Notification provides for its Scope of Application. It states that foreign ships over 1,000 gross tonnage can be permitted entry into or exit from the ports of China only if they hold the Certificate of Insurance or Other Financial Security in Respect of Civil Liability for Bunker Oil Pollution Damage (the 'Certificate') issued by the Administration or other authorized organizations of the Contracting States. Chinese-flag ships over 1,000 gross tonnes that engage in international transport service can apply for the Certificate from the Maritime Safety Administration (MSA) (of a relevant Province), directly under MSA of the PRC (the 'Administration') since the Notification has been promulgated. These ships must have held the Certificate before 9 March 2009 that was issued by the Administration. Certificates issued by other Contracting States that were still in effect before 9 March 2009 and the Certificate issued by the Administration should be held once the original Certificate lapses. Ships engaged in Chinese coastal transport service must have held the Certificate before 1 July 2009 that was issued by the Administration.

Ships carrying permanent oil in bulk to which the International Convention on Civil Liability for Oil Pollution Damage, 1992 ('CLC 1992') is applicable are not obliged to have the Certificate.

Clause 2 provides the Scope of Liability of the Notification and states that Chinese-flag ships over 1,000 gross tonnes engaged in international and coastal lines service must have insurance or other financial security. The insurance amount of the coastal transport ships cannot be lower than the ships' civil liability limitation stated in China Maritime Code (CMC). The insurance coverage of ships engaged in international transport service is determined by the managing courses and the laws of Contracting States into which they enter. This will ensure that ships cannot be refused entry into ports or be detained because of the breach of laws of relevant Contacting States. However, the insurance amount cannot be lower than the ship's civil liability limitation as stated in the CMC.

CLC 1992, the Bunker Convention 2001, and the Hazardous and Noxious Substances (HNS) Convention 1996, which is not yet effective, all require compulsory insurance or financial security. It has significant meaning in respect to reducing pollution risks. Pollution from ships might cause huge losses. Compulsory insurance can diversify the risks and guarantee the victims' compensation arising from oil pollution. Ships are disallowed in or out of other countries' ports unless they have the certificate of compulsory insurance or financial security, which is required under the conventions.

Clauses 1 and 2 of the Notification execute Articles 6 and 7 of Bunker Convention 2001. These articles read as follows:

> Article 6 Limitation of liability
>
> Nothing in this Convention shall affect the right of the ship owner and the person or persons providing insurance or other financial security to limit liability under any applicable national or international regime, such as the Convention on Limitation of Liability for Maritime Claims, 1976, as amended.

Article 7 Compulsory insurance or financial security

1. The registered owner of a ship having a gross tonnage greater than 1000 registered in a State Party shall be required to maintain insurance or other financial security, such as the guarantee of a bank or similar financial institution, to cover the liability of the registered owner for pollution damage in an amount equal to the limits of liability under the applicable national or international limitation regime, but in all cases, not exceeding an amount calculated in accordance with the Convention on Limitation of Liability for Maritime Claims, 1976, as amended.

From these two articles, one can see that the amount of compulsory insurance or financial security provided by shipowners is in fact the amount stipulated in domestic laws or in the Convention on Limitation of Liability for Maritime Claims 1976 ('LLMC 1976') or its protocols that shall be applied.

2 SHIPOWNERS' LIMITATION OF LIABILITY STATED IN DOMESTIC LAWS

According to Article 6 of the Bunker Convention 2001, there are no independent rules on limitation of liability for bunker oil pollution compensation. The limitation of liability shall be determined according to the relevant rules of domestic or international laws. The international laws refer to LLMC 1976 and its protocols. China is not a Contracting Party to LLMC 1976 and its protocols. In regard to liability of bunker oil pollution of Chinese-flag ships, except for those regulated by CLC 1992, shipowners should provide compulsory insurance or financial security in accordance with domestic laws. Once a ship is engaged in international transport, and concerned states impose special requirements or conditions, those requirements should be fulfilled accordingly.

2.1 When CLC 1992 Will Apply

This matter concerns the relationships between CLC 1992 and the Bunker Convention 2001.

The system of limitation of liability for oil pollution from ships results from the CLC 1969. Before CLC 1969, the general limitation of liability for maritime claims applied to oil pollution damage. With the development of international transport of oil by sea, the possibility of oil pollution incidents has increased enormously. The damages are also more serious than ever before. The oil pollution incident involving the *Torrey Canyon* of 1967 made the world aware that the system of global liability for maritime claims could not adapt to the changes. Then, CLC 1969 introduced a new system of civil liability for oil pollution damage, which departs from the traditional system of maritime compensation. The limitation of liability for oil pollution damage compensation is the core of the system. The amount of limitation of liability under CLC 1969 is doubled compared to the limit of International Convention relating to the Limitation of the Liability for Owners of Sea-going Ships, 1957 ('Limitation Convention for Ship owners 1957'). Given that there is a special convention relating to oil pollution, the LLMC 1976 stated in paragraph 2 of Article 3 that:

> The rules of this Convention shall not apply to claims for oil pollution damage within the meaning of the International Convention on Civil Liability for Oil Pollution Damage, dated 29 November 1969 or of any amendment or Protocol thereto which is in force.

In order to clarify the difference in limitation of liability between the two conventions, the global limitation shall not apply to the oil pollution damage regulated in CLC system and vice versa.

Whether oil pollution damage falls under the scope of application of CLC depends on the type of ship, type of oil and geographical area.[1] In this chapter, only the ships and oil are discussed.

2.1.1 Ships under the Scope of the Conventions

Article 2.1 of CLC 1992 states that:

> Ship means any sea-going vessel and seaborne craft of any type whatsoever constructed or adapted for the carriage of oil in bulk as cargo, provided that a ship capable of carrying oil and other cargoes shall be regarded as a ship only when it is actually carrying oil in bulk as cargo and during any voyage following such carriage unless it is proved that it has no residues of such carriage of oil in bulk aboard.

According to the rules, in principle, CLC 1992 applies to oil tankers and includes ships capable of carrying oil and other cargoes. However, we should pay attention to what happens when there are residues of bulk cargo oil onboard the ship. The oil pollution damage caused by the residues, oily sewage or bunker oil are all governed by the CLC.

Other ships such as inland waterway ships, lake ships and so on are excluded from the scope of application of CLC 1992. If a ship capable of carrying oil and other cargoes does not actually carry oil in bulk as cargo or has no residues of such

1. Article 2 of CLC 1992 stated that: 'This Convention shall apply exclusively: (a) to pollution damage caused: (i) in the territory, including the territorial sea, of a Contracting State, and (ii) in the exclusive economic zone of a Contracting State, established in accordance with international law, or, if a Contracting State has not established such a zone, in an area beyond and adjacent to the territorial sea of that State determined by that State in accordance with international law and extending not more than 200 nautical miles from the baselines from which the breadth of its territorial sea is measured; (b) to preventive measures, wherever taken, to prevent or minimize such damage.'

carriage of oil in bulk aboard, CLC 1992 shall not apply to the damage caused by its bunker oil.

Article 1.1 of Bunker Convention 2001 states that, 'Ship means any seagoing vessel and seaborne craft of any type whatsoever.' There is a difference in the scope of application of ships in the two conventions. The Bunker Convention 2001 does not apply to ships under CLC, and the Bunker Convention 2001 does not apply to non-seagoing ships or boats, including inland waterway boats, lake boats, offshore drilling platforms, oil storage equipment and so on.

2.1.2 Oil under the Scope of the Conventions

Article 1.5 of the CLC 1992 states that:

> Oil means any persistent hydrocarbon mineral oil such as crude oil, bunker oil, heavy diesel oil and lubricating oil, whether carried on board a ship as cargo or in the bunkers of such a ship.

Accordingly, CLC 1992 applies to the compensation of pollution damage caused by cargoes of persistent hydrocarbon mineral oil and persistent hydrocarbon mineral bunker oil. CLC 1992 does not apply to the compensation of pollution damage caused by non-persistent hydrocarbon mineral oil and bunker oil of ships that are out of the scope of the convention.

The Bunker Convention 2001 applies to all kinds of seagoing ships, seagoing vessels and fuel oil used on them. 'Bunker oil' means any hydrocarbon mineral oil, including lubricating oil, used or intended to be used for the operation or propulsion of ships, and any residues of such oil. Bunker oil stated in Bunker Convention 2001 refers not only to the bunker oil used to supply the power for the main engine and the auxiliary engine but also hydrocarbon mineral oil, including lubricating oil that keep boilers and other machines working. The said hydrocarbon mineral oil after use (and therefore useless oil) and sewage are also considered to be fuel oil.

Because there is overlap in the scope of ships and oils between the two conventions, when ships falling under the scope of CLC 1992 cause pollution damage arising from a spill of bunker oil, the question arises as to which convention shall apply, CLC 1992 or the Bunker Convention 2001. Article 4.1 of the Bunker Convention 2001 states that:

> This Convention shall not apply to pollution damage as defined in the Civil Liability Convention, whether or not compensation is payable in respect of it under that Convention.

Therefore, the CLC 1992 shall apply in priority to the damage caused by bunker oil, governed by CLC 1992, that is, the pollution damage caused by persistent hydrocarbon mineral oil. Otherwise, the Bunker Convention 2001 shall apply, that is, when the damage caused by the bunker oil is beyond the scope stated in the CLC 1992 or caused by non-persistent hydrocarbon mineral oil.

2.2 The Application of Domestic Law: CMC and the Provisions Promulgated by Ministry of Transport of the PRC

As discussed earlier, as a Contracting State, China shall apply domestic laws to the limitation of liability of bunker oil pollution damage governed by the Bunker Convention 2001. However, there are no special provisions concerning compensation of oil pollution damage in CMC. Chapter 11 of CMC – the Limitation of Liability for Maritime Claims noticed the compensation for oil pollution damage, that is, paragraph 2 of Article 208 states that:

> The provisions of this Chapter shall not be applicable to the following claims: Claims for oil pollution damage under the International Convention on Civil Liability for Oil Pollution Damage to which the People's Republic of China is a party.

Obviously, if the claims for compensation for oil pollution damage are within the scope of CLC 1992, which China has acceded to, the responsible parties cannot limit their liability according to Chapter 11 of CMC but the convention. If claims are not in the scope of compensation for bunker oil pollution damage regulated in CLC 1992, parties might still be able to limit their liability according to Chapter 11 of CMC if the claims are within the scope of paragraph 3 of Article 207 of CMC, which states that:

> Claims in respect of other loss resulting from infringement of rights other than contractual rights occurring in direct connection with the operation of the ship or salvage operations.

Consider also paragraph 2 of Article 210 of CMC, which states that:

> The limitation of liability for ships with a gross tonnage not exceeding 300 tons and those engaging in transport services between the ports of the People's Republic of China as well as those for other coastal works shall be worked out by the competent authorities of transport and communications under the State Council and implemented after its being submitted to and approved by the State Council.

According to the Article, the Pre-Ministry of Communication of the PRC promulgated The Provisions Concerning the Limitation of Liability for Maritime Claims for Ships with a Gross Tonnage not Exceeding 300 Tonnes and Those Engaging in Coastal Transport Services as well as Those for Other Coastal Operations ('The Provisions'). The Provisions have been approved by the State Council and have been in effect since 1 January 1994. Therefore, in China, there are three possibilities to determine the limitation of liability for claims concerning bunker oil pollution damage: (1) the CLC 1992 shall apply when the damage is within the scope of this convention. (2) Chapter 11 of the CMC shall apply when the damage is caused by ships engaging in international service with gross tonnage over 300 tonnes and is not within the scope of the CLC 1992. (3) The Provisions shall apply when the damage is caused by bunker oil of ships engaging in international transport services with a gross tonnage not exceeding 300 tonnes and those engaging in coastal transport services, as well as those for other coastal work.

2.3 The Amount of the Limitation of Liability in Domestic Laws

2.3.1 Relevant Provisions in Chapter 11 of the CMC

According to Article 210 of the CMC, the limitation of liability for maritime claims, except as otherwise provided for in Article 211 of this Code[2] shall be calculated as demonstrated in Table 9.1.

Table 9.1 Calculation of Liability Limitation

<table>
<tr><th>Gross Tonnage</th><th>Loss of Life or Personal Injury (Special Drawing Right)</th><th>Damage to Property (Special Drawing Right)</th></tr>
<tr><td>300–500</td><td>333,000</td><td>167,000</td></tr>
<tr><td colspan="3">Amounts in Addition For Each Tonne (SDR)</td></tr>
<tr><td>501–3,000</td><td>500</td><td rowspan="2">167</td></tr>
<tr><td>3,001–30,000</td><td>333</td></tr>
<tr><td>30,001–70,000</td><td>250</td><td>125</td></tr>
<tr><td>>70,000</td><td>167</td><td>83</td></tr>
</table>

2.3.2 The Provisions Promulgated by the Pre-Ministry of Communication[3]

The main three articles concerning the amount of limitation of liability of The Provisions are examined in the following discussion.

2. Article 211 of the CMC is the regulation of limitation of liability for claims for loss of life or personal injury to passengers carried by sea.
3. The name of the Ministry of Communication has now been changed to 'Ministry of Transport'.

Article 2 of the Provisions shall apply to ships with a gross tonnage exceeding 20 tonnes and not exceeding 300 tonnes and those with a gross tonnage exceeding 300 tonnes engaging in transport services between the ports of the PRC, as well as those for other coastal operations.

In Article 3 the limitation of liability for maritime claims for ships with a gross tonnage not exceeding 300 tonnes, except as otherwise provided for in Article 4 of the Provisions, shall be calculated as follows:

(1) In respect of claims for loss of life or physical injury:
   (a) 54000 Units of Account for a ship with a gross tonnage in excess of 20 tonnes and less than 21 tonnes.
   (b) For a ship with a gross tonnage in excess of 21 tonnes, 1000 Units of Account shall be added for each ton in excess of 21 tonnes.
(2) In respect of claims other than that for loss of life or physical injury:
   (a) 27500 Units of Account for a ship with a gross tonnage in excess of 20 tonnes and less than 21 tonnes;
   (b) For a ship with a gross tonnage in excess of 21 tonnes 500 Units of Account shall be added for each ton in excess of 21 tonnes.

According to Article 4, the limitation of liability for maritime claims for ships with a gross tonnage not exceeding 300 tonnes engaging in carriage of goods between the ports of the PRC, as well as those for other coastal operations, shall be calculated on the basis of 50% of the limitation of liability provided for in Article 3 of the Provisions, and that for ships with a gross tonnage exceeding 300 tonnes shall be calculated on the basis of 50% of the limitation of liability provided for in sub-paragraph (1) of Article 210 of 'Maritime Code of the People's Republic of China'.

By analysing these provisions, we can reach the following conclusions:

(a) The limitation amount for bunker oil pollution damage, caused by ships with a gross tonnage over 20 tonnes but not exceeding 300 tonnes engaging in international transport service, is equal to the amount provided in Article 3 of the Provisions.
(b) The limitation amount for bunker oil pollution damage, caused by ships with a gross tonnage over 20 tonnes but not exceeding 300 tonnes engaging in transport services between the ports of the PRC as well as those for other coastal operations, is 50% of the amount provided in Article 3 of the Provisions.
(c) The limitation amount for bunker oil pollution damage, caused by ships with a gross tonnage over 300 tonnes engaging in transport services between the ports of the PRC as well as those for other coastal operations, is 50% of the amount regulated in paragraph 1 of Article 210 of CMC.
(d) The limitation amount for bunker oil pollution damage, caused by ships with a gross tonnage over 300 tonnes engaging in international transport services, is equal to the amount provided in paragraph 1 of Article 210 of CMC. However, if the bunker oil pollution belongs to

the scope of application of CLC 1992, the limitation amount of the latter shall apply.

3 THE LIMITATION REGULATED IN THE 1976 LLMC AND ITS PROTOCOLS

The LLMC 1976 came into force on 1 December 1986. Until February 2009 there were fifty Contracting States, including Hong Kong. The convention divides the limitation of liability into two kinds, that is, the limitation of liability of loss of life or personal injury and the limitation of liability of damage to property (such as the damage to other ships, properties or ports). According to the convention, a responsible party shall not be entitled to limit his liability if the loss resulted from his intentional act or omission.

The Protocol 1996 of the LLMC 1976 came into force on 13 May 2004. There are thirty-three Contracting States[4] as of February 2009. The amounts for limitation of liability for loss of life or personal injury and for damage to property of this Protocol have been raised substantially.

3.1 Ships under the Scope of the Convention

Ships are defined as sea-going vessels in the convention. The convention does not apply to hovercrafts, floating platforms constructed for the purpose of exploring or exploiting the natural resources of the seabed or the subsoil thereof. There are two particular regulations involved; first, the Contracting States can regulate in their domestic law for ships with a gross tonnage not exceeding 300 tonnes engaging in transport services in inland waterways. Second, the Convention shall not apply to the drilling vessel when the limitation of liability regulated in the domestic laws of Contracting State is higher than in the Convention.

3.2 The Amount for Limitation of Liability

The amount for limitation of liability provided for in Article 6 of the LLMC 1976 is shown in Table 9.2.

4. The 33 Contracting States are: Albania, Australia, Bulgaria, Canada, Cook Island, Croatia, Cyprus, Denmark, Finland, France, Germany, Hungary, Iceland, Jamaica, Japan, Latvia, Liberia, Lithuania, Luxembourg, Malaysia, Malta, Marshall Islands, Norway, Romania, Russian Federation, Saint Lucia, Samoa, Sierra Leone, Spain, Sweden, Syrian Arab Republic, Tonga, and United Kingdom.

Table 9.2 Liability Limitation in Article 6 of LLMC 1976

Gross Tonnage	*Loss of Life or Personal Injury (Special Drawing Right)*	*Damage to Property (Special Drawing Right)*
1–500	333,000	167,000
Amounts in Addition For Each Tonne (SDR)		
501–3,000	500	167
3,001–30,000	333	
30,001–70,000	250	125
>70,000	167	83

The amount for limitation of liability provided for in Article 6 of the Protocol 1996 has been increased as shown in Table 9.3.

Table 9.3 Liability Limitation Increase in Article 6 of Protocol 1996

Gross Tonnage	*Loss of Life or Personal Injury (Special Drawing Right)*	*Damage to Property (Special Drawing Right)*
Not exceeding 2,000	2,000,000	1,000,000
Amounts in Addition For Each Tonne (SDR)		
2,001–30,000	800	400
30,001–70,000	600	300
Over 70,000	400	200

Concerning the exchange rate of Special Drawing Right (SDR) and the national currency of the State, it is provided in Article 8.1 of LLMC 1976 that:

> The value of a national currency in terms of the Special Drawing Right, of a State Party which is a member of the International Monetary Fund, shall be calculated in accordance with the method of valuation applied by the International Monetary Fund in effect at the date in question for its operations and transactions.

4 THE ESTABLISHMENT OF A LIMITATION FUND FOR COMPENSATION FOR BUNKER OIL POLLUTION DAMAGE

There is no independent regulation of limitation for liability in Bunker Convention 2001 where national law or international convention applies. In the case of accidents at sea leading to bunker oil pollution and damage to other properties, the following matters should be considered: the shipowner should establish only one fund of limitation of liability for the victims of bunker oil pollution damage and other property damage to distribute or establish two funds of limitation of liability for the two kinds of damage compensation. It is said that during the drafting of Bunker Convention 2001, all the countries had a tacit mutual understanding in the distribution of the fund; that is, the limitation of compensation of bunker oil pollution should be limited within the liability limit of LLMC 1976. However, there is no unanimity concerning the interpretation of the Convention. Some claim that Article 7 of Bunker Convention holds that the amount of compulsory insurance or financial security is determined according to the liability limit of national or international law because the compulsory insurance or financial security is specifically regulated for compensation for damage of bunker oil pollution but not for other limited maritime claims in the LLMC 1976. The reasoning for this is to ensure that the limitation of liability pursuant to LLMC 1976 does not become compulsory insurance, which would not reflect the original intention of the legislation. Therefore, in accordance with LLMC 1976 or the national law applied, there should be a fund of limitation of liability of compensation for bulk oil pollution damage besides the fund for other maritime claims.[5] In Chinese judicial practice, the position of the plaintiff and the defendant is opposing. The ship owner of the leaking oil ship argues that the fund is established for all limited maritime claims, which includes general maritime claims and claims for bulk oil pollution damage, whereas the victims' claim that the shipowner has no right of limitation of liability for bulk oil pollution damage or the fund should be established individually. Generally, the maritime court considers that the claims that are not in the application scope of CLC (which China has entered into) but in the application scope of paragraph 3 of Article 207 of CMC, which states that:

> claims in respect of other loss resulting from infringement of rights other than contractual rights occurring in direct connection with the operation of the ship or salvage operations,

the shipowners have the right of global maritime limitation of liability and only need to establish one limitation fund for bunker oil pollution damage and other property damage.

---

5. Si. Yuzhuo, *Maritime Law* (Beijing: Law Press, 2007), 358–359.

In the authors' view, the said judgment of the maritime court is correct until the Bunker Convention applies to China, because this view is in accordance with the original intention of the international and national legislation. Before CLC 1969 the claim for compensation for oil pollution damage belonged to general maritime claim and within the scope of the limited maritime claims regulated in Limitation Convention for Ship Owners 1957. According to the discussion earlier, in accordance with Article 207 of CMC, the claim for bunker oil pollution damage that falls outside the scope of application of CLC 1992 can be considered as general maritime claims and the owner is entitled to limit his liability according to the global limitation of liability for maritime claims. Once the Bunker Convention 2001 came into force in China, because the Convention has regulated compensation systems especially for bunker oil pollution damage, compared with the understanding of CLC, there should be an independent limitation for bunker oil pollution damage separated from the limit of other property damage. But unlike CLC, the Bunker Convention does not regulate independent limitation amounts calculated according to ship tonnage, but owners can limit their liability according to national law or international limitation of liability system. Therefore, when the Bunker Convention comes into force, if bunker oil pollution damage and other property damage occurred in one accident, two separate funds of limitation of liability are necessary.

5 COMPULSORY INSURANCE OR FINANCIAL SECURITY

5.1 Ships Applied

Article 7 of Bunker Convention 2001 states that the registered owner of a ship having a gross tonnage greater than 1,000 registered in a State Party shall be required to maintain insurance or other financial security. Provisions of this Convention shall not apply to warships, naval auxiliary or other ships owned or operated by a State and used, for the time being, only on Government non-commercial service only if it shall notify the Secretary-General thereof specifying the terms and conditions of such application. Insurance or other financial security is not maintained in respect to a ship owned by a State Party, but the ship shall carry a certificate issued by the appropriate authority of the State of the ship's registry stating that the ship is owned by that State Party and that the ship's liability is covered within the limit prescribed in accordance with paragraph 1.

The Convention provides that a State Party may declare that this article is not to apply to ships operating exclusively within the area of that State; that is, the Contracting State may regulate that ships over 1,000 tonnes can operate transport service within the area of the State without compulsory insurance or other financial security. According to the Notification, China did not make such a reservation for such ships upon acceding to the convention.

5.2 The Amount of Compulsory Insurance or Financial Security

Clause 2 of the Notification provides for the Scope of Liability of the Notification. It provides that Chinese-flag ships over 1,000 gross tonnes navigating international and coastal service must have insurance or other financial security. The insurance amount of the coastal transport ships cannot be lower than the ship's civil liability limitation as stated in CMC. The insurance amount of ships engaging in international navigation is determined by the managing courses and the laws of Contracting States into whose territory they have entered. However, the insurance amount cannot be lower than the ship's civil liability limitation stated in CMC.

According to Article 4 of the Provisions, the limitation of liability for maritime claims for ships engaging in coastal transport services, as well as those for other coastal operations with a gross tonnage exceeding 300 tonnes, shall be calculated on the basis of 50% of the limitation of liability provided for in sub-paragraph 1 of Article 210 of CMC. Therefore, except for ships covered by CLC 1992, the insurance amount of ships over 1,000 tonnes engaging in coastal transport service shall be calculated on the basis of 50% of the limitation of liability provided for in sub-paragraph 1 of Article 210 of CMC. The insurance amount of ships over 1,000 tonnes engaging in international transport service shall be calculated according to sub-paragraph 1 of Article 210 of CMC.

In shipping practice, there are some oil tankers between 1,000 and 2,000 tonnes in China that are engaging in transport service between China and Japan. Although they are not required to have compulsory insurance according to CLC 1992, they should have the certificate of CLC 1992 but not the certificate of Bunker Convention 2001 considering the difference between the risks and protection of victims. The amount of compulsory insurance or financial security for such ships should be decided according to CLC 1992.

5.3 The Name List of the Insurers

For the improvement of reliability of compulsory insurance and financial security to ensure the compensation for bunker oil pollution damage, China Maritime Safety Administration in January 2009 promulgated the name list of insurers qualified to provide insurance or security of civil liability for oil pollution for Chinese ships. The eighteen guarantors are China Ship owners Mutual Assurance Association, Assuranceforeningen Gard Gjensidig, The Steamship Mutual Underwriting Association (Bermuda) Limited, PICC Property and Casualty Company Limited, Ping An Insurance Company of China Limited, The United Kingdom Mutual Steam Ship Assurance Association, The Britannia Steam Ship Insurance Association Limited, The London Steam Ship Owners Mutual Insurance Association Limited, Assuranceforeningen SKULD (Gjensidig), West of England Insurance Services (Luxembourg) S.A., The Swedish Club, North of England P&I Association Limited, China Pacific Insurance Company Limited, China Continent Property and Casualty Insurance Company Limited, American Steamship Owners

Mutual Protection and Indemnity Association, China Life Property and Casualty Insurance Company Limited, Sunshine Insurance Group Corporation Limited, and Taiping General Insurance Co. Ltd

Chinese-flag ships should obtain compulsory insurance or financial security from the above eighteen insurers; otherwise, they will not be ratified by China Maritime Safety Administration and cannot obtain the certificate of compulsory insurance or financial security issued by such authority.

6 CONCLUSION

Since China acceded to CLC 1992 and Bunker Convention 2001, the regulations of the scope of application on ships, kind of oils and the limitation of liability of the two conventions are quite different, and the regulations of limitation of liability for ships with different routes and tonnage in Chinese domestic law are also different. This leads to the complexity of limitation of liability in the event of bunker oil pollution in the dominated sea area of China. The amounts for compulsory insurance and financial security also remain complicated. This chapter analysis concludes that there are various cases of limitation of liability and compulsory insurance in accordance with the Chinese domestic law and international conventions.

Reference

Si, Yuzhuo. *Maritime Law*. Beijing: Law Press, 2007, 358–359.

Part III
Criminal Law

Chapter 10

Criminal Liability for Oil Pollution Damage: An Economic Analysis

*Michael Faure**

1 INTRODUCTION

For many years a significant number of international conventions have been developed through the International Maritime Organization (IMO). Many aim to provide compensation for victims of oil pollution damage.[1] Other international conventions, more particularly the International Convention for the Prevention of Pollution from Ships (MARPOL) Convention and the United Nations Convention on the Law of the Sea (UNCLOS), aim at providing international regulations to prevent oil pollution damage. These international conventions are subsequently implemented by the signatory states within their own legal system.

It is rare that an international convention would also indicate how the national legislation implementing the convention would have to be enforced. Many problems have arisen in that respect. First, in some cases the non-compliance with international conventions was merely sanctioned through fines; imprisonment was

* Maastricht University and Erasmus University Rotterdam, the Netherlands.

1. For an overview see A. Verheij, 'Shifts in Governance: Oil Pollution', in *Shifts in Compensation for Environmental Damage*, ed. M. Faure & A. Verheij (Vienna: Springer, 2007), 133–195; and H. Wang, 'Shifts in Governance in the International Regime of Marine Oil Pollution Compensation: A Legal History Perspective', in *Shifts in Compensation for Environmental Damage*, ed. M. Faure & A. Verheij (Vienna: Springer, 2007), 197–241.

Michael G. Faure, Han Lixin & Shan Hongjun, *Maritime Pollution Liability and Policy: China, Europe and the US*, 161–192.

not always possible, and in the countries where it was it was rarely applied. Moreover, many problems arose as well as far as the application of sanctions was concerned when violations occurred outside of the territorial scope of the criminal law of national states. Traditionally, flag states had jurisdiction to enforce regulations upon ships. Their incentives to seriously enforce were, however, very limited given the practice of so-called 'flags of convenience'. To some extent, coastal state jurisdiction would be a solution for that problem, but the danger could exist that coastal states would be too severe with ships passing through their waters, thus potentially limiting the possibilities of free navigation. UNCLOS is now moving towards port state jurisdiction. The advantage is that port states can both have shipping interests as well as environmental interests because they may also be victimized by pollution incidents.[2]

Notwithstanding these shifts, many problems arise, because it is often felt that sanctions applicable to the violation of rules aiming at the prevention of marine pollution are not sufficiently severe. This has led to an increased interest in the role of criminal law, more particularly in the European Union (EU). In 2005, the EU launched a Council Framework Decision to strengthen the criminal law framework for the enforcement of the law against ship-source pollution.[3] Shortly later, a directive was accepted on ship-source pollution and on the introduction of penalties for infringements.[4] Even though the framework decision was later annulled by a decision of the Court of Justice of 23 October 2007 for a lack of competence,[5] it shows the increasing interest in using criminal law to enforce regulations aiming at the prevention of oil pollution damage.

This paper aims at using economic analysis of law to examine this tendency towards an increased criminal liability for oil pollution damage. It seems useful to call on this economic approach for a variety of reasons: first, economic analysis has clearly indicated that under particular circumstances criminal law may have an added value compared to other (civil, administrative) legal instruments. Second, economic analysis treats potential oil polluters as rational criminals and hence also has useful indications on the types and magnitude of sanctions to be imposed for optimal deterrence of oil pollution. Although economic analysis of law has been widely applied to environmental law[6] it has not been applied yet to the specific case of criminal liability for oil pollution damage. The aim of this paper is therefore to show how economic analysis can contribute to explaining the role of the criminal law in remedying damage caused by oil pollution. The focus of this paper will

2. See on this evolution H. Wang, 'Transboundary Vessel-Source Marine Pollution – International Legal Framework and Its Application to China', in *China and International Environmental Liability. Legal Remedies for Transboundary Pollution*, ed. M. Faure & Y. Song (Cheltenham: Edward Elgar, 2008), 64–106.
3. Council Framework Decision 2005/667 of 12 Jul. 2005, OJ L255/164 of 30 Sep. 2005.
4. Directive 2005/35 of 7 Sep. 2005 on ship-source pollution and on the introduction of penalties for infringement, OJ L255/11 of 30 Sep. 2005.
5. This decision will be further discussed *infra*.
6. For a summary of the literature see M. Faure & G. Skogh, *Economic Analysis of Environmental Policy and Law: An Introduction* (Cheltenham: Edward Elgar, 2003).

therefore be much more on the theory provided by economic analysis and less on actual legal policy. However, some examples of actual policy and cases will be given. Moreover, even if the approach is mainly theoretical, economic analysis allows for clear indications of the circumstances under which criminal law may contribute to remedying oil pollution damage to be considered. These insights obviously have practical value as well.

The remainder of this chapter is structured as follows: first, the question will be asked why criminal law would be needed at all to remedy oil pollution damage. Next, the economic approach to the decision of the rational criminal as introduced by Gary Becker will be used to explain how optimal sanctions for oil pollution damage could be determined. The question will also be asked whether criminal sanctions should be merely applied to individual actors, to corporations or to both. The insights of economic analysis then will be applied to a few specific problems arising in legal policy.

2 WHY CRIMINAL LAW FOR OIL POLLUTION DAMAGE?

2.1 Arguments in Favour of Public Enforcement

There is a rich law and economics literature describing under what kind of circumstances criminal law may be more warranted than other legal instruments.[7] However, much of the literature in this respect in fact provides arguments in favour of public enforcement of regulation rather than merely in favour of criminal law. This is also understandable – criminal law is after all used to enforce public regulations. Hence, if regulations were not necessary and enforcement was possible, for example, via private law remedies, criminal law would not be needed either.

The principal argument in favour of public enforcement is that private law remedies often will not sufficiently deter environmental pollution.[8] The arguments in favour of public enforcement have been presented in the law and economics literature in the well-known studies of Steven Shavell.[9] The arguments of Shavell in favour of public enforcement also apply to oil pollution damage – oil pollution

7. See, e.g., R. Bowles, M. Faure & N. Garoupa, 'The Scope of Criminal Law and Criminal Sanctions: An Economic View and Policy Implications', *Journal of Law and Society* 35, no. 3 (2008): 389–416; M. Faure & M. Visser, 'Law and Economics of Environmental Crime', in *New Perspectives on Economic Crimes*, ed. H. Sjögren & G. Skogh (Cheltenham: Edward Elgar, 2004), 57–75; and A.I. Ogus, 'Enforcing Regulation: Do We Need the Criminal Law?', in *New Perspectives on Economic Crimes*, ed. H. Sjögren & G. Skogh (Cheltenham: Edward Elgar, 2004), 42–56.
8. See, e.g., M. Wilde, *Civil Liability for Environmental Damage: A Comparative Analysis of Law and Policy in Europe and the United States* (The Hague/London: Kluwer Law International, 2002), 307–310.
9. See S. Shavell, 'Liability for Harm versus Regulation of Safety', *Journal of Legal Studies* (1984): 357–374; and S. Shavell, 'A Model of the Optimal Use of Liability and Safety Regulation', *RAND Journal of Economics* (1984): 271–281.

damage often has no individual victim that could file a liability suit (this is especially the case with damage caused on the high seas that does not directly affect coastal areas), and a causal relationship may also be difficult to prove with a particular tanker (even though detection technologies have greatly improved recently). Moreover, there is a likelihood that the amount of the damage caused can be substantially larger than the individual wealth of an injurer. Hence, an insolvency problem may arise.

Although many have stressed that for oil pollution damage civil liability can, provided it is structured correctly, well provide sufficient incentives to deter oil pollution,[10] there is also a great likelihood that private law rules (and more particularly tort law) may not provide a sufficient deterrent. A serious problem is the fact that tanker owners often organize their own insolvency, for example, by creating so-called single ship legal entities, as a result of which the liability is limited to the value of the ship. The total magnitude of the damage caused by marine pollution can be substantially larger; thus, civil liability alone will result in under-deterrence.

These are powerful arguments in favour of public regulation (in addition to using tort law) to deter marine oil pollution and to use public enforcement to induce compliance with public regulation.

2.2 Low Probability of Detection

An important reason advanced in economic theory in favour of public regulation, but also to explain the use of the criminal law, is the low probability of detection.[11] To understand why this is the case I have to introduce the theory of Gary Becker concerning the decision of a potential criminal to commit a crime or not.[12] Becker holds that a potential criminal will weigh the benefits and costs of his action and only commit the violation when the benefits outweigh the costs. The benefits for a potential polluter in case of marine pollution are the savings from spilling the oil instead of discharging it legally, including any costs of avoidance of spills. The costs associated with the oil pollution can be broken down into two components. The first part is the probability that the offender will be convicted, and the second consists of the size of the penalty that is imposed in the event of a conviction. The probability itself can be divided into the probability of being caught and the probability of being convicted. The expected cost to the offender is thus the

10. See J. Boyd, 'Compensation for Oil Pollution Damages: The American Oil Pollution Act as an Example for Global Solutions?', in *Prevention and Compensation of Marine Pollution Damage: Recent Developments in Europe, China and the US*, ed. M.G. Faure & J. Hu (The Hague: Kluwer Law International, 2006), 137–163.
11. See R. Posner, 'An Economic Theory of the Criminal Law', *Columbia Law Review* 85 (1985): 1193–1209 and S. Shavell, 'Criminal Law and the Optimal Use of Non-monetary Sanctions as a Deterrent', *Columbia Law Review* 85 (1985): 1232–1265.
12. See G.S. Becker, 'Irrational Behavior and Economic Theory', *Journal of Political Economy* (1962): 1–13 and G.S. Becker, 'Crime and Punishment: An Economic Approach', *Journal of Political Economy* (1968): 169–216.

combination of the two probabilities multiplied by the penalty imposed under a conviction. The problem of the criminal is then to compare the benefits of the oil spill with the expected costs of the spill. Assuming that the rational offender is risk neutral, he will pollute if the benefits outweigh the expected costs of the spill. Assume for example, that the benefits of an oil spill would be United States Dollar (USD) 100,000, that the fine to be imposed by a court were USD 1,000,000 and the probability of being sentenced to such a fine were 1%. In that case the expected cost of the oil spill to the potential offender would be 1% × USD 1,000,000 = USD 10,000. Because these costs are lower than the expected benefit of USD 100,000, a risk-neutral potential offender would decide to pollute.

In many cases of marine pollution the probability of being caught may not be 100%, but in fact much lower. The result is that according to the Becker model, the efficient sanction for deterring the potential polluter should be correspondingly higher. In general, this effect cannot be achieved with tort law, because tort law in principle only forces the injurer to compensate the victim the amount of damage suffered and not more. This again shows the inappropriateness of civil law in cases in which the probability of detection is less than 100%. For optimal deterrence a higher sanction has to be imposed in order to compensate for this low detection rate. This cannot be provided through private law, which explains the need for public sanctions that allow compensating for the lower detection rate.[13]

The low probability of detecting oil pollution damage[14] is hence an important argument against merely using civil liability to deter oil pollution damage. To continue with the example presented earlier, civil liability would only force the polluter to compensate the damage, but would not take into account the low probability of detection. Assuming that the damage caused to society would also be USD 1,000,000 (and hence equal to the fine), under a probability of detection of 1% the expected cost for the potential offender would still be 1% × USD 1,000,000 = USD 10,000. Because we assume the benefit from pollution to be USD 100,000, a rational risk neutral offender would still commit the pollution.

This simple example shows that sanctions are needed that are substantially higher than the benefit of pollution to the offender in order to outweigh the low probability of detection. Under traditional tort law the compensation due by the polluter is equal to the damage caused; however, private law cannot reach that goal.

---

13. G. Skogh & C. Stuart, 'An Economic Analysis of Crime Rates, Punishment and the Social Consequences of Crime', *Public Choice* (1982): 171–179 and G. Skogh, 'A Note on Gary Becker's Crime and Punishment: An Economic Approach', *Swedish Journal of Economics* (1973): 305–311.
14. For a summary of the literature on detection of oil pollution see M.A. Cohen, 'Oil Pollution Prevention and Enforcement Measures and Their Effectiveness: A Survey of Empirical Research from the US', in *Prevention and Compensation of Marine Pollution Damage: Recent Developments in Europe, China and the US*, ed. M.G. Faure & J. Hu (The Hague: Kluwer Law International, 2006), 25–39.

2.3 Administrative or Criminal Law?

Although the low probability of detection may be a reason in favour of public enforcement, it is not necessarily an argument in favour of using criminal law. Indeed, the low probability of detection could well be compensated by imposing a high fine on the polluter.

Fines have always been considered the preferred sanction in economic theory for the simple reason that the costs of imposition are low and fines in fact only generate money for the public budget.[15] Monetary sanctions can in principle have both a criminal and an administrative nature. All things being equal the administrative procedure has the major advantage that it is far less costly than the criminal procedure. Administrative fines can within what is sometimes referred to as 'administrative penal law'[16] be imposed by administrative authorities after a relatively simple procedure and require usually a relatively low threshold of proof. Compared with criminal law, the costs of the administrative procedure are substantially lower. All things being equal, it can therefore be argued that if optimal deterrence can be achieved through fines, it seems desirable to use the less costly administrative law instead of the relatively more costly criminal procedure. This has led many scholars to argue that the imposition of relatively modest fines through the criminal procedure is inefficient because a similar result could be achieved at lower costs through administrative law. More particularly, Ogus and Abbot have therefore argued that in the UK more use should be made of administrative fines (and other administrative sanctions for that matter) to enforce violations of environmental regulations.[17] A clear, normative conclusion from this literature is therefore that in many more instances than is the case today, administrative law could be used to deter environmental pollution, especially when the optimal penalties consist of relatively low fines or other (not too infringing) administrative sanctions.

However, there are two important reasons why not all efficient penalties necessary to deter marine oil pollution can be imposed through administrative law and why criminal law therefore remains necessary. The first reason is that because the probability of detection of marine pollution can in practice often be very low, the optimal sanction to deter pollution may become very high as well. The likelihood that this optimal fine might outweigh the individual wealth of an offender is relatively high. Marine polluters are often organized as corporate

15. S. Shavell, *supra* n. 11 and R. Posner, 'Optimal Sentences for White-Collar Criminals', *American Criminal Law Review* (1980): 400–418.
16. This expression 'administrative penal law' may be confusing to some who consider 'penal' synonymous with 'criminal'. In the literature this notion is used to refer to a system whereby administrative authorities impose penalties.
17. A. Ogus & C. Abbot, 'Pollution and Penalties', in *An Introduction to the Law and Economics of Environmental Policy: Issues in Institutional Design*, ed. T. Swanson (The Hague: Elsevier, 2002), 493–516 and A. Ogus & C. Abbot, 'Sanctions for Pollution: Do We Have the Right Regime?', *Journal of Environmental Law* 14 (2002): 283–300.

entities that benefit from limited liability.[18] Hence, there is always a risk that marine pollution may cause costs that are higher than the assets of the firm or, in the criminal law context, that the optimal fine (to outweigh a low detection rate) will be much higher than the assets of the firm. Indeed, the optimal monetary sanction required for deterrence so frequently exceeds the offenders' assets that non-monetary sanctions, such as imprisonment, are necessary. The major advantage of the fine (lower administrative costs) therefore leads to favouring this type of sanction only when the risk for insolvency can be controlled. It should also be recalled that the probability of an administrative fine being imposed will be much higher (given a lower procedural threshold) than that of a criminal fine. As a result the administrative fine should not necessarily be nearly as large as the criminal fine. This can again reduce the insolvency problem.

There is a clear economic reason why society does not want to impose very stringent sanctions, such as imprisonment (but also high fines) through an administrative proceeding. The reason is that the costs of the administrative proceedings may be lower than the costs of the criminal proceedings, but the accuracy of the latter (where the investigations are often undertaken by professional lawyers) may be a lot higher as well.

This aspect is also important because a task of criminal law is not only to apply optimal sanctions to the guilty, but also to avoid punishing the innocent. This is referred to as the goal of reduction of error costs.[19] The error cost is obviously a lot higher when very serious sanctions, such as imprisonment, may be imposed, rather than monetary sanctions only. It is therefore understandable that less costly administrative proceedings are chosen in all cases in which the consequences (and thus the error cost) will not be too high in the event of a wrongful conviction. Arguably also, a goal of the administrative procedure is to avoid punishing the innocent (and thus reducing error costs), although they operate at a lower standard. That explains why administrative law (and the corresponding administrative procedure) will be reserved for cases in which relatively low penalties can suffice to provide deterrence.

The policy lesson from this economic literature is therefore rather straightforward: in cases in which optimal deterrence of marine polluters can be achieved through relatively modest sanctions (e.g., not excessively high administrative fines or other administrative sanctions) the use of the less costly administrative penal law may be warranted. However, in cases in which the probability of detection is relatively low, social harm and the potential gain to the polluter are high and thus a more severe sanction is needed, it may be warranted to use the more costly criminal procedure in order to reduce error costs. This is certainly the case when the optimal fine would reach the insolvency limit and non-monetary

---

18. See generally on this problem posed by the limited liability of corporations: H. Hansmann & R.H. Kraakman, 'Toward Unlimited Shareholder Liability for Corporate Torts', *Yale Law Journal* 100 (1991): 1879; and W.H.F.M. Cortenraad, *The Corporate Paradox: Economic Realities of the Corporate Form of Organization* (Boston: Kluwer Academic, 1999).
19. T. Miceli, 'Optimal Prosecution of Defendants Whose Guilt Is Uncertain', *Journal of Law, Economics and Organization* 6 (1990): 189–201.

sanctions are thus needed for deterrence, but equally where for the same reason very high administrative fines would have to be imposed. In reality in many legal systems there are now possibilities to use either the criminal law or administrative penal law to particular environmental offences or in some cases even a combination of those.[20]

2.4 ALTERNATIVES?

This economic reasoning explaining the use of the criminal law also makes clear that criminal law is needed to back up public regulation and that public enforcement is needed because of weaknesses in the private law system. By presenting criminal law thus as an instrument that is necessary to remedy weaknesses in private law it may equally be clear that one way towards decriminalization is to improve the effectiveness of private law. There are good reasons to think about this because, as discussed earlier, criminal law is a highly costly instrument to apply. That is why it should be reserved for those cases in which deterrence cannot be achieved through other (lower cost) alternatives.

If, for example, a problem with tort law is that it limits the compensation for the harm done to the victim, one way of providing a counterweight for the low detection rate in oil pollution cases to increase the amount of compensation payable by the injurer under tort law. This is exactly the idea behind the concept of 'punitive damages'.[21] However, punitive damages may not be a solution when the amount of the damages would make the polluter judgment proof. In that case, non-monetary sanctions will still be needed to achieve deterrence.[22]

A second example could relate to the insolvency problem under tort law. I indicated that the inability of the potential polluter to compensate the harm under tort law may be an important reason in favour of public enforcement and of the criminal law. Given the practice of so-called single ship entities, this insolvency problem is a realistic one in oil pollution cases. However, to some extent this can be remedied by introducing compulsory insurance. By forcing a potential polluter to take out insurance coverage, some of the underterrence caused through insolvency could be removed.[23] Article VII of the Civil Liability Convention (CLC) forces the

20. For a discussion on the optimal use of criminal sanction in addition to administrative penalties see N. Garoupa & F. Gomez-Pomar, 'Punish Once or Punish Twice: A Theory of the Use of Criminal Sanctions in Addition to Regulatory Penalties', *American Law and Economics Review* 6 (2004): 410–433.
21. See R.D. Cooter, 'Economic Analysis of Punitive Damages', *Southern California Law Review* (1982): 79–101 and W. Landes & R. Posner, 'An Economic Theory of Intentional Tort', *International Review of Law and Economics* (1981): 127–154.
22. That may explain why one could notice more particularly in the United States both the application of punitive damages (more particularly to oil pollution cases) and criminalization of marine pollution.
23. For a more detailed analysis see M. Faure & H. Wang, 'Economic Analysis of Compensation for Oil Pollution Damage', *Journal of Maritime Law and Commerce* 37 (2006): 179–217.

tanker owner to provide 'insurance or other financial security, such as the guarantee of a bank or a certificate delivered by an international compensation fund'.

Another example, more specifically relating to the issue at hand, refers to the argument that civil law will often not be useful as a deterrent against environmental pollution because the damage may be widespread, the whole community victimized or the damage caused to collective goods that are not owned by one individual. In such cases, no one will bring a suit in tort. If, in these cases, one wishes to increase the effectiveness of civil law, one way of doing so is to allow the use of civil law also when no individual victim is harmed, but where a (high) level of social harm is caused by the tortuous act. One way of achieving this is by allowing the state or other public authorities to act on behalf of the environment and to let them use tort law; another way is to grant non-governmental organizations or individuals the right to institute actions in tort, for example, to ask for an injunction to stop environmental pollution from occurring. Also, class actions brought on behalf of a group of victims could be considered a way to improve the functioning of private law, thus reducing the need for criminalization.

2.5 Criminal Law for Marine Pollution?

Applying the criteria sketched in the earlier discussion to the case of marine pollution damage it is not difficult to make a case in favour of public enforcement and more particularly of the criminal law. It is, given Shavell's criteria for safety regulation, no surprise that increasingly the risk for pollution incidents is controlled through regulation aiming at the design of tankers in such a way that pollution incidents can be prevented. The phasing out of single hull tankers, state inspections by port states and controls by classification societies are just some of the regulations that have received increasing attention in recent years. Of course, increased safety design will lead to (a small) increase in oil prices, which has been calculated in various empirical studies.[24] This is not the place to discuss the efficiency and effectiveness of these tanker design regulations.[25] More interesting is the question as to whether in addition to this regulation, any additional deterrent effect can be expected from liability rules. It is, lacking empirical evidence, not possible to provide hard data on the effectiveness of liability rules in supplementing safety regulation. There are, however, some indications in the literature. For instance Wood argues that,

24. For an early one see J.L. Pedrick, 'Tank Ship Design Regulation and Its Economic Effect on Oil Consumers', *Journal of Maritime Law and Commerce* (1978): 377–395.
25. For an overview of developments in this respect at the EU level see H. Wang, 'The EU Marine Oil Pollution Regime – Recent Developments', *European Environmental Law Review* (2004): 292–303.

> Civil liability can also serve a prophylactic function. A properly structured system for civil liability would exert a powerful influence to discourage polluting discharges of oil.[26]

Cummins et al examine explicitly the effectiveness of liability versus regulation in oil tanker pollution control.[27] They argue strongly in favour of liability rules and against regulation, arguing that liability for major oil spills would leave it up to the tanker owner to choose the least cost method of pollution control in case of oil spills.[28] They conclude that imposing design standards might inhibit innovation and would be much too rigid compared to a liability system.[29] Of course, one could state in rebuttal to these authors that thirty years of experience after they published their papers has indicated that the liability regime has not been able to prevent major oil spills such as occurred with the *Erika* or the *Prestige*. However, this would hardly be fair given the fact that the liability system as it was installed in the 1969. The CLC did not provide adequate incentives, given the channelling of liability and the low limits on the liability of the tanker owner.

Turning then to the question of criminal law it also seems clear that most cases of marine pollution are those in which the probability of detection is (notwithstanding improved detection technology) still relatively low. The potential gain (saved costs) to polluters can still be high, and the social harm caused by major oil spills is obviously spectacularly high. For those cases it can easily be argued that given the low probability of detection and high potential gains, a high sanction would be needed to deter. Moreover, given the practice of single ship entities, monetary sanctions are not likely to suffice. Hence, non-monetary sanctions may be warranted and given high error costs involved, this warrants the application of the criminal procedure, which can provide higher guarantees of quality in truth finding.

This does not imply that all stipulations in regulations aiming at the prevention of oil pollution damage should necessarily be enforced through the criminal law. I have discussed here merely the deterrence of major spills. However, increasingly, regulation in the area of marine pollution also involves many administrative duties, for example, to keep registers, file specific declarations and so forth. In those cases, one can argue that the non-respect of such an administrative duty does not lead to high gains to the offender. Moreover, when administrative duties are violated (e.g., a tanker owner does not report to the port authorities where this would be required) the probability of detection is obviously also a lot higher than, for example, with an oil spill on the high seas. These types of administrative offences could hence be deterred through administrative fines because violation of these

26. L.D. Wood, 'An Integrated International and Domestic Approach to Civil Liability for Vessel-Source Oil Pollution', *Journal of Maritime Law and Commerce* (1975): 2.
27. Ph.A. Cummins et al., 'Oil Tanker Pollution Control: Design Criteria vs. Effective Liability Assessment', *Journal of Maritime Law and Commerce* (1975): 169–206.
28. *Ibid.*, 175.
29. *Ibid.*, 204–206.

administrative duties usually does not lead to high social harm. To enforce those administrative duties administrative fines may well suffice.

The result is therefore that it may be optimal to enforce regulations aiming at the prevention of marine pollution through a combination of administrative fines in the case of violations of administrative duties and criminal (non-monetary) sanctions in case actual spills took place.

3 OPTIMAL SANCTIONS FOR MARINE POLLUTION

3.1 Fines or Imprisonment?

As far as the type of sanction is concerned, the previous section mentioned that in some cases administrative fines may suffice (for minor administrative violations), whereas for others criminal non-monetary sanctions should be applied.

When optimal deterrence can be achieved equally through fines and prison sanctions, fines are preferred because they are less costly to impose than prison sanctions.[30] As indicated earlier, many economists have for that reason shown themselves to be opponents of prison sanctions simply because the costs for the implementation of those sanctions are much higher than the costs for imposing fines. That is why the fine has been advanced as the ideal sanction in the case of corporate crimes such as environmental crime.[31] This was, however, criticized by Coffee[32] as well as by Shavell,[33] who argued that fines can work as an effective deterrent only if there is no insolvency problem. As argued earlier, in the case of a low probability of detection and an insolvency risk, non-monetary sanctions will have to be applied.[34] This lesson from economic theory seems to be followed in legal practice as well: the available criminology studies report that the fine is the most commonly used sanction; the application of prison sanctions for environmental crime would be rare and would be limited to cases of fraud or serious environmental harm.[35]

30. M. Polinsky & S. Shavell, 'A Note on Optimal Fines When Wealth Varies among Individuals', *American Economic Review* 81 (1991): 618; A.M. Polinsky & S. Shavell, 'The Optimal Trade Off between the Probability and the Magnitude of Fines', *American Economic Review* (1979): 880–891 and for a summary R. Bowles, *Law and the Economy* (Oxford: Martin Robertson, 1982).
31. Posner, *supra* n. 15.
32. J.S. Coffee, 'Corporate Crime and Punishment: A Non-Chicago View of the Economics of Criminal Sanctions', *American Criminal Law Review* (1980): 419–476.
33. Shavell, *supra* n. 11.
34. Polinsky & Shavell, *supra* n. 30, 880–891 and K. Segerson & T. Tietenberg, 'Defining Efficient Sanctions', in *Innovation in Environmental Policy*, ed. T.H. Tietenberg (Cheltenham: Edward Elgar, 1992), 63–65.
35. See for the situation in Europe generally G. Heine & V. Meinberg, *Empfehlen sich Änderungen im Strafrechtlichen Umweltschutz, Ins Besondere in Verbindung mit dem Verwaltungsrecht? Gutachten für den 57. Deutschen Juristentag* (München: Beck, 1988); and V. Meinberg, 'Empirische Erkenntnisse zum Vollzug des Umweltstrafrechts', *Zeitschrift für die Gesamte*

3.2 Determining the Optimal Sanction

It may be recalled that on the basis of the economic theory sanctions should be set in such a way that they optimally deter potential polluters. Ideally, the sanction should be of such a magnitude that, taking into account the probability of detection and sanction, it creates expected costs for the potential polluter that are higher than his gains. The result of applying economic theory is that optimal sanctions in the case of environmental crimes depend upon a variety of elements, such as the probability of detection, the harm caused, available assets of the polluter, the corresponding insolvency risk and the costs of the sanctions to be imposed. Optimal sentencing policy would consist of applying fines as long as possible (in the light of the insolvency risk) and combining them in a differentiated manner with non-monetary sanctions where necessary.[36] In practice, the probability of detection is given to the judge (in the sense that it cannot be influenced by the judge when deciding on the case). The judge then can only fix the sanction taking into account the probability of detection and the gain of the oil spill to the polluter. If this would lead to very high monetary sanctions (which are in principle preferred), for example, because the probability of detection was low and the gain to the polluter was high, non-monetary sanctions may have to be applied.

However, in reality, sanctioning and sentencing do not always correspond with the previously mentioned criteria based on economic analysis. This is more particularly the case in the United States, where strict sentencing guidelines apply, to which the judge in principle should comply. Differently from the economic model, the sentencing guidelines also provide quite often for incarceration as the main penalty in the case of violation of environmental regulations. Frank Easterbrook has strongly opposed this idea of strict sentencing guidelines, arguing that 'there is no correct price for crime in the same way that there is no correct price for apples'. He holds that the price of (environmental) crime depends upon a variety of elements, indicated earlier, that enable the judge to determine the optimal penalty in a differentiated way.[37] Easterbrook therefore supports discretionary sanctioning powers of the judge, arguing that this will allow the judge to differentiate the sanctions in a more precise manner, taking into account relevant economic criteria. The economic lessons as formulated by Easterbrook seem to be followed in Europe, where, if sentencing guidelines exist at all, they seem to provide sufficient flexibility to the judge to differentiate sanctions according to the economic criteria. Recently, the Department for the Environment, Food and Rural Affairs (DEFRA) in the United Kingdom published a Review of Enforcement in Environmental Regulation that equally provides for sentencing guidelines, but at the same time holds that this should

---

Strafrechtswissenschaften (1988): 112–157. For the situation of Belgium see C.M. Billiet & S. Rousseau, 'De hoogte van strafrechtelijke boetes. Een rechtseconomische analyse van milieurechtspraak (1990–2000) van het Hof van Beroep te Gent', *Tijdschrift voor Milieurecht* (2003): 120–134.

36. R.D. Cooter & T. Ulen, *Law and Economics*, 4th edn (Boston: Addison Wesley, 2004), 477.
37. F. Easterbrook, 'Criminal Procedure as a Market System', *Journal of Legal Studies* 12 (1983): 295, 325.

not impose a rigid tariff but should merely provide a more systematic framework for more effective and consistent sentencing decisions.[38] Also, recent empirical evidence seems to support Easterbrook's criticism on the sentencing guidelines in the United States. Barrett concluded in a study on the application of the sentencing guidelines for environmental crime that,

> 'The sentences imposed in the majority of cases reflected the reluctance of judges to impose significant incarceration for violations of environmental laws'.[39]

Also, others have recently held that the sentencing guidelines lead to negative outcomes as far as enforcement and deterrence is concerned. Recent research by Babbitt, Cory and Kruchek showed that a practice of lenient sentencing of environmental criminals, as a refusal to apply the guidelines, was well documented. Because the judiciary feels that the guidelines 'over criminalize' environmental violations, they may systematically sentence (also significant) violations too leniently, which may undermine the deterrent value of environmental enforcement and may trivialize environmental law itself.[40]

3.3 Non-Monetary Sanctions

3.3.1 Imprisonment or Restoration?

The non-monetary sanction I discussed so far was the imprisonment. However, imprisonment also has its limitations: it cannot be applied against corporate defendants (for the simple reason that one cannot put a legal entity in jail), and imprisonment is generally disliked by economists because it is very costly. There are, however, a variety of non-monetary sanctions that can be found in modern environmental statutes. Some of these measures force the convicted polluter to restore the harm committed or prevent future harm. In oil pollution cases an order, for example, to restore soils polluted by oil, can be quite effective. Even though these measures are usually treated as non-monetary in legal doctrine, many of those arguably also have monetary implications as well. The simple reason is that it may be extremely costly for the offender to restore, for example, a beach polluted by oil in its status quo ante.

Imposing specific measures and duties upon a convicted polluter has on the one hand the advantage that it can lead to additional deterrence (given the high costs involved) in the execution of the measure. On the other hand it has the

38. N. Parpworth, 'Enforcement in Environmental Regulation: The Defra Review', *Environmental Liability* (2007): 21.
39. J.J. Barrett, 'Sentencing Environmental Crimes under the United States Sentencing Guidelines: A Sentencing Lottery', *Environmental Law* 22 (1992): 1421.
40. Ch.J. Babbitt, D.C. Cory & B.L. Kruchek, 'Discretion and Criminalization of Environmental Law', *Duke Environmental Law and Policy Forum* 15 (2004): 63.

advantage that at least the environmental problem caused by the offence will also be solved. Indeed, merely sending for example, a captain to jail will not guarantee that the polluted environment is also restored. In some legal systems these orders (e.g., to clean up) can also be enforced by means of a penalty payment. This is an amount due by the offender if the duties imposed are not fulfilled within the time limit laid down in the judgment.

3.3.2 Publicity

In some cases the legislator also provides the power to the judge to order the publication of a judgment to the mass media. For oil pollution damage such a publication can have a deterrent effect because the loss of reputation caused by publication can have financial consequences for the company involved. Of course the effect may be stronger when the defendant is an oil company with a reputation at stake than when it would be a company merely transporting the oil, where the reputational losses from publication could be lower.

3.3.3 Forfeiture of Illegal Gains

One particular sanction that increasingly gains popularity in the fight against white-collar crime generally, but also in the area of environmental crime, is the forfeiture of illegal gains. From a legal perspective, this forfeiture of illegal gains rests on the notion that 'crime should not pay'. The gains obtained by the crime should be removed from the criminal in order to put him back in the status quo ante.[41] Bowles, Faure and Garoupa showed that the forfeiture of illegal gain can be efficient in order to provide additional deterrence in the case of a low detection rate and low effective fines.[42] However, early empirical studies by German criminologists showed that the risk for removal of illegal gain had hardly any effect on potential criminals in their decision whether to commit the crime.[43] Meanwhile, the possibilities to remove illegal gain as well as the practical application in the environmental area have spectacularly increased since these early studies of

41. For an overview of these legal foundations see M. Faure, 'Gewinnabschöpfung und Verfall auf ökonomischen Prüfstand', in *Menschengerechtes Strafrecht. Festschrift für Albin Eser zum 70. Geburtstag*, ed. J. Arnold et al. (München: Beck, 2005), 1311–1330 and for a comparative overview see M. Kilchling (ed.), *Die Praxis der Gewinnabschöpfung in Europa* (Freiburg im Breisgau: Max Planck Institute for Foreign and International Criminal Law, 2002); and M. Kilchling, 'Comparative Perspectives on Forfeiture Legislation in Europe and the United States', *European Journal of Crime, Criminal Law and Criminal Justice* 5 (1997): 342–361.
42. R. Bowles, M. Faure & N. Garoupa, 'Economic Analysis of the Removal of Illegal Gains', *International Review of Law and Economics* (2000): 537–549; and R. Bowles, M. Faure & N. Garoupa, 'Forfeiture of Illegal Gain: An Economic Perspective', *Oxford Journal of Legal Studies* 25 (2005): 275–295.
43. See J.R. Smettan, *Kriminelle Bereicherung in Abhängigkeit von Gewinnen, Risiken, Strafen und Moral* (Freiburg im Breisgau: Max Planck Institute for Foreign and International Criminal Law, 1992); and W. Perron, 'Vermögensstrafe und Erweiterte Verfall', *Juristenzeitung* (1993): 919–920.

1992–1993.[44] The additional deterrent effect and therewith the effectiveness of the removal of illegal gain in case of environmental crime may thus well have substantially increased.

3.3.4 Confiscation

Finally, confiscation also could be considered as an alternative sanction. In this respect I am not referring to the confiscation of illegal gains obtained by the oil pollution, but by a confiscation of the instruments with which the crime has been committed, for example, the tanker. This is a sanction that has already been applied more particularly by the Criminal Court of Hamburg (Germany).[45] In one case, an Egyptian oil tanker was not seaworthy. This was, according to the court, known by the tanker owner. As a result the court ordered the confiscation of the tanker because it was considered an instrument with which the crime had been committed. This can in some cases be quite an effective sanction because it directly hurts the tanker owner and may therefore give incentives to avoid oil pollution in the future. Also, the advantage may be that the tanker that caused the pollution is still available in the port where the offence took place. In that case, collection costs should not be too high. Of course this sanction may not be useful in cases in which the tanker itself should be considered as lost (such as in the cases of *Erika* and *Prestige*) and where the costs to salvage the tanker may in some cases even be higher than what may be left of the ship. In some legal systems confiscation may only be possible if the property belongs to the persons actually convicted. Because the tanker usually does not belong to the master or crew member who is actually prosecuted for oil pollution, confiscation would then be problematic, unless the legal entity that owns the tanker is also prosecuted. Moreover, recent changes in legislation have enlarged possibilities of confiscation to include confiscation of objects that do not belong to the convicted.

4 CORPORATE CRIMINAL LIABILITY FOR OIL POLLUTION?

4.1 The Economic Case for Corporate Criminal Liability

Marine oil pollution is often not (only) committed just by one individual actor, but usually by several persons acting together and working on behalf of a company, a

44. For an overview of the practice in Belgium and the Netherlands see M. Faure & Th. De Roos (eds), *De berekening van het wederrechterlijk verkregen voordeel uit milieudelicten* (The Hague: SDU Uitgevers, 1998).
45. Amtsgericht Hamburg, Judgment of 30 Oct. 1985, Nr. 142a-1420/85 discussed by M. Faure & G. Heine, 'The Insurance of Fines: The Case of Oil Pollution', *The Geneva Papers on Risk and Insurance* 16 (1991): 54–55.

legal entity. Many of the serious cases of marine pollution have been committed within corporate entities. Therefore, unavoidably the question arises as to whether the corporation, the individual or both should be held liable. There is currently a general tendency towards criminal liability also for corporate entities and also for marine pollution.[46] Within the scope of this Chapter I will examine whether there are economic arguments in favour of imposing criminal liability on the corporate entity.

From an economic point of view, designating the liable party is unimportant as long as the sanctions are freely transferable and the parties are fully informed. With transferable sanctions, either the corporation charges to the liable employee the fine that it paid, or the employee asks the corporation for reimbursement of the fine that he paid. According to this line of reasoning, it is unimportant whether the fine is imposed on the corporation or the individual because the contractual relationship between the individual and the corporation governs these matters. This, obviously, is an application of the Coase-theorem.[47]

However, this Coasian solution to penalties within corporate entities may not always work in practice. One problem is that monetary sanctions are not always freely transferable between the employer (the corporation) and the employee, sometimes because the law prohibits paying someone else's fine. This is, for example, the case in Germany.[48] A contractual transfer of the fine may thus not always be possible. Therefore, in practice it does matter to whom the penalty is directed.[49]

The reason in favour of holding the employer liable is related to the insolvency problem mentioned earlier. When the corporation is held liable, the less costly fines can be applied in reaction to environmental crime; the corporation (employer) can in turn apply sanctions to the employee, such as refusing promotion or termination of the contract.[50] Also, in the area of environmental crime it is held that when the employee cannot bear the full burden of the optimal penalty, the government might still be able to impose the optimal penalty on the company.[51]

46. For a detailed discussion see G. Heine, 'Marine (Oil) Pollution: Prevention and Protection by Criminal Law – International Perspectives, Corporate and/or Individual Criminal Liability', in *Prevention and Compensation of Marine Pollution Damage: Recent Developments in Europe, China and the US*, ed. M.G. Faure & J. Hu (The Hague: Kluwer Law International, 2006), 41–59.
47. R.H. Coase, 'The Problem of Social Cost', *Journal of Law and Economics* (1960): 1–44.
48. See Faure & Heine, *supra* n. 45, 43–44.
49. Segerson & Tietenberg, *supra* n. 34, 61–63.
50. Compare with the principle of *respondeat superior* in tort law, W. Landes & R. Posner, 'The Positive Economic Theory of Tort Law', *Georgia Law Review* 15 (1981): 914.
51. M.A. Cohen, 'Criminal Law as an Instrument of Environmental Policy: Theory and Empirics', in *Law and Economics of the Environment*, ed. A. Heyes (Cheltenham: Edward Elgar, 2001), 208–209.

4.2 . . . AND ON INDIVIDUALS?

However, the literature equally indicates that despite the strong arguments for corporate criminal liability[52] there are reasons to apply criminal liability to individuals within the corporation as well. We have shown that non-monetary sanctions may have to be applied to environmental crime as well. Indeed, corporate entities may also be equally unable to pay for the damage caused by their pollution. A polluting corporation may escape a liability suit because the damage to the environment is widely dispersed and therefore difficult to trace. Monetary sanctions may exceed the corporation's assets and thus not be an effective deterrent. Because non-monetary sanctions are hard to apply to corporations (this is obvious for incarceration), they must also be applied to individual employees as well. Polinsky and Shavell have made a strong economic argument in favour of the individual criminal liability of the employee. One of the arguments they make is that monitoring by firms may often be imperfect: another problem is that if employees only face fines, they may not be induced to exercise socially optimal levels of care. Therefore, they argue that non-monetary sanctions on employees (imprisonment) should be used in addition to fines applied on the corporations.[53]

4.3 LIMITS OF CORPORATE CRIMINAL LIABILITY

This leads to the conclusion that corporate liability (leading to fines and non-monetary sanctions applicable to corporations) should be combined with the individual criminal liability of employees. However, some criticism has been formulated against corporate criminal liability, stressing even more the need to have individual liability on the part of the employee. Wheeler has argued that, in general, criminal statutes should not be used to regulate product safety.[54] His arguments against criminal law are mainly arguments against punishing the corporation and are, to a certain extent, interesting for environmental crime as well. One of his points is that a civil lawsuit may be as much of a deterrent as criminal sanctions.[55] Another is that criminal law may lead to costly over-deterrence and that the costs of detecting, investigating, prosecuting and sentencing are much

52. L. Friedman, 'In Defence of Corporate Criminal Liability', *Harvard Journal of Law and Public Policy* 23 (2000): 833; and D.M. Kahan, 'Social Meaning and the Economic Analysis of Crime', *Journal of Legal Studies* 27 (1998): 609.
53. A.M. Polinsky & S. Shavell, 'Should Employees Be Subject to Fines and Imprisonment Given the Existence of Corporate Liability?', *International Review of Law and Economics* 13 (1993): 239–257 and L.A. Kornhauser, 'An Economic Analysis of the Choice between Enterprise and Personal Liability for Accidents', *California Law Review* 70 (1982): 1345–1392.
54. N. Wheeler, 'The Use of Criminal Statutes to Regulate Product Safety', *Journal of Legal Studies* (1984): 593–622.
55. Powerful counter arguments have been produced by Shavell (Shavell, *supra* n. 9, 357–374 and Shavell, *supra* n. 9, 271–281) who showed that in specific circumstances, tort law may lack a deterrent effect.

higher than the benefits of marginal deterrence. His argument that the existing market forces in place, such as private tort actions and civil regulatory systems, can better regulate product safety than criminal law might hold true for the field of product liability. However, it is obviously less convincing with regard to the area of environmental crime, precisely because this area requires *ex ante* regulation, without which potential polluters might otherwise escape liability.

Another criticism of the use of corporate criminal liability has been formulated by Jennifer Arlen. She argues that the traditional argument in favour of what is referred to as vicarious liability is seen as an indirect means of sanctioning wrongful employees, assuming that corporations subject to criminal liability will in turn sanction the wrongful agency. This regime can, however, according to Arlen, lead to potentially perverse incentives for the following reason: if a corporation is able to monitor its employees optimally, it will have to increase its level of corporate enforcement expenditures. This might reduce the number of agents who commit crime by increasing the probability of detection and thus reducing the costs of crime. On the other hand, the increased enforcement expenditures may also increase the probability that the government will detect those crimes, whereby the corporations' expected criminal liability for those crimes is increased. This means that additional enforcement by the firm only increases the firms' expected criminal liability. Thus vicarious liability could lead to the perverse incentive that the corporation will reduce the monitoring of its employees in order to avoid the detection of corporate (environmental) crime.[56]

5 CRIMINAL LIABILITY FOR OIL POLLUTION DAMAGE: LEGAL PRACTICE

Having discussed the application of economic theory to the criminal liability for oil pollution damage, I will now briefly discuss a few issues relating to legal policy and practice in this domain. Economic analysis of law as presented earlier will be used to discuss a few of these policy issues.

5.1 Insurance of Fines

An interesting decision of the Hamburg Criminal Court of 24 October 1985 shows a particular problem that arises in the case of enforcement of regulations aiming at the prevention of oil pollution damage, being that fines are often insured by protection and indemnity (P&I) clubs.[57] The case dealt with the following facts: a

56. J. Arlen, 'The Potentially Perverse Effects of Corporate Criminal Liability', *Journal of Legal Studies* (1994): 833–867.

57. Amtsgericht Hamburg, Judgment of 24 Oct. 1985, Nr. 1452b-1108/25. For a discussion see G. Heine, 'Erkennung und Verfolgung von Umweltstraftaten im Europäischen Rechtsraum', *Umwelt- und Planungsrecht* (1987): 289 ff.

Brazilian cargo ship, the MS *Tapajos*, arrived in the harbour of Hamburg. The defendant Milton da Silva was responsible for the operation of the engines of the *Tapajos*, which was run on oil. The water police discovered that the water was polluted with an oil slick of two sea miles in length and 200 m in width. The defendant was found guilty of intentional water pollution (paragraph 324 of the German Criminal Code).

The court held that the behaviour of the defendant caused substantial pollution to the environment. Therefore, the court decided that a severe punishment should be imposed. Given the preventive function of criminal law, the court held that to deter other potential wrongdoers, imprisonment should be imposed, especially because the criminal sanctions that courts actually impose tend to become rapidly known in the maritime world.

Furthermore, the court held that a fine should not be imposed, because this would have no influence on the defendant. In this respect it was noted that it has been known for some time that ocean ships have had a wide range of insurance coverage through P&I clubs for all kinds of risks. In particular, this insurance would also cover fines that would be imposed upon members of the crew. According to the court, even fines for intentional criminal acts have been reimbursed under the P&I insurance coverage.[58] The facts were affirmed by a P&I representative who participated in the court hearing as a witness. For these reasons, the court decided to impose a prison sentence of six months.

In addition, the court had to fix the amount of the Bußgeld.[59] The court again took into consideration the fact that the fines would be covered under the P&I insurance. Therefore, the court decided that the amount of the fine should be high enough to deter the shipowners from polluting the environment. The fine should be higher than the costs of the legal oil removal (oil clean-up or avoidance of spills). If the P&I club would, as a consequence, raise the premiums for insurance against fines for illegal spills, the avoidance of spills would become financially attractive.

---

58. Looking at the various statutes and rules of P&I clubs, applicable at the time of the Hamburg case, one can conclude that the Hamburg Court had a point: full insurance coverage was provided for the fines that the insured has to pay for oil pollution. For instance, rule 14 of the 1985 Statutes and Rules of Gard, a large Scandinavian P&I club, states: 'The association shall cover liability, fines and loss arising in consequence of the discharge or escape from the ship of oil or any other substance or the threat of such discharge or escape.' In addition to this explicit provision for fines as a result of oil pollution, rule 23 had an even broader scope: 'The association shall cover: 1. fines as set out below whether imposed upon the member or upon a member of the crew whom the member may be legally liable to reimburse or reimburse with the approval of the association:

    a. immigration and custom fines;
    b. fines incurred as a result of the conduct of the crew.'

    The rules of Gard were no exception since similar provisions could be found in the rules of other P&I clubs.
59. A *Bußgeld* can be imposed according to the German *Ordnungswidrigkeitengesetz* and is not a criminal sanction but an administrative fine.

The amount of the actual Bußgeld in this case was, however, not mentioned in the decision.

This case is of course highly interesting from an economic perspective. It shows once more the limits of monetary sanctions to be applied to marine oil pollution: given the low probability of detection and the fact that P&I clubs are groups of shipowners, one can assume that a suboptimal control of moral hazard by the insurance company took place. The low detection rate and low fines reduce all incentives of the insurer to control moral hazard. The approach followed by the Hamburg Court to move to the imposition of a non-monetary (prison) sentence can hence be understood, because in this case fines would probably not have led to the desired deterrent effect.[60] The case also shows that it may be important for criminal courts to have adequate information on whether someone else (*in casu* an insurance company) is paying the fine. Theoretically there could still be a sufficient deterrence if fines were set high enough, because this would equally lead to an increase of insurance premiums. However, given the gains obtained with oil pollution and the low probability of detection optimal fines may be spectacularly high. It may run counter to the proportionality principle in criminal law to set these very high fines. Lower fines in combination with the low probability of detection may thus lead to a situation in which insurance amounts to under-deterrence. In that case the application of non-monetary sanctions (imprisonment as in this case or the confiscation of the vessel as was the case in another decision of the Hamburg Court) may then prove to be more effective.

5.2 Ship Source Pollution in the EU

Following the *Erika* incident, the EU has increasingly become involved with ship-source pollution and more particularly with the introduction of criminal sanctions.[61] These regulations have led to a few interesting decisions concerning the power of the European institutions to force Member States to use criminal law to sanction legislation implementing European directives.[62] The starting point was a Council Framework Decision of 12 July 2005 to strengthen the criminal law framework for the enforcement of the law against ship-source pollution.[63] This Council Framework Decision, taken within the framework of the intergovernmental cooperation (the so-called third pillar in the EU) described that Member States had to consider specific infringements as a criminal offence. Moreover, in Article 4

60. For a more detailed analysis see Faure & Heine, *supra* n. 45, 39–58.
61. For an overview see F. Zia-Mansoor, 'International Regime and the EU Developments for Preventing and Controlling Vessel-Source Oil Pollution', *European Environmental Law Review* (2005): 165–173.
62. See on these evolutions M. Faure, 'European Environmental Criminal Law: Do We Really Need It?', *European Environmental Law Review* (2004): 18–29; and M. Faure, 'The Continuing Story of European Environmental Criminal Law in Europe after 23 October 2007', *European Energy and Environmental Law Review* (2008): 68–75.
63. OJ L255/164 of 30 Sep. 2005.

of this Framework Decision, specific penalties were also prescribed. For example, Article 4(1) prescribed that criminal penalties would be of a maximum of at least one to three years of imprisonment. For intentionally committed offences, more severe punishment would apply. Article 5 also stipulated that each Member State shall take the measures necessary to ensure that legal persons can be held liable for offences, and Article 6 provided for specific penalties against legal persons.[64]

Shortly after the pronunciation of this Council Framework Decision, on 7 September 2005, a directive was also accepted on ship-source pollution and on the introduction of penalties for infringements.[65] This directive had as far as the contents is concerned the same goal as the mentioned Framework Decision. The reason that both a directive and a framework decision were promulgated (and published on the same day in the *Official Journal*) is that it was held before that within the so-called first pillar, directives did not have the possibility to provide rules on criminal offences and penalties, whereas this could be done within the powers accorded to the Council in the third pillar. The Directive of 7 September 2005 refers explicitly to the Council Framework Decision in Recital 6.[66] In turn, the framework decision also refers explicitly to the directive in Recital 4. Both should hence be read together. The cooperation was the result of a complicated division of competences (probably not understandable to any non-European lawyer!) whereby on the one hand directives were issued within the first pillar, but the possibilities to issue decisions on security (criminal law) belonged to the third pillar.

This complicated division of labour somewhat changed with a Decision of the European Court of Justice of 13 September 2005. In the judgment in Case 176/03 the court found that although 'as a general rule, neither criminal law nor the rules of criminal procedure fall within the Community competence', 'the last mentioned finding does not prevent the Community legislature, when the application of effective, proportionate and dissuasive criminal penalties by the competent national authorities is an essential measure for combating serious environmental offences, from taking measures that relate to the criminal law of the Member States that it considers necessary to ensure that the rules it lays down on environmental protection are fully effective'.[67] In a communication from the Commission after the judgment, it made clear that the Commission is now competent also to propose provisions concerning the criminal law if this is required for an effective implementation of Community law.[68] In that same communication the Commission also held that as a result of the Court's judgment, many framework decisions were entirely or partly incorrect because they were adopted on the wrong legal basis.

64. Note that an earlier draft of this Framework Decision specified explicitly that fines would not be insurable. See Heine, *supra* n. 46, 53–54.
65. OJ L255/11 of 30 Sep. 2005.
66. It mentions 'This directive is supplemented by detailed rules on criminal offences and penalties as well as other provisions set out in Council Framework Decision 2005/667/JHA of 12 July 2005.'
67. Paragraph 48 of the Decision of 13 Sep. 2005.
68. COM (2005) 583 final.

Therefore, the Commission held that in these cases it will launch appeals wherever this is possible (if the deadline has not expired). That is why the Commission argued on 23 November 2005 that it would appeal to the European Court of Justice for the annulment of the Framework Decision 2005/66 of 12 July 2005 to strengthen the criminal law framework for the enforcement of the law against ship-source pollution.[69] The simple reason was that the Council would have no competence to issue such a framework decision, which would belong to the Commission within the framework of the directive. In a Decision of the European Court of Justice of 23 October 2007 (Case C-440/05) the court indeed annulled Framework Decision 2005/667 for the following reason:

> Since Articles 2, 3 and 5 of Framework Decision 2005/667 are designed to ensure the efficacy of the rules adopted in the field of maritime safety, non-compliance with which may have serious environmental consequences, by requiring Member States to apply criminal penalties to certain forms of conduct, those articles must be regarded as being essentially aimed at improving maritime safety, as well as environmental protection, and could have been validly adopted on the basis of Article 80(2) EC.

However, the European Court of Justice continues with an interesting observation:

> 'By contrast, and contrary to the submission of the Commission, the determination of the type and level of the criminal penalties to be applied does not fall within the Community's sphere of competence'.
>
> 'It follows that the Community legislature may not adopt provisions such as Articles 4 and 6 of the Framework Decision 2005/667, since those articles relate to the type and level of the applicable criminal penalties'.

The consequence of this (admittedly rather complicated) case law is therefore that the Commission can issue directives whereby it forces Member States to use criminal sanctions to enforce regulations, for example concerning ship-source pollution. However, it is not allowed to also prescribe the type and level of penalties, as was the case in Council Framework Decision 2005/667, which was meanwhile annulled. This resulted *inter alia* in a Directive 2008/99 of 19 November 2008 with respect to the protection of the environment through the criminal law.[70] That directive prescribes the use and introduction of criminal sanctions in particular cases, but does not touch upon the type and level of penalties, which is left to the Member States.

Logically the European Commission has after the decision of the European Court of Justice of 23 October 2007 taken the initiative regarding a new directive with respect to ship-source pollution. It is Directive 2009/123/EC of 21 October 2009[71] amends the old directive 2005/35/EC. The directive now provides that each Member State shall take the necessary measures to ensure that the offences referred

69. Paragraph 15 of the Communication of 24 Nov. 2005.
70. OJ L328/28 of 6 Dec. 2008.
71. OJ L280/52 of 27 Nov. 2009.

to are punishable by effective, proportionate and dissuasive criminal penalties, without mentioning the amount of the punishment.

5.3 *Erika*

On 11 December 1999 the tanker *Erika* broke down off the coast of Brittany (France) causing substantial damage to the coast of the French *département* of Brittany. Damage was caused to many victims, including various enterprises and communities, and severe environmental losses were also suffered.

The case was in the first instance dealt with before the Tribunal de Grande Instance de Paris, 11th Chamber, which resulted in a judgment of 16 January 2008 of no less than 278 pages. Even though an appeal has been launched against this judgment it is worthwhile presenting, albeit briefly, a few of the elements of the case as an example of how in legal practice criminal law deals with these major cases of oil spills in Europe.

The public ministry launched criminal prosecutions against various individuals and legal entities; in addition under French law it is also possible that victims constitute themselves as so-called civil parties and claim compensation within the criminal court. Given the focus in this chapter on criminal law, I will not discuss the civil aspects of the case.

It is first of all interesting to mention that the court discusses a great number of persons and legal entities involved in the disaster. The judgment discusses in detail the contribution of all persons involved that eventually led to the disaster. A first individual involved is M. Savarese who, with his company Tevere Shipping, is the owner of the tanker. The tribunal is critical of the fact that M. Savarese created for each of the four tankers that he owned several legal entities. All that Tevere Shipping hence possessed was the tanker *Erika*. The court criticizes this practice of creating a 'single ship company' arguing that it is as such not illegal but could of course create an insolvency problem.

The second main player and individual involved was M. Pollara. He had been contacted by the ship owner Savarese to execute work on the *Erika*.

A third party involved is a legal entity, being the classification society Rina. According to an expert that examined the ship, the tanker did not conform to the conditions and requirements imposed by Rina and should hence not have been accredited.

The final entity involved was the well-known French oil company Total, which was, to put it simply, the owner of the cargo.[72]

The judgment held that all parties involved had in some way or another contributed to the *Erika* disaster. M. Savarese obtained certificates concerning the *Erika* that did not correspond with the real quality of the ship; he had, moreover, ordered changes to the *Erika* that caused difficulties for an older ship that was

72. Obviously the main facts of the judgment, which consists of 278 pages, are summarized here.

already heavily rusted.[73] Moreover, Savarese knew that reparation was still necessary and nevertheless he decided still to use the ship. The court is less severe of M. Pollara. M. Pollara is criticized for allowing the *Erika* to sail even though it was in bad condition. On the other hand it is acknowledged that even if M. Pollara would have told M. Savarese not to use the tanker anymore, this would probably not have changed the course of events. The classification society Rina is blamed for the fact of providing an International Safety Management (ISN) certificate for Erika even tough it was known that there were technical shortcomings. A new certificate was issued without verifying whether these shortcomings had been repaired.[74] Finally Total is criticized for accepting a ship of bad quality like the *Erika* for transporting its cargo (oil). Moreover, Total is particularly criticized by the court because Total possessed significant expertise that could have enabled the quality of the ship to be verified by Total.[75]

On the basis of this analysis of the involvement of all these various parties the court held that each of these individuals and entities committed faults that led to their criminal liability. The court more particularly applies the Act Nr. 83–583 of 5 July 1983, which creates criminal liability for pollution following an accident at sea that results from a fault. This act imposes criminal liability on all of those (individuals or legal persons) who have contributed to this event. The court holds that M. Savarese clearly had management power and control over the ship. The same was, according to the court, the case for M. Pollara, given the work he executed on the *Erika*. The classification society Rina executed yearly inspections of the *Erika* and hence had a clear power of control over the ship. The company Total inspected the tanker *Erika*. Hence, even though Total did not have formal powers of control over *Erika* (in the legal sense) it effectively executed those powers. As a result, all four parties were declared criminally liable for the offence of pollution committed at the occasion of the *Erika* disaster.[76]

This accumulation of criminal liability of natural persons and legal entities corresponds with the principles mentioned earlier concerning corporate criminal liability.[77] The Court clearly shows that each of the four persons involved had in some way the power to take care to avoid the accident and omitted to do so.

The penalties imposed seem, however, relatively modest from an economic perspective. The legal entities Rina and Total are both ordered to pay a fine of Euro (EUR) 375,000 each and the individuals (Savarese and Pollara) to pay a fine of EUR 75,000.[78] However, the Court can hardly be blamed because it indicates that in both cases this is the maximum amount of the fine that can be imposed. Obviously, given the principles of deterrence explained previously,[79] one can wonder whether these fines are sufficiently deterrent. On the one hand the probability of

73. *Erika* judgment, 208.
74. *Erika* judgment, 210.
75. *Erika* judgment, 214–218.
76. *Erika* judgment, 226–228.
77. See *supra* s. 4.
78. *Erika* judgment, 233.
79. See s. 3.

detection was of course large *ex ante* (after the accident), but before the accident occurred the probability to detect that a ship of inadequate quality was put into circulation was not that high at all. The potential gains to all persons involved (who all exercised their professions) may be substantial, and moreover the damage to society was huge. Hence, the maximum fines fixed in the French legislation are from a deterrence perspective probably low.

An interesting aspect is, however, that also substantial amounts of compensation are awarded to many of the victims. The four defendants are held jointly and separately liable to pay these amounts. For example, to the French State more than EUR 153,000,000 had to be paid and also substantial amounts to many other individuals involved. These liability actions may hence provide substantial additional deterrence, provided of course that no insolvency problem arises. For Total that may not be the case, but it may be different for the other agents involved. An interesting aspect that arose during the civil case is that some victims had asked for the publication of the judgment to be ordered. This was, however, rejected by the Court, arguing that the *Erika* disaster had already received sufficient attention in the media.

In sum, this case shows that although it may be possible to apply criminal sanctions to many responsible parties involved (both natural persons and legal entities), it is not certain that the criminal sanction in this particular case did indeed have a substantial deterrent effect given the low penalties (fines) available in French legislation. The complementary application of liability law may thus be required to generate the necessary deterrent effect.

5.4 Empirical Evidence

Quite a bit of empirical literature has meanwhile been executed concerning the effectiveness of the regulation and its enforcement concerning marine pollution. Economists have always been interested in the question of how accidental or voluntary marine oil pollution by tankers could be prevented optimally either by investments in the safety of the tankers (in case of accidental pollution) or by increasing detection (in case of voluntary discharge).[80] The problem of detection of oil spills has been modelled by Epple and Visscher.[81] They show how vessel size, the price of oil, the enforcement of pollution control regulations and the risk associated with variance in spill size affect the oil transporters' decisions concerning expenditures on measures for spill prevention. They provide empirical data to support their theoretical analysis. Cohen[82] has followed up on their work by

80. See, e.g., P. Burrows, C.K. Rowley & D. Owen, 'The Economics of Accidental Oil Pollution by Tankers in Coastal Waters', *Journal of Public Economics* 3 (1974): 251–268.
81. D. Epple & M. Visscher, 'Environmental Pollution: Modeling Occurrence, Detection, and Deterrence', *Journal of Law and Economics* 27 (1984): 29–60.
82. M.A. Cohen, 'Optimal Enforcement Strategy to Prevent Oil Spills: An Application of a Principal-Agent Model with Moral Hazard', *Journal of Law and Economics* 30 (1987): 23–51.

providing an optimal enforcement strategy to prevent oil spills. We can also point to a paper by Dunford,[83] addressing the natural resource damages from oil spills, addressing the question of under what kind of conditions can there be liability under the US Oil Pollution Act for natural resource damages. The recovery for economic loss following the *Exxon Valdez* oil spill has been examined by Goldberg.[84]

In a recent paper, Cohen also provides an overview of the available empirical literature on the deterrent effect of monetary penalties on oil spills.[85] He discusses *inter alia* a study by Weber and Crew showing that not only may more severe penalties have a deterrent effect but also that the speed of punishment may reduce oil spillage as well.[86] They conclude, just as Cohen did in another study[87] that the statutory maximum penalty in the United States of USD 5,000 is too small compared to the optimal penalty required.

Hendriks also provides an empirical analysis of the effectiveness of regulation of maritime pollution (both regulation as well as civil penalties).[88] Hendriks uses data from the Oil Spill Intelligence Report (OSIR) which provides an oil spill database. This provides *inter alia* information on the number of oil spills of between 10,000 and 100,000 gallons. Hendriks argues that the data show that whenever a new piece of legislation concerning oil spills is enacted or whenever a safer transport technology becomes available (such as the enactment of the 1990 US Oil Pollution Act), this had a strong influence on the number of spills, not only in 1991, but also in subsequent years.[89] He shows that, especially for the United States, there has been since 1990 a significant downward trend.

Less positive news, however, is that for a higher proportion of spills the polluter cannot be identified. Put cynically, Hendriks argues that as a result of stricter liability polluters might have improved their skill of concealing their identity.[90] Different from what one would hence expect (that increased detection technology would lead to a higher probability of detection), the reverse seems to be the case. He also shows that a higher proportion of polluters are unable to pay for the damages they are liable for and that hence a larger proportion of costs is born by the International Oil Pollution Compensation Fund (IOPC).

83. R. Dunford, 'Natural Resource Damages from Oil Spills', in *Innovation in Environmental Policy, Economic and Legal Aspects of Recent Developments in Environmental Enforcement and Liability*, ed. T.H. Tietenberg (Brookfield: Edward Elgar, 1992), 165–193.
84. V.P. Goldberg, 'Recovery for Economic Loss Following the Exxon Valdez Oil Spill', *Journal of Legal Studies* 23 (1994): 1–39.
85. Cohen, *supra* n. 14, 25–39.
86. J.M. Weber & R.E. Crew, 'Deterrence Theory and Marine Oil Spills: Do Coastguard Civil Penalties Deter Pollution?', *Journal of Environmental Management* 58 (2000): 161.
87. M.A. Cohen, 'The Costs and Benefits of Oil Spill Prevention and Enforcement', *Journal of Environmental Economics and Management* 13 (1986): 167–188.
88. R. Hendriks, 'Maritime Oil Pollution: An Empirical Analysis', in *Shifts in Compensation for Environmental Damage*, ed. M. Faure & A. Verheij (Vienna: Springer, 2007), 243–260.
89. *Ibid.*, 248.
90. *Ibid.*, 257.

In summarizing some of this empirical evidence one is mindful that on the one hand empirical evidence shows that increased monetary penalties and increased detection activities (e.g., of the Coast Guard) lead to increased deterrence, but that on the other hand the maximum penalties available in statutes may be too low and increased civil liabilities for oil pollution damage may lead to the perverse effect of polluters increasing their skill in concealing their identity, thus lowering the probability of detection.

6 CONCLUSION

The goal of this contribution was to show the usefulness of the economic analysis of law in analyzing the application of the criminal law to marine oil pollution. Of course within the scope of this contribution it was not possible to provide details on the applicable rules concerning criminal liability. Still, the advantage of presenting the economic analysis of law was that a few indications could be provided on how criminal law could be shaped if an optimal use of this legal instrument were to be made.

The belief in the criminal law seems to be strong, especially when large-scale pollution disasters occur. Oil pollution disasters that receive a lot of media attention often prompt calls in the media for the application of the criminal law. These calls have apparently also reached the EU, which recently launched various proposals to introduce criminal penalties for ship-source pollution. The main argument underlying this trend is that a similar deterrent effect could not be reached with either civil or administrative penalties.

This contribution started by precisely addressing that question: why would society need the criminal law to enforce regulation aiming at the prevention of marine oil pollution? Given the fact that gains from oil spills can be large, that harm to society can be large as well and that the probability of detection may be low, the case for the criminal law is strong. Indeed, because civil sanctions do not have the possibility to provide a remedy high enough to outweigh for the low probability of detection, civil penalties alone will not provide a sufficient deterrent effect. Administrative fines may fulfil this function, but they may reach an insolvency limit, in which case non-monetary sanctions would have to be applied.

A brief look at legal policy and practice, however, shows that still much is left to be desired in that respect. Recent empirical evidence shows that the number of unidentified oil spills is increasing rather than decreasing (pointing at an even lowering of the probability of detection) and the available maximum sanctions seem to be low. For example, the *Erika* case showed that even when judges apply the maximum statutory penalty it is doubtful that this will provide sufficient deterrence. Hence, non-monetary sanctions will have to be applied. The original decisions of the Hamburg Court showed what is possible in that respect: where fines were insured by the P&I club, the court refused to impose a fine and condemned a person liable of pollution to a prison sanction. In another case, confiscation of the vessel was ordered. These types of non-monetary sanctions (e.g., confiscation of illegal gain and a duty to clean up damage caused by the oil pollution) may in some cases be more effective than either a fine or the imposition of a prison sanction. They may serve the environmental interests better and at the same time (because they are costly) also provide deterrence.

The *Erika* case also showed the importance of addressing the criminal action against all those involved in an oil pollution case, for example, who contributed to the accident. In that respect it is important to address the criminal sanction against both individuals and against legal entities. In the latter case, the penalties need of course to be adapted and again sufficiently high to provide adequate deterrence.

This chapter provides an introduction to the importance of economic analysis for criminal law with respect to marine oil pollution. Many other interesting aspects could be studied, such as the question of which jurisdiction may be optimal from an economic perspective (flag state, coastal state or port state control). Also, the optimal combination of civil penalties and criminal liability (which was also at stake in the *Erika* case) may need further research as well. Much still remains for further study in this interesting domain of economic analysis of the criminal law applicable to marine oil pollution.

References

Arlen, J. 'The Potentially Perverse Effects of Corporate Criminal Liability'. *Journal of Legal Studies* 23 (1994): 833–867.

Babbitt, Ch.J., D.C. Cory & B.L. Kruchek. 'Discretion and Criminalization of Environmental Law'. *Duke Environmental Law and Policy Forum* 15 (2004): 1–64.

Barrett, J.J. 'Sentencing Environmental Crimes under the United States Sentencing Guidelines: A Sentencing Lottery'. *Environmental Law* 22 (1992): 1421–1449.

Becker, G.S. 'Irrational Behaviour and Economic Theory'. *Journal of Political Economy* 70 (1962): 1–13

Becker, G.S. 'Crime and Punishment: An Economic Approach'. *Journal of Political Economy* 76 (1968): 169–216.

Billiet, C.M. & S. Rousseau. 'De hoogte van strafrechtelijke boetes. Een rechtseconomische analyse van milieurechtspraak (1990–2000) van het Hof van Beroep te Gent'. *Tijdschrift voor Milieurecht* 2 (2003): 120–134.

Bowles, R. *Law and the Economy.* Oxford: Martin Robertson, 1982.

Bowles, R., M. Faure & N. Garoupa. 'Economic Analysis of the Removal of Illegal Gains'. *International Review of Law and Economics* 20 (2000): 537–549.

Bowles, R., M. Faure & N. Garoupa. 'Forfeiture of Illegal Gain: An Economic Perspective'. *Oxford Journal of Legal Studies* 25 (2005): 275–295.

Bowles, R., M. Faure & N. Garoupa. 'The Scope of Criminal Law and Criminal Sanctions: An Economic View and Policy Implications'. *Journal of Law and Society* 35, no. 3 (2008): 389–416.

Boyd, J. 'Compensation for Oil Pollution Damages: The American Oil Pollution Act as an Example for Global Solutions?. In *Prevention and Compensation of*

Marine Pollution Damage: Recent Developments in Europe, China and the US, edited by M.G. Faure & J. Hu. The Hague: Kluwer Law International, 2006.

Burrows, P., C.K. Rowley & D. Owen, 'The Economics of Accidental Oil Pollution by Tankers in Coastal Waters'. *Journal of Public Economics* 3 (1974): 251–268.

Coase, R.H. 'The Problem of Social Cost'. *Journal of Law and Economics* 3 (1960): 1–44.

Coffee, J.S. 'Corporate Crime and Punishment: A Non-Chicago View of the Economics of Criminal Sanctions'. *American Criminal Law Review* 17 (1980): 419–476.

Cohen, M.A. 'Criminal Law as an Instrument of Environmental Policy: Theory and Empirics'. In *Law and Economics of the Environment*, edited by A. Heyes. Cheltenham: Edward Elgar, 2001, 208–209.

Cohen, M.A. 'Optimal Enforcement Strategy to Prevent Oil Spills: An Application of a Principal-Agent Model with Moral Hazard'. *Journal of Law and Economics* 30 (1987): 23–51.

Cohen, M.A. 'The Costs and Benefits of Oil Spill Prevention and Enforcement'. *Journal of Environmental Economics and Management* 13 (1986): 167–188.

Cohen, M.A. 'Oil Pollution Prevention and Enforcement Measures and Their Effectiveness: A Survey of Empirical Research from the US'. In *Prevention and Compensation of Marine Pollution Damage: Recent Developments in Europe, China and the US*, edited by M.G. Faure & J. Hu. The Hague: Kluwer Law International, 2006.

Cooter, R.D. 'Economic Analysis of Punitive Damages'. *Southern California Law Review* 56 (1982): 79–101.

Cooter, R.D. & T. Ulen. *Law and Economics*, 4th edn. Boston: Addison Wesley, 2004.

Cortenraad, W.H.F.M. *The Corporate Paradox: Economic Realities of the Corporate Form of Organization.* Boston: Kluwer Academic, 1999.

Cummins Ph.A., et al. 'Oil Tanker Pollution Control: Design Criteria vs. Effective Liability Assessment'. *Journal of Maritime Law and Commerce* (1975): 169–206.

Dunford, R. 'Natural Resource Damages from Oil Spills'. In *Innovation in Environmental Policy, Economic and Legal Aspects of Recent Developments in Environmental Enforcement and Liability*, edited by T.H. Tietenberg. Brookfield, VT: Edward Elgar, 1992.

Easterbrook, F. 'Criminal Procedure as a Market System'. *Journal of Legal Studies* 12 (1983): 295, 325.

Epple, D. & M. Visscher, 'Environmental Pollution: Modelling Occurrence, Detection, and Deterrence'. *Journal of Law and Economics* 27 (1984): 29–60.

Faure, M. 'European Environmental Criminal Law: Do We Really Need It?'. *European Environmental Law Review* 18 (2004): 18–29.

Faure, M. 'Gewinnabschöpfung und Verfall auf ökonomischen Prüfstand'. In *Menschengerechtes Strafrecht. Festschrift für Albin Eser zum 70. Geburtstag*, edited by J. Arnold et al. München: Beck, 2005.
Faure, M. 'The Continuing Story of European Environmental Criminal Law in Europe after 23 October 2007'. *European Energy and Environmental Law Review* (2008): 68–75.
Faure, M. & Th. De Roos (eds). *De berekening van het wederrechterlijk verkregen voordeel uit milieudelicten*. The Hague: SDU Uitgevers, 1998.
Faure, M. & G. Heine. 'The Insurance of Fines: The Case of Oil Pollution'. *The Geneva Papers on Risk and Insurance* 16 (1991): 54–55.
Faure, M. & G. Skogh. *Economic Analysis of Environmental Policy and Law: An Introduction*. Cheltenham: Edward Elgar, 2003.
Faure, M. & M. Visser. 'Law and Economics of Environmental Crime'. In *New Perspectives on Economic Crimes*, edited by H. Sjögren & G. Skogh. Cheltenham: Edward Elgar, 2004).
Faure, M. & H. Wang. 'Economic Analysis of Compensation for Oil Pollution Damage'. *Journal of Maritime Law and Commerce* 37 (2006): 179–217.
Friedman, L. 'In Defence of Corporate Criminal Liability'. *Harvard Journal of Law and Public Policy* 23 (2000): 833–859.
Garoupa, N. & F. Gomez-Pomar. 'Punish Once or Punish Twice: A Theory of the Use of Criminal Sanctions in Addition to Regulatory Penalties'. *American Law and Economics Review* 6 (2004): 410–433.
Goldberg, V.P. 'Recovery for Economic Loss Following the Exxon Valdez Oil Spill'. *Journal of Legal Studies* 23 (1994): 1–39.
Hansmann, H. & R. Kraakman. 'Toward Unlimited Shareholder Liability for Corporate Torts'. *Yale Law Journal* 100 (1991): 1879–1934.
Heine, G. 'Erkennung und Verfolgung von Umweltstraftaten im Europäischen Rechtsraum'. *Umwelt- und Planungsrecht* 2 (1987): 89 ff.
Heine, G. 'Marine (Oil) Pollution: Prevention and Protection by Criminal Law – International Perspectives, Corporate and/or Individual Criminal Liability'. In *Prevention and Compensation of Marine Pollution Damage: Recent Developments in Europe, China and the US*, edited by M.G. Faure & J. Hu. The Hague: Kluwer Law International, 2006.
Heine, G. & V. Meinberg. *Empfehlen sich Änderungen im Strafrechtlichen Umweltschutz, Ins Besondere in Verbindung mit dem Verwaltungsrecht? Gutachten für den 57. Deutschen Juristentag*. München: Beck, 1988.
Hendriks, R. 'Maritime Oil Pollution: An Empirical Analysis'. In *Shifts in Compensation for Environmental Damage*, edited by M. Faure & A. Verheij. Vienna: Springer, 2007.
Heyes, A. (ed.). *Law and Economics of the Environment*. Cheltenham: Edward Elgar, 2001.
Kahan, D.M. 'Social Meaning and the Economic Analysis of Crime'. *Journal of Legal Studies* 27 (1998): 609–622.

Kilchling, M. 'Comparative Perspectives on Forfeiture Legislation in Europe and the United States'. *European Journal of Crime, Criminal Law and Criminal Justice* 5 (1997): 342–361.

Kilchling, M. (ed.). *Die Praxis der Gewinnabschöpfung in Europa.* Freiburg im Breisgau: Max Planck Institute for Foreign and International Criminal Law, 2002.

Kornhauser, L.A. 'An Economic Analysis of the Choice between Enterprise and Personal Liability for Accidents'. *California Law Review* 70 (1982): 1345–1392.

Landes, W. & R. Posner. 'An Economic Theory of Intentional Tort'. *International Review of Law and Economics* 1 (1981): 127–154.

Landes, W. & R. Posner. 'The Positive Economic Theory of Tort Law'. *Georgia Law Review* 15 (1981): 914.

Meinberg, V. 'Empirische Erkenntnisse zum Vollzug des Umweltstrafrechts'. *Zeitschrift für die Gesamte Strafrechtswissenschaften* (1988): 112–157.

Miceli, T. 'Optimal Prosecution of Defendants Whose Guilt Is Uncertain'. *Journal of Law, Economics and Organisation* 6 (1990): 189–201.

Ogus, A.I. 'Enforcing Regulation: Do We Need the Criminal Law?' In *New Perspectives on Economic Crimes*, edited by H. Sjögren & G. Skogh. Cheltenham: Edward Elgar, 2004.

Ogus, A. & C. Abbot. 'Pollution and Penalties'. In *An Introduction to the Law and Economics of Environmental Policy: Issues in Institutional Design*, edited by T. Swanson. The Hague: Elsevier, 2002.

Ogus, A. & C. Abbot. 'Sanctions for Pollution: Do We Have the Right Regime?'. *Journal of Environmental Law* 14 (2002): 283–300.

Parpworth, N. 'Enforcement in Environmental Regulation: The Defra Review'. *Environmental Liability* (2007): 15–26.

Pedrick, J.L. 'Tank Ship Design Regulation and Its Economic Effect on Oil Consumers'. *Journal of Maritime Law and Commerce* 9 (1978): 377–395.

Perron, W. 'Vermögensstrafe und Erweiterte Verfall'. *Juristenzeitung* (1993): 919–920.

Polinsky, M. & S. Shavell. 'A Note on Optimal Fines When Wealth Varies among Individuals'. *American Economic Review* 81 (1991): 618–621.

Polinsky, A.M. & S. Shavell. 'Should Employees Be Subject to Fines and Imprisonment Given the Existence of Corporate Liability?'. *International Review of Law and Economics* 13 (1993): 239–257.

Polinsky, A.M. & S. Shavell. 'The Optimal Trade Off between the Probability and the Magnitude of Fines'. *American Economic Review* 69 (1979): 880–891.

Posner, R. 'An Economic Theory of the Criminal Law'. *Columbia Law Review* 85 (1985): 1193–1209.

Posner, R. 'Optimal Sentences for White-Collar Criminals'. *American Criminal Law Review* (1980): 400–418.

Segerson, K. & T. Tietenberg. 'Defining Efficient Sanctions'. In *Innovation in Environmental Policy*, edited by T.H. Tietenberg. Cheltenham: Edward Elgar, 1992.

Shavell, S. 'A Model of the Optimal Use of Liability and Safety Regulation'. *RAND Journal of Economics* 15 (1984): 271–280.

Shavell, S. 'Criminal Law and the Optimal Use of Non-monetary Sanctions as a Deterrent'. *Columbia Law Review* 85 (1985): 1232–1265.

Shavell, S. 'Liability for Harm versus Regulation of Safety'. *Journal of Legal Studies* 13 (1984): 357–374.

Skogh, G. 'A Note on Gary Becker's Crime and Punishment: An Economic Approach'. *Swedish Journal of Economics* 75 (1973): 305–311.

Skogh, G. & C. Stuart. 'An Economic Analysis of Crime Rates, Punishment and the Social Consequences of Crime'. *Public Choice* 38 (1982): 171–179.

Smettan, J.R. *Kriminelle Bereicherung in Abhängigkeit von Gewinnen, Risiken, Strafen und Moral*. Freiburg im Breisgau: Max Planck Institute for Foreign and International Criminal Law, 1992.

Verheij, A. 'Shifts in Governance: Oil Pollution'. In *Shifts in Compensation for Environmental Damage*, edited by M. Faure & A. Verheij. Vienna: Springer, 2007.

Wang, H. 'The EU Marine Oil Pollution Regime – Recent Developments'. *European Environmental Law Review* 15 (2004): 292–303.

Wang, H. 'Transboundary Vessel-Source Marine Pollution – International Legal Framework and Its Application to China'. In *China and International Environmental Liability. Legal Remedies for Transboundary Pollution*, edited by M. Faure & Y. Song. Cheltenham: Edward Elgar, 2008.

Wang, H. 'Shifts in Governance in the International Regime of Marine Oil Pollution Compensation: A Legal History Perspective'. In *Shifts in Compensation for Environmental Damage*, edited by M. Faure & A. Verheij. Vienna: Springer, 2007.

Weber, J.M. & R.E. Crew. 'Deterrence Theory and Marine Oil Spills: Do Coastguard Civil Penalties Deter Pollution?'. *Journal of Environmental Management* 58 (2000): 161–168.

Wheeler, N. 'The Use of Criminal Statutes to Regulate Product Safety'. *Journal of Legal Studies* 13 (1984): 593–622.

Wilde, M. *Civil Liability for Environmental Damage: A Comparative Analysis of Law and Policy in Europe and the United States*. The Hague: Kluwer Law International, 2002.

Wood, L.D. 'An Integrated International and Domestic Approach to Civil Liability for Vessel-Source Oil Pollution'. *Journal of Maritime Law and Commerce* (1975): 2.

Zia-Mansoor, F. 'International Regime and the EU Developments for Preventing and Controlling Vessel-Source Oil Pollution'. *European Environmental Law Review* (2005): 165–173.

Chapter 11

Criminal Liability and Vessel-Source Pollution in the European Union and the United States: Inspiration for the Prevention of Vessel-Source Pollution in China?

Liu Nengye & Frank Maes***

1 INTRODUCTION

Vessel-source pollution accounts for 12% of all marine pollution. However, this pollution often affects more than one state alone. Furthermore, shipping is an activity with intense communication between different states and individuals and is regulated by international conventions, including, for example, the United Nations Convention on the Law of the Sea (UNCLOS).[1] Following particularly serious disasters such as the *Torrey Canyon* (1967), the *Amoco Cadiz* (1987), the *Exxon Valdez* (1989), the *Erika* (1999) and the *Prestige* (2002), a series of

* Leading author; PhD Candidate, Department of Public International Law/Maritime Institute, Faculty of Law, Ghent University Postal Address: Universiteitstraat 4, B9000 Gent, Belgium. Tel.: +32(0)9 264 8441, Fax: +32 (0)9 264 6989, e-mail: nengye.liu@ugent.be.

** Professor, Department of Public International Law/Maritime Institute, Faculty of Law, Ghent University.

1. E.J. Molenaar, *Coastal State Jurisdiction over Vessel-Source Pollution* (The Hague: Kluwer Law International, 1998), 18–19.

Michael G. Faure, Han Lixin & Shan Hongjun, *Maritime Pollution Liability and Policy: China, Europe and the US*, 193–214.

international, regional and national regulations were established as remedies to promote the prevention of marine pollution caused by accidents and as a political reaction to public anger. However, the most obvious weakness of the regulatory system appears to be its failure to ensure effective enforcement of and compliance with the relevant rules and standards.[2] In general, compliance with treaty obligations can be seen as a function of three factors – legal authority, capacity and incentives.[3] It is known that the peculiar features of the shipping industry have led to a general lack of incentives for compliance with safety and pollution control rules.[4] In addition to preventive measures, civil and criminal liability are considered a strong incentive on the part of interested parties to put in place controls and carry out measures with a view to prevent pollution.[5]

As 'A Constitution for the Oceans', part of the UNCLOS balances rights and obligations of states for the protection of the marine environment. It sets a minimum standard for flag States (whose own laws and regulations 'shall at least have the same effect') under Article 211(2) of UNCLOS.[6] Articles 94 (general terms), 217 (enforcement obligation) and 228 (enforcement jurisdiction of flag state preceding over coastal state) also show that the UNCLOS places the main duty on flag states to prevent vessel-source pollution. Meanwhile, Article 218 introduces a universal enforcement jurisdiction of port states in the case of breaches of international rules and standards. Because ports lie wholly within a state's territory and therefore fall under its territorial sovereignty, customary international law acknowledges a port state's wide discretion in exercising jurisdiction in its port.[7] The International Court of Justice in the Nicaragua case confirmed this,[8] as do Articles 25(2), 211(3) and 255 of UNCLOS. A port state has jurisdiction over any discharge from a vessel, even if the discharge occurred outside its internal waters, territorial sea or exclusive economic zone (EEZ) (Article 218(1)). Therefore, discharges are covered on the high seas, irrespective of any damage, whereas the coastal state's jurisdiction is restricted to its territorial sea and EEZ. What is required is that applicable international rules and standards, such as the

---

2. P.S. Dempsey, 'Compliance and Enforcement in International Law – Oil Pollution of the Marine Environment by Ocean Vessels', *Nw. Journal of International Law and Business* 6 (1984): 459, 541.
3. R.B. Mitchell, *International Oil Pollution at Sea – Environmental Policy and Treaty Compliance* (Cambridge: MIT Press, 1994), 11.
4. A. Khee-jin Tan, *Vessel Source Marine Pollution: The Law and Politics of International Regulation* (Cambridge: Cambridge University Press, 2006), 5–6.
5. G.M. Gauci, 'Protection of the Marine Environment through the International Ship-Source Oil Pollution Compensation Regimes', *Review of European Community & International Environmental Law* 8 (1999): 30.
6. P.W. Birnie & A. Boyle, *International Law and the Environment*, 2nd edn (Oxford: Oxford University Press, 1992), 273.
7. E.J. Molenaar, 'Port State Jurisdiction: Toward Comprehensive, Mandatory and Global Coverage', *Ocean Development & International Law* 38 (2007): 227.
8. Case concerning Military and Paramilitary Activities in and against Nicaragua (*Nicaragua v. United States of America*), [1986] I.C.J. Rep., at. 111, para. 123. For details about the case, see <en.wikipedia.org/wiki/Nicaragua_v._United_States#Judgment> 29 Oct. 2008.

International Convention for the Prevention of Pollution from Ships 1973, as modified by the Protocol of 1978 (MARPOL) have been violated.[9] In addition to port state jurisdiction, coastal state jurisdiction is also an important part of the safety net for the protection of the marine environment,[10] which also varies in different maritime zones. Enforcement by the coastal state is governed largely by Article 220. However, proceedings instituted by the coastal state should be subject to section 7 of Part XII. For example, section 7 of Article 230 provides that penalties other than monetary ones can be imposed only in case of a wilful and serious act of pollution in the territorial sea.

Article 4(4) MARPOL states that penalties specified under the law of a party pursuant to the present article are to discourage violations of the Convention. The penalties are to be adequate in severity and equally severe, irrespective of where the violations occur. Although MARPOL itself is silent on criminal liability and leaves it to the discretion of Member States, there is no doubt that criminal punishment is contemplated by MARPOL against polluters, especially for intentional discharge.[11] According to Annex I (Chapter II, Regulation 9, 10, 11) and Annex II (Regulation 5) of MARPOL, there is a distinction between operational and accidental discharges. Although operational discharges are deliberate discharges of fuel oil, oily mixtures, oil waste, noxious liquid substances and so on, accidental discharges are discharges resulting from accidents such as collision, rounding, and so forth. Regulation 11 of Annex I and Regulation 6 of Annex II[12] allow for discharges in four cases: (1) operational discharges that meet the discharge requirements of MARPOL; (2) a discharge in case of necessity or force majeure and necessary to save vessels of crew; (3) a discharge approved by the Administration for purposes of combating pollution and (4) a discharge as a result of an unexpected damage to ship and equipment when suitable precautions are subsequently taken and there was no intent or recklessness on the part of the owner, the master or other crew members of the vessel.[13]

---

9. M.G. Faure & J. Hu, *Prevention and Compensation of Marine Pollution Damage: Recent Development in Europe, China and the US* (Alphen aan den Rijn: Kluwer Law International, 2006), 46.
10. The 'coastal states' are shoreline states that seek to regulate the movement and activities of foreign vessels in or near the waters adjacent to their coasts. It should be noted that the expressions 'coastal state' or 'maritime state' are largely terms of convenience, not mutually exclusive today. See Gauci, *supra* n. 5, 30–31.
11. O.G. Anthony, 'Criminalization of Seafarers for Accidental Discharge of Oil: Is There Jurisdiction in International Law for Criminal Sanction for Negligent or Accidental Pollution of the Sea?', *Journal of Maritime law and Commerce* 37 (2006): 230.
12. Exceptions: (a) Discharges necessary for the purpose of securing the safety of a ship or to save a life at sea. (b) Discharge resulting from damage to a ship or its equipment: (i) provided that all reasonable precautions have been taken after the occurrence of the damage or discovery of the discharge for the purpose of preventing or minimizing the discharge. (ii) Except if the owner or the master acted either with intent to cause damage, or recklessly and with knowledge that damage probably result'. (c) Discharges approved by the Administration, precisely for purposes of combating pollution.
13. J.-M.M. Osante, 'Competition and the European Union Directive on Criminal Penalties for Ship-Source Pollution', *Journal of International Maritime Law* 14 (2008): 419, 420.

Ensuring compliance by members of the international community with their international environmental obligations continues to be a matter of increasing concern.[14] As major economies of the world, the European Union (EU), the United States and China all have essential maritime interests. Not only does maritime transport play a leading role in foreign trade and energy supply, the protection of the marine environment is also crucial for their sustainable development. This paper explores state practice in both the EU and the United States on imposing criminal liability to create incentives for the prevention of vessel-source pollution. Are there any differences between them and can their practice inspire China to act accordingly?

2 PRACTICE OF THE EUROPEAN UNION

2.1 BACKGROUND

The Community action to combat marine pollution started with a Council Resolution of 26 June 1978, which was adopted in response to the oil spill caused by the *Amoco Cadiz* incident in 1978. The reduction of pollution from shipping has formed an integral part of the EU maritime safety policy from its beginning in 1993.[15] Indeed, around half of the more than forty EU regulatory instruments (directives and regulations) dealing with maritime safety have an explicit objective to protect the environment, and almost all maritime safety measures contribute at least indirectly to that objective.[16]

At the present time, the EU is the only international organization that is a party to UNCLOS. However, all Member States are signatories to the MARPOL and the EU is not, because membership of International Maritime Organization (IMO) and IMO conventions is exclusively reserved to States.[17] The EU Member States are generally viewed as being supportive of IMO, but the EU is sometimes accused of regional activism and unilateralism.[18] This cannot be properly understood without an appreciation of the political dynamics of the EU institutions. The political philosophy of the EU is the division of competence between Member States on the one hand and Community institutions on the other. With the ever closer integration being forged among the EU Member States, supra-national institutions like the European Commission are increasingly assuming an activist stance, seeking greater competence to initiate legislation for the Union.[19] The EU can impose

14. P. Sands, *Principles of International Environmental Law* (Cambridge: Cambridge University Press, 2003), 171.
15. COM (1993) 66 final.
16. EMSA leaflet: Preventing Pollution from Ships <extranet.emsa.europa.eu/index.php?option=com_docman&task=cat_view&gid=148&Itemid=193> 30 Oct. 2008.
17. Former IMO Secretary-General, W. O'Neil, at MEPC 50, 4 Dec. 2003.
18. See A. Boyle, 'EU Unilateralism and the Law of the Sea', *International Journal of Marine and Coastal Law* 21 (2006): 15 et seq.
19. See Khee-jin Tan, *supra* n. 4, 86–87.

binding rules on its Member States through its institutions. Moreover, the Commission is entrusted to ensure application of Community law (Article 211, EC Treaty). If the Commission considers that a Member State has failed to fulfil an obligation under the Treaty, it shall deliver a reasoned opinion on the matter after giving the State concerned the opportunity to submit its observations. In the case of continuous disagreement, the Commission can start judicial proceedings against a Member State (Article 226) for non-compliance with Community law. Judicial proceedings against a member State can be brought by another Member State (Article 227). Not complying with European Court of Justice (ECJ) judgments can be fined by the ECJ upon request of the Commission (Article 228). Therefore, for the implementation of international law, the EU certainly has a better chance for establishing more exhaustive, effective and rapid legislation on European level and to enforce this legislation toward Member States.[20]

It is well known that maritime casualties act as catalysts for the development of new or improvement of existing international and regional legislation.[21] In the aftermath of the *Erika* (1999) and the *Prestige* (2002) oil pollution disasters a series of legal issues were raised, such as limited jurisdiction of coastal states, lack of flag state implementation and so on.[22] The European Commission thus issued two Communications calling for measures to be adopted to strengthen maritime safety: a Communication on 'Improving Safety at Sea as a Result of the Prestige Accident'[23] and a Communication 'Towards a Strategy to Protect and Conserve the Marine Environment'.[24] Some measures include stronger rules for the phasing-out of single hull tankers,[25] an improved monitoring of the performance of classification societies,[26] harmonizing port state control[27] and the establishment of a European Maritime Safety Agency.[28] Before *Erika* and *Prestige*, the EU normally

20. See also H. Wang, 'The EU Marine Oil Pollution Prevention Regime – Recent Development', *European Environmental Law Review* (2004): 293.
21. I.C. Varotsi, 'Recent Developments in the EC Legal Framework on Ship-Source Pollution: The Ambivalence of the EC's Penal Approach', *Transportation Law Journal* 4 (2007): 374.
22. V. Frank, 'Consequences of the *Prestige* Sinking for European and International Law', *International Journal of Marine and Coastal Law* 20 (2005): 1 et seq.
23. COM (2002) 681 final.
24. COM (2002) 539 final.
25. Regulation (EC) No. 1726/2003 of the European Parliament and of the Council of 22 Jul. 2003 amending Regulation EC No. 417/2002 on the accelerating of phasing in of double hull or equivalent requirements for single hull oil tankers, OJ L249/1, 1 Oct. 2003.
26. Directive 2001/105/EC of the European Parliament and of the Council of 19 Dec. 2001 amending Council Directive 94/57/EC on common rules and standards for ship inspection and survey organizations and for the relevant activities of maritime administrations, OJ L19/16, 22 Jan. 2002.
27. Directive 2001/106/EC (as part of the Erika I package) of the European Parliament and of the Council of 19 Dec. 2001 amending Council Directive 95/21/EC concerning the enforcement, in respect of shipping using Community ports and sailing in the waters under the jurisdiction of the Member States, of international standards for ship safety, pollution prevention and shipboard living and working conditions (port State control), OJ L19/17, 22 Jan. 2002.
28. Regulation (EC) 1406/2002 of the European Parliament and Council Establishing a European Maritime Agency, O J L208/1, 5 Aug. 2002.

relied heavily on international *fora* like the IMO to address protection of the marine environment; to some extent, the three 'Erika packages', as proposed by the Commission, may be seen as representing a shift in focus of EU maritime policy.[29] Directive 2005/35/EC (the Directive)[30] and Decision 2005/667/JHA[31] created a legal framework for imposing criminal penalties in the event of discharges of oil and other noxious substances from ships in Community waters.

2.2 Relevant Cases

2.2.1 Operational Discharges

The EU noted that in spite of international efforts and efforts on the part of EU, the problem of operational discharges remains still widespread. For example, since 1988 up to 2003 a total of 7,216 confirmed vessel-source pollution discharges have been detected in the Baltic Sea, 596 were detected in the North Sea in 2001 alone and 1,638 such discharges were reported for the Mediterranean Sea in 1999.[32] Some recent cases are as follows.[33]

In February 2005, a British shipping company Norbulk Shipping Ltd cooperated with Danish authorities and pleaded guilty to causing pollution in Danish waters. The fine was 50,000 Danish Kroner (around DKK 6,700), and cleanup costs were DKK 150,000 (around DKK 20,000).

In September 2004, the French Coast Guard observed a visible hydrocarbon wake 11 kilometres long and forty meters wide emanating from the Panama-flagged *Atlantic Hero*. The vessel was detained by French authorities and ordered to pay a fine of Euro (EUR) 300,000.

In June 2004, French authorities detained the United Arab Emirates–flagged container ship, *Khaled Ibn Al Waleed*, after customs investigators photographed an oil slick 18 kilometres long by 50 meters wide behind the vessel in France's Mediterranean waters. A EUR 500,000 fine was imposed. A twelve-month prison sentence for the ship's captain was deferred.

---

29. M.H. Nordquist & J.N. Moore (eds), *Current Marine Environmental Issue and the International Tribunal for the Law of Sea* (The Hague: Kluwer Law International, 2001), 281.
30. Directive 2005/35/EC of the European Parliament and of the Council of 7 Sep. 2005 on ship-source pollution and on the introduction of penalties for infringements, OJ L255/11, 30 Sep. 2005.
31. Council Framework Decision 2005/667/JHA of 12 Jul. 2005 to strengthen the criminal-law framework for the enforcement of the law against ship-source pollution, OJ L255/164 of 30 Sep. 2005.
32. See Explanatory Memorandum, COM (2003) 92 final, 2 <http://eur-lex.europa.eu/LexUriServ/LexUriServ.do?uri=COM:2003:0092:FIN:EN:PDF> 1 Dec. 2008.
33. J.M. Grasso & A. Fennell, 'Criminal Prosecutions and the Maritime Industry – A World-Wide Trend', <www.iosc.org/papers/IOSC%202005%20a159.pdf>, 15 Nov. 2008.

2.2.2 Accidental Discharges

On 10 December 1999, a twenty-five–year–old, single hull oil tanker, *Erika*, caused the leakage of more than 20,000 tonnes of fuel oil, polluting 400 km of France's coastal area and killing tens of thousands of seabirds. Within days of the sinking, the investigating French judge had the ship's Indian Master, Capt. Karun Mathur behind bars on charges of endangering human life and causing marine pollution. The prosecution in the *Erika* trial has called for maximum sentences of a year in prison and EUR 75,000 (United States Dollar [USD] 101,500) fines for Giuseppe Savarese and Antonio Pollara, respectively, the owner and technical manager. In a ruling on 16 January 2008, the Criminal Court of Paris condemned the world's fourth largest oil group, Total SA, to a fine of EUR 375,000, the maximum allowable penalty for maritime pollution – claiming 'ecological prejudice' caused by the sinking of the *Erika.* The ship's Italian owner Guiseppe Saverese and its Italian manager Antonio Pollara were also found guilty and each fined EUR 75,000 – the maximum penalty for physical persons. According to the Court, the two men had committed an unlawful act by cutting down maintenance work on the *Erika* in order to save money, despite heavy corrosion of the ship's structures. However, eleven other accused parties, including the ship's Indian captain and Total's former security chief, Bertrand Thouillin, were let off the hook. Accusations of 'putting other people's lives in danger', which could have led to prison sentences, were dropped against all parties.[34]

On 13 November 2002, the vessel *Prestige* broke up off the coast of Spain on its way to Singapore, with 77,000 metric tonnes of oil leaking into the Spanish waters from the wreckage. From an outside perspective, the disaster was caused by the Spanish government itself for not allowing the vessel entry into the port.[35] Capt. Apostolos Mangouras, the captain of the *Prestige*, was held in detention in Spain. In issuing the arrest order for the captain, the Spanish judge indicated that the captain may have shared some responsibility for the accident and could have faced up to nine years in jail.

In the context of vessel-source pollution, the legal framework for criminal sanctions is left to the discretion of states.[36] Why does the EU remain so interested in it? For one thing, political pressure and public opinion after dreadful casualties such as *Erika* and *Prestige* in European waters have prompted the EU to highlight the need to tighten the net in relation to vessel-source pollution and eliminate the discrepancies in the implementation of MARPOL 73/78 among EU Member States. The Directive and Decision 2005/667/JHA were also introduced in such a context.[37] Moreover, it is noted that European institutions are extending their

34. 'Erika oil-slick trial sets "ecological prejudice" precedent' <www.euractiv.com/en/environment> search 'erika', 11 Nov. 2008.
35. The flag state investigation into the loss of the *Prestige* undertaken by the Bahamas Maritime Authority (BMA) does not attribute any blame at all to the Master of the ship for the incident.
36. See Varotsi, *supra* n. 21, 380.
37. The adoption of them reflects the compromise between EU institutions: the Directive defines specific 'infringements' under Art. 4 which, if committed with intent, recklessness or by serious

competence of maritime safety toward Member States by the European law making process.

2.3 The European Court of Justice Case C-308/06 Concerning the Validity of EU Directive 2005/35/EC on Ship Source Pollution

Despite opposition from some Member States, in particular Greece, Malta and Cyprus, which have strong ship-related interests, the Directive was adopted. The European Commission was of the opinion that the introduction of adequate sanctions for pollution offences is particularly important in relation to pollution by shipping, because the international civil liability regimes for vessel-source pollution accidents have significant shortcomings with respect to their dissuasive effects.[38] The EU agreed that although MARPOL has detailed standards and strict conditions for discharge of pollution substances at sea, the applicable pollution standards lack implementation and enforcement. Moreover, the EU noted that the violation of MARPOL is not fully covered by EU law and that the implementation of MARPOL by Member States shows variations, both in practice and in law.[39] Hence, it is believed by the EU that a specific Directive on vessel-source pollution discharges would have the dual benefit of completing and clarifying Community law in this area, and harmonize the enforcement of the rules.[40]

According to Article 3 (1) of the Criminal Sanction Directive, it applies to discharges of polluting substances from any ship, irrespective of its flag, in (1) internal waters, including ports of a Member State; (2) the territorial sea of a Member State; (3) straits used for international navigation subject to the regime of transit passage; (4) the EEZ or equivalent zone of a Member State and (5) the high seas. Article 4 describes that vessel-source discharges of polluting substances into any of the areas referred to in Article 3(1) are regarded as infringements if committed with intent, recklessly or by serious negligence. These infringements are regarded as criminal offences. Furthermore, Article 5 specifically lays down the exceptions.

Compared with MARPOL, the Directive does not make a distinction between accidental or operational discharge. It is not certain whether 'serious negligence' covers cases of accidental discharges. For the chairman of the Greek Shipping

---

negligence, must be criminalized by Member States under the conditions established under the Framework Decision. In the ECJ case C-440/05 – *Commission v. Council*, the Court in following the opinion of the Advocate General ruled that the Framework Decision be annulled. See Judgment of the Court (Grand Chamber), <eur-lex.europa.eu/LexUriServ/LexUriServ.do?uri=CELEX:62005J0440:EN:HTML> 1 Dec. 2008. See also R. Pereira, 'On the Legality of the Ship-Source Pollution 2005/35/EC Directive – The Intertanko Case and Selected Others', *European Energy and Environmental Law Review* (2008): 374.

38. See COM (2003) 92 final, 6.
39. See COM (2003) 92 final, 4.
40. See Anthony, *supra* n. 11, 226.

Cooperation Committee, the term 'serious negligence' is vague, subjective and ill defined.[41] An impartial, but devastating analysis of the compatibility of the Directive with MAROL and UNCLOS was provided by Judge Thomas Mensah, formerly Presiding Judge of the International Tribunal on the Law of the Sea, in his Cadwallader Lecture of 2005 on 'Sovereign Rights in Legislation of Member States under UNCLOS and MARPOL'.

> The UNCLOS permitted a coastal State to impose more stringent requirements regarding pollution only for vessels flying its own flag, since imposing such requirements on other vessels would have the effect of hindering innocent passage of foreign ships through its territorial sea. Article 19 of UNCLOS permitting the coastal State to deny passage to a foreign ship if in the territorial sea it is engaged in activities prejudicial to the peace, good order or security of the coastal State, covered any act of wilful and serious pollution contrary to this Convention, but did not cover discharges resulting from 'serious negligence'. The Directive by imposing liability for such discharges in the territorial sea – a lower criteria not contained in UNCLOS or MARPOL – would therefore hamper passage of foreign vessels through the territorial sea of States applying it.[42]

According to Article 5(1), discharges into areas referred to in Article 3(1) ('areas') are legal if they are in compliance with the conditions established in Regulations 9 and 10 of Annex I (oil) and in Regulation 5 of Annex II (hazardous noxious substances in liquid form). Discharges into areas where necessary for the purpose of securing the safety of the ship or saving human lives are also legal. This coincides with MARPOL. However, according to MARPOL, accidental discharges are legal when suitable precautions are subsequently taken and there is no intent or recklessness on the part of the owner, the master or other crew member of the vessel. Although article 5(2) of the Directive has a similar rule,[43] it applies only in international straits, the EEZ of a Member State or the high seas. It is unavailable in the case of discharge in the territorial sea. Moreover, Article 5(2) confines the application of the exception to ship owners, masters and crew, meaning that there is nothing to prevent the discharges being considered as an infringement by other parties, for example, the cargo owner, the classification society, the salvage firm, and so on.[44]

---

41. E. Embiricos, The Extra-Territorial Jurisdiction in Criminalization Cases: Sovereign Rights in Legislation and New Risks for the Shipping Industry, 6, <www.london-shipping-law.com/pdfs/Presentation-Mr Embiricos.pdf>, 25 Dec. 2008.
42. The full text of the lecture by T. Mensah, <www.london-shiing-law.com/pdfs/Eighth%20Cadwallader%20Transcriptm.pdf>, 13 Jul. 2009. See also E. Denza, 'A Note on Intertanko', *European Law Review* 33 (2008): 877.
43. It reads: 'A discharge of polluting substances into the straits, EEZ and the high seas shall not be regarded as an infringement for the owner, the master or the crew when acting under the master's responsibility if it satisfies the conditions set out in Annex I, Regulation 11(b) of MARPOL 73/78.'
44. See Osante, *supra* n. 13, 421.

Fearing a heavier burden in European waters, a coalition including the International Association of Independent Tanker Owners (INTERTANKO), the International Association of Dry Cargo Ship owners (INTERCARGO), the Greek Shipping Cooperation Committee, Lloyd's Register and the International Salvage Union representing the shipping industry was formed and started legal proceedings in December 2005 in the High Court of Justice in London. The Case was brought against the UK's Secretary of State for Transport as a technical defendant. Applicants tried to challenge the legality of the Directive within the territorial sea and EEZ of the Member States as well as on the high seas, in particular whether it is contrary to the right of innocent passage and whether the term 'serious negligence' is consistent with legal certainty.[45] Justice Hodge requested the ECJ to make a preliminary ruling on the following issues: (1) whether it is lawful for the EU to impose criminal liability in respect to discharges from foreign ships on the high seas or the EEZ, and to limit MARPOL defences in such cases; (2) whether it is lawful for the EU to exclude MARPOL defences for discharges in the territorial sea; (3) whether the imposition of criminal liability for discharges caused by 'serious negligence' hampers the right of innocent passage; and (4) whether the standard of liability in the Directive of 'serious negligence' satisfies the requirement of legal certainty. The ECJ heard oral arguments on 25 September 2007.[46] On 20 November 2007, Advocate General Kokott delivered her opinion to the Court. She suggested that although the validity of a Community act could not depend on an international agreement of which the EU was not a party,[47] the concept of serious negligence could be interpreted to mean 'recklessness in the knowledge that damage will probably occur'[48], as embodied in MARPOL.[49] Although the opinion of Advocate General was followed, on 3 June 2008, the Court (Grand Chamber) ruled that the validity of the Directive cannot be assessed by reference to the international conventions, MARPOL and UNCLOS. It was held that the Directive satisfies the general principle of legal certainty under EU law.[50]

In the judgment given by the Grand Chamber of the ECJ, it upheld the Commission's argument that because the Community itself (unlike its Member States) is not a party to MARPOL, it is not bound by the Convention. The Court has taken the view that although the Community is a party to UNCLOS, the Convention does

45. EU Ship Source Pollution Directive – High Court Ruling, <www.Intertanko.com/templates/Page.aspx?id=36080>, 11 Dec. 2008.
46. Aside for hearing, UK, Greece, Cyprus, Malta, Italy, Spain, France and Sweden addressed the Court. Moreover, nations outside the EU, like Hong Kong (China), the Marshall Islands and Liberia also had their concerns about the Directive. It was clear that the cases had raised very important questions that it was in the interests of Member States and industry alike to have the Court answer.
47. Kokott AG, para. 37, <www.intertanko.com/upload/15345/AG's%20opinion%20-%2020.11.07.pdf> 25 Dec. 2008.
48. Kokott, para. 109.
49. Kokott, para. 102–112.
50. Industry coalition statement on ECJ ruling on EU SSP Directive: <www.intertanko.com> Press Releases: Year 2008, 12 Jan. 2009.

not give individuals rights or freedoms on which they can rely against states. It also held that the use of the term 'serious negligence' does not infringe the requirement of certainty in Community legislation. In reaching the conclusion it gave guidance on the interpretation of the phrase, holding that it can refer only to a patent breach of a duty of care. By ruling that the Community is not bound by the MARPOL 73/78 and that UNCLOS is not directly effective in the Community legal order, the ECJ was able to avoid the difficult question of establishing the compatibility of the Directive with international law.[51] However, because the purpose of the Directive is to harmonize the implementation of MARPOL in its Member States, it is arguable that the validity of the Directive cannot be assessed in light of MARPOL, to which all Member States are signatories. As the coalition's solicitor, Colin de la Rue from Ince & Co said:

> The ruling appears to mean that if there are concerns about the compatibility of a Community instrument with international maritime law, these cannot be tested in the European Court and can be examined only if flag States outside the Community refer them to other international tribunals.[52]

2.4 Conclusions

Reasons why the Court upheld arguments of the Commission can be analysed as follows.

First, from an economic point of view, shipping interests are not that important in the EU, especially for main powers such as United Kingdom, France and Germany. Although certain Member States such as Greece, Malta and Cyprus still have strong interest in shipping, they were not able to block the adoption of the Directive under the qualified majority voting system applicable for the adoption of a transport measure based on Article 80(2) EC.[53] The judgment proves again that the EU decision-making is now pro-coastal.

Second, although the European States have a stronger tradition of upholding multilateral decision-making processes at IMO, such as the acceleration of phasing out single hull tankers, both the Directive and the Judgment also show that impatience with the IMO system has grown in Europe, particularly in the wake of several major vessel pollution incidents.[54]

Since the adoption of the Directive, the EU is now seeking to criminalize illegal discharge, whether operational or accidental, caused by any ship in its waters and even on the high seas. There is no clear interpretation of 'serious negligence' from the Community, and the right to interpret still belongs to Member States. If Member States do not give a broader interpretation, the EU will not be in conflict with the international law in practice. However, because of EU's significant economic power, nobody doubts the EU's ability to act unilaterally.

---

51. Pereira, *supra* n. 37, 382.
52. Press Release: Year 2008, <www.intertanko.com> 15 Jan. 2009.
53. See Pereira, *supra* n. 37, 374.
54. See Khee-jin Tan, *supra* n. 4, 88.

3 PRACTICE OF THE UNITED STATES

3.1 Background

Being called the strongest comprehensive environmental treaty in existence,[55] the major part of the UNCLOS had been supported by the US Administration, beginning with President Reagan, by fulfilling US interests in having a comprehensive legal framework relating to competing uses of the world's ocean.[56] However, because of concerns that the United States will abandon its sovereignty if it were to join the new international regime,[57] to date, it is still not a full member of UNCLOS. The United States claims that certain principles found in UNCLOS inure to its benefit and apply to other nations, not as treaty provisions per se, but rather as 'reflection' of customary international law.[58]

Despite remaining 'on the outside looking in' with regard to UNCLOS, the United States has accepted other major marine pollution conventions, such as MARPOL, along with Annexes I, II, III, V and VI.[59] Specifically, MARPOL is implemented in the United States by the Act to Prevent Pollution from Ships (APPS), under the lead of the US Coast Guard.

3.2 Domestic Legal Regime

Under US law, criminal conduct proscribed in pollution laws may be separated into two main areas: reckless or grossly negligent carriage of oil cargos in violation of existing regulations and false statements or obstruction of justice in responding to an oil spill.[60]

APPS make it a crime for any person who knowingly violates MARPOL, APPS, or the federal regulations promulgated under APPS 33 U. S. C. 1908(a). Generally, to be guilty of a crime, a person must act with criminal intent or *mens rea*. These regulations apply to all commercial vessels, including vessels operating under the authority of a country other than the United States, when these vessels are

55. R. Stevenson & B.H. Oxman, 'The Future of the United Nations Convention on the Law of the Sea', *American Journal of International Law* 88 (1994): 496.
56. CRS Issue Brief for Congress: The Law of Sea Convention and US Policy <www.climateactionproject.com>, 10 Feb. 2009.
57. D.J. Bederman, 'The Old Isolationism and the New Law of the Sea: Reflections on Advice and Consent for UNCLOS', *Harvard International Journal Online* (2008): 21.
58. J.A. Duff, 'The United States and the Law of the Sea Convention: Sliding back from Accession and Ratification', *Ocean and Coastal Law Journal* (2005–2006): 6.
59. The most recent US action concerning MARPOL occurred in April 2006, when the US Senate approved Annex VI, which regulates air pollution (Treaty Doc. 108-7, Exec. Rept. 109-13).
60. V.J. Foley & C.R. Nolan, 'The Erika Judgment-Environmental liability and Places of Refuge: A Sea Change in Civil and Criminal Responsibility that the Maritime Community Must Heed', *Tulane Maritime Law Journal* 33 (2008): 53.

operating in US waters or while at a port or terminal under US jurisdiction. The Clean Water Act (CWA) provides criminal penalties for pollution of navigable waterways. It requires negligent or knowing violations for criminal conviction. Besides, for health and public welfare violations, strict liability applies. It does not require the government to prove intent or negligence. Instead, the government merely has to prove that the prohibited act has occurred. It does not matter that the oil spill or pollution incident was accidental.[61] The Refuse Act and the Migratory Bird Treaty Act, though marginally related to oil spill, are potentially applicable.

APPS gives the US Coast Guard (USCG) authority to issue 'regulations implementing the requirements of MARPOL'.[62] The USCG has the authority to board and inspect any vessel while that vessel is in US waters or at a US port under APPS.[63] It regularly inspects the Oil Record Book (ORB) during port state inspections, determines compliance with US law and MARPOL. When a vessel violates its obligations, the USCG is empowered to deny that vessel's entry to a US port or detain it.[64] If the USCG or another agency, such as the Environmental Protection Agency (EPA),[65] determines that illegal discharges have been made in US waters, those discharges may be prosecuted as substantive violations of APPS, CWA or any other applicable domestic environmental law.[66] However, criminal offences must be brought by the US Department of Justice (DOJ) through both the Environmental Crimes Section and US Attorney General.[67] The USCG is also required to report all suspected violations of any US laws to the US Attorney General's office for the District in which the vessel is inspected. A broad array of criminal sanctions are available to prosecutors against crew members, owning and managing corporations and individuals in such corporations, including fines and duration of sentences. It is interesting to find in the APPS that half of a criminal penalty assessed to a violator may be awarded to a person ('whistleblower') who provides information leading to a conviction.

---

61. Environmental Criminal Liability in the United States: A Handbook for the Marine Industry <www.marisec.org/resources/EnvironmentalCriminalLiabilityInTheUS.pdf>, 20 Feb. 2009.
62. *United States v. Abrogar*, 459 F.3d at 432, 2006 AMC at 2156 (citing 33 USC 1903, 1907).
63. E. Franckx (ed.), *Vessel-Source Pollution and Coastal State Jurisdiction: The Work of the ILA Committee on Coastal State Jurisdiction Relating to Marine Pollution (1991–2000)* (The Hague: Kluwer Law International, 2001), 13.
64. 33 CFR 151.07 (b).
65. Because the EPA has separate authority to enforce the CWA and many other environmental laws that affect the marine environment, these agents are more frequently becoming involved in investigating vessel-source pollution incidents. In addition, these agents will often work with the Coast Guard and other federal or state enforcement agencies in conducing criminal investigations of pollution. See *supra* n. 61, 14.
66. A.W. Homer, 'Red Sky at Morning: The Horizon for Corporations, Crew Members, and Corporate Officers as the United States Continues Aggressive Criminal Prosecution of Intentional Pollution from Ships', *Tulane Maritime Law Journal* (2007): 152.
67. See Faure & Hu, *supra* n. 9, 26.

3.3 Relevant Cases

3.3.1 Operational Discharges

In the late 1990s, the DOJ began to fine-tune its methods for prosecuting violations of MARPOL/APPS in a way that allowed them to reach beyond US waters and apply serious criminal penalties, despite basic protections for individual defendants built into UNCLOS and recognized as customary international law.[68] The very first major cases dealing with these issues were the Royal Caribbean cases.[69]

On 25 October 1994, a USCG aircraft observed a Norwegian-flag cruise ship, the *Sovereign of the Seas*, owned and operated by Royal Caribbean Cruises, Ltd (RCCL), discharging approximately thirty gallons of oil off the coast of Puerto Rico. When the vessel docked in San Juan that day, the USCG boarded the vessel to investigate the discharge. After the investigation, a federal grand jury in Puerto Rico returned a ten-count indictment against the RCCL and additional indictments against the chief engineer and first engineer on the vessel. It included counts for violations of APPS, False Statement Act (FSA) and conspiracy. However, the court held that although MARPOL and UNCLOS did not preclude the DOJ from bringing criminal charges against the defendants for acts of pollution committed in US territorial waters,[70] the UNCLOS limited DOJ to pursuing only monetary penalties for cases of wilful and serious act of pollution.[71] Finding no serious pollution, the court finally sentenced the accused monetary penalties and the individuals were not incarcerated.

On 1 February 1993, a USCG aircraft on patrol in the Caribbean observed a Liberian-flag cruise ship, also owned by RCCL, discharging oil in Bahamian waters while en route to the United States. When the vessel arrived at Miami, Florida, the USCG boarded and examined the ORB. The matter was first referred to the flag State, Liberia. But no prosecution was initiated by the flag State. In response, the DOJ indicted the vessel for making false statements through its employees by presenting a falsified ORB to the USCG. Under the plain meaning of MARPOL/APPS, the United States is powerless to prosecute intentional illegal discharges occurring outside of its waters.[72] However, the court held that MARPOL/APPS complement the FSA so as to maximize pollution enforcement efforts in both the domestic and international arenas. Whereas MARPOL/APPS establish a comprehensive, well-balanced international regulatory scheme for limiting discharges of pollutants on the high seas, FSA establishes a framework through which

68. See Homer, *supra* n. 66, 153.
69. For details of two Royal Caribbean Cases, please see *supra* n. 68, 156–158 and D.G. Dickman, 'Recent Development in the Criminal Enforcement of Maritime Environmental Law', *Tulane Maritime Law Journal* (1999): 27–32.
70. 24 F. Supp. 2d 159, 1998 AMC at 1845.
71. *Ibid.*, 1998 AMC at 1846 (citing UNCLOSUNCLOS Art. 230 (2)).
72. See Homer, *supra* n. 66, 157.

to prosecute false writings or statements that are made knowingly and wilfully to an agency of the United States.[73] Hence, even there is no clear intention in US domestic legislation to codify Article 218 of UNCLOS concerning port state enforcement for discharges outside national jurisdiction of coastal states[74] the US court acknowledged in both cases the right to impose criminal penalties for illegal discharges of foreign vessels occurring outside US waters, based on port state enforcement jurisdiction. With the endorsement of the district courts in the Royal Caribbean cases, the DOJ began to step up its enforcement of MARPOL/APPS violations occurring on the high seas using a variety of domestic criminal charges.[75]

Since the Royal Caribbean cases, a series of criminal cases against crew members and corporations have taken place. The most recent is *United States v. Kun Yun Jho* before the Fifth Circuit Court of Appeals 30 June 2008. The government obtained information from an engineer on the MIT *Pacific Ruby* (whistleblower) about illegal discharges and tampering with the vessel's equipment. Ten charges were brought against the vessel owner, Overseas Shipholding Group, Inc. (OSG), including entering a US port with a knowingly inaccurate ORB. However, the District Court considered that the ORB violations occurred aboard a foreign-flag ship while outside US waters and that international law would be violated by prosecuting offences in such circumstances. The Court of Appeal, in contrast, held that such offences occurred entirely within the ports of the United States on the eight occasions that the vessel entered them, irrespective of whether the alleged false ORB entries were made in international waters. It upheld the right of the United States to prosecute a foreign-flag shipowner (and its engineer) for falsification of an ORB, even when such falsification took place outside of US ports or navigable waters, as a breach of the legal duty to maintain an ORB under US law.[76]

3.3.2 Accidental Discharges

On 24 March 1989, the *Exxon Valdez* oil spill occurred in Prince William Sound, Alaska. The vessel spilled 10.8 million US gallons (about 40 million litres) of Prudhoe Bay crude oil into the sea, and the oil covered 11,000 square miles (28,000 km^2) of ocean.[77] In the wake of the *Exxon Valdez* spill the US Congress passed the Oil Pollution Act (OPA 90), in which for the first time in maritime history a country imposed unilateral regulations that required compliance from both domestic and foreign-flag tankers carrying oil in US waters.[78]

73. 11 F. Supp. 2d at 1368, 1998 AMC at 1829.
74. See Franckx, *supra* n. 63, 364.
75. See Homer, *supra* n. 66, 158.
76. U. S. Criminal Jurisdiction over MARPOL Offences <www.simsl.com/OSG0908.html> 28 Feb. 2009.
77. *Exxon Valdez* Oil Spill, <en.wikipedia.org/wiki/Exxon_Valdez_oil_spill>, 1 Mar. 2009.
78. K.W. Ketkar, 'The Oil Pollution Act of 1990: A Decade Later', *Spill Science and Technology Bulletin* (2002): 46.

After the disaster, an indictment was filed against Exxon Corporation and Exxon Shipping Company for violations of the CWA, the Refuse Act, the Migratory Bird Treaty Act (MBTA), the Ports and Waterways Safety Act and the Dangerous Cargo Act. On 8 October 1991, Exxon Corporation pleaded guilty to one count under the Migratory Bird Treaty Act, while Exxon Shipping Company pled guilty to violations of the CWA, the Refuse Act and MBTA. The Corporations were sentenced to pay a criminal fine in the amount of USD 125 million, with USD 100 million remitted based on restitutionary payments to the United States and the State of Alaska.[79]

3.4 Conclusions

Although the United States has not ratified UNCLOS, at least some US courts believe that because the United States is actively considering accession to UNCLOS, it has, in their view, an obligation 'to refrain from acts which would defeat the objective and purposes of the treaty'. Based on port state jurisdiction, there is a long-standing history of criminal prosecutions of illegal discharges, both operational and accidental. As a DOJ press release stated: 'The aggressive enforcement of federal criminal environmental laws is a priority' and 'there will be no safe harbour for those who intentionally pollute',[80] the United States is fully aware of imposing criminal liability to create incentives for the prevention of vessel-source pollution. One interesting measure is to give half of a criminal fine to a person who provides information leading to a conviction, the so called 'whistleblower'. This may be more effective than just imposing strict criminal sanctions, because to what extent can seamen, as employees, refuse to obey their employers, or implicate them by reporting them to an authority that they know in the end may likely detain the ship, which is their instrument of contract and livelihood with that employer?[81]

4 WHAT ABOUT CHINA?

4.1 Background

China is a major coastal State with a continental coastline of some 18,000 km facing east and another 14,000 km of coastline around its 6,500 islands. Along China's coastlines are hundreds of ports and harbours that connect China to the rest of the world.[82] Thus, the marine environment plays a crucial role for the

79. See Anthony, *supra* n. 11, 229.
80. Press Release: Tanker Company Sentenced for Conspiring to Conceal Hazardous Condition and Oil Pollution in Baltimore Harbor. <www.usdoj.gov/opa>, 3 Mar. 2009.
81. See Anthony, *supra* n. 11, 232.
82. See Franckx, *supra* n. 63, 201.

development of China and is addressed by the Government. On the other hand, China is also a major maritime nation and has long been cross-pressured by its own maritime interests. Rapid economic development has had extreme consequences for foreign trade, energy production and energy consumption. According to the National Report on Social and Economic Development, China imported 178,880,000 tonnes of crude oil and 38,850,000 tonnes of refined oil in 2008.[83] Ninety-five per cent of the import is carried by maritime transportation.

Therefore, it is not wise to conclude whether China is pro-coastal or pro-maritime. Concerning prevention of vessel-source pollution, China is like other developing countries whose motives have less to do with domestic public interest pressures but more with the need to exercise control over resources in adjacent seas and, along with that, foreign vessels engaged in pollution.[84] China is a supporter of the international legal regime. It has ratified UNCLOS, MARPOL and most other relevant international conventions, as well as adopted domestic laws and regulations regarding implementation.

4.2 Domestic Legal Regime

In 1982, the Marine Environmental Protection Law (MEPL) was approved, which is the first specific and most important legislation addressing problems of pollution of the marine environment.[85] It not only applies to China's internal waters, contiguous zone, EEZ and continental shelf but also to discharges beyond if pollution affects sea areas within the jurisdiction of China.[86] In order to implement the 1982 MEPL, a number of regulations and detailed measures were laid down. Of particular importance for vessel-source pollution is the Regulation on Preventing Pollution of Sea Areas from Ships (RPPSAS). It is applicable to all Chinese and foreign vessels as well as to all shipowners and other individuals within the sea areas and sea ports under China's jurisdiction.[87] All are required to comply with the 1982 MEPL and regulations regarding the discharge of oil, oily mixtures, wastes and other poisonous or harmful substances.[88] Even not specifically implementing MARPOL, many provisions and standards of MARPOL were taken into account and some have been incorporated into the RPPSAS.

On 25 September 1999, MEPL was revised in order to harmonize with the entry into force of UNCLOS and to deal with a greater pressure on the marine environment resulting from the economic boom. Chapter 9 of the MEPL, 'Legal Liability' was expanded from 4 articles to 22 articles, with the intent to make

83. National Bureau of Statistics of China <www.stats.gov.cn/tjgb/ndtjgb/qgndtjgb/t20090226_402540710.htm>, 13 Jul. 2009.
84. See Khee-jin Tan, *supra* n. 4, 73.
85. See Franckx, *supra* n. 63, 206.
86. See Art. 2, MEPL, revised in 1999 and became effective since 1 Apr. 2000.
87. See Art. 2, RPPSAS.
88. See Art. 4, RPPSAS.

enforcement easier.[89] In Chapter 9, most of the sanctions that can be used for the prevention of marine pollution are administrative measures. Article 91, paragraph 3 is the only junction between the MEPL and criminal law.[90] According to the explanation of the Standing Committee of the National People's Congress, this article is added in accordance with Crimes of Impairing the Protection of Environment and Resources in the Penal Law.[91] It provides that in the case of significant pollution of the marine environment resulting in major damage (great loss of public or private property/bodily injury or death of another person), criminal liability shall be imposed according to applicable law.[92]

In the Penal Law of China, approved in 1979 and modified in 1997, there is no specific offence of polluting the sea. It seems that Article 338, 'Considerable Accident with Environmental Pollution' is the most likely offence that fits with MEPL.

Article 338 is mentioned in section 6 'Crimes of Impairing the Protection of Environment and Resources' of Chapter 6. It reads that a person who, in violation of the state's regulations, discharges, dumps or disposes radioactive wastes, wastes carrying infectious pathogens, poisonous substances or any other dangerous substances to land, waters or air, thus causing serious environmental pollution, shall be sentenced to fixed-term imprisonment of no more than three years or criminal detention and concurrently or independently, to a fine, if a great loss of public or private property or bodily injury or death of another person is caused. If the result is more significant, a fixed-term imprisonment of not less than three years and not more than seven years, concurrent to a fine, can be imposed.[93] The article applies to both individuals and legal persons. It does not differentiate recklessness and intention for the application. Concerned with accidental discharges, one author's opinion is that the transported crude oil cannot be considered as waste, but is one of the most important raw materials for industry.[94] However, in the recent ECJ case C – 188/07, *Commune de Mesquer v. Total France SA* and Total International Ltd, the judgment of the Court (Grand Chamber) rules that hydrocarbons accidentally spilled at sea following a shipwreck, mixed with water and sediment and drifting along the coast of a Member State until being washed up on that coast, constitute waste in the meaning of Article 1(a) of Directive 75/442, as amended by

89. The Content of Revision of Marine Environmental Protection Law (.........), Environment Protection and Resources Conservation Committee, Standing Committee of National People's Congress(.........), <www.people.com.cn/zgrdxw/news/200004/04/lf30302.html>, 13 Mar. 2009.
90. See *supra* n. 9, 68.
91. See *supra* n. 89.
92. It is noted that MEPL uses the same terms as the Chinese version of UNCLOS, such as 'significant pollution' (重大污染) and 'major damage' (重大损害). For the Chinese version of UNCLOS, see <www.un.org/chinese/law/sea/>, 15 Mar. 2009.
93. In the Panel law, there is no specific range of amount of the fine that can be imposed by courts, so does any Judicial Interpretation by the Supreme Court of China. This may be a loophole because the court in fact has discretion of the amount of the fine, although in practice until now there is no criminal case on vessel-source pollution.
94. See Faure & Hu, *supra* n. 9, 70.

Decision 96/350.[95] The Chinese courts may also be interested in considering this. When it comes to operational discharges, Article 338 may be applied. Nevertheless, the premise is whether dangerous discharges cause serious consequences for public or private property or human casualties.

With respect to enforcement, the Maritime Safety Agency (MSA) is responsible for the supervision and administration of pollution of the marine environment caused by non-military vessels inside the port waters and non-fishery vessels and non-military vessels in waters under jurisdiction of China. The MSA is responsible for the investigation and handling of pollution accidents.[96] In the event of a pollution caused by a foreign vessel navigating, berthing and anchoring and operating in the sea area under the jurisdiction of the People's Republic of China, officers in charge shall board the vessel in question to examine and handle the case.

4.3 CHALLENGES

According to the 2008 National Report on the Marine Environment issued by the State Oceanic Administration, coastal areas of China are still seriously polluted. In 2008, the whole polluted area was about 137, 000 km^2.[97] Although no serious tanker spill has been reported to date, the huge amount of freight and oil import makes the risk for both operational and accidental pollution much higher in the future. Stringent rules on vessels in the US and European waters force more substandard vessel flow into Asia, especially East Asia. The recent *Hebei Spirit oil* spill disaster,[98] the most serious oil spill in the history of South Korea, draws special attention to this problem again. Moreover, the current administrative sanctions seem not enough for preventing serious pollution. The maximum fine for committing an illegal discharge is modestly Chinese Yuan (CNY) 200,000.[99] However, until now, there is still no case that imposed criminal liability.[100] Possible reasons are as follows.

First and not only for imposing criminal liability, great emphasis is placed on the distribution of powers as well as on cooperation, and less emphasis is placed on coordination among the administrative departments.[101] For example, the

95. See Judgment of case C-188/07, *Commune de Mesquer v. Total France SA and Total International Ltd*, <http://eur-lex.europa.eu/LexUriServ/LexUriServ.do?uri=CELEX:62007J0188:EN:HTML>, 6 Apr. 2009.
96. See Art. 5, MEPL.
97. State Oceanic Administration of P. R. China, 2008 National Communiqué of Marine Environment (《中国海洋环境质量公报》, 2008年), <www.soa.gov.cn/hyjww/ml/news/ba/webinfo/2009/02/1225332550960914.htm>, 6 Mar. 2009.
98. For detail information, see <en.wikipedia.org/wiki/2007_Korea_oil_spill>, 22 Dec. 2007.
99. See Art. 73, MEPL. It is about EUR 22,000 according to the exchange rate of 17 Mar. 2000.
100. Thousands of Cases Violating Ocean Law without Criminal Penalties, the System Need to be Improved <www.chinaorg.cn>, 7 Mar. 2009.
101. See Faure & Hu, *supra* n. 9, 295.

MSA is authorized to enforce port state jurisdiction and coastal state jurisdiction on vessel-source pollution. But the China Sea Supervision, the most important marine surveillance power in China, is affiliated with the State Oceanic Administration.

Second, problems exist of junction between the MEPL and Penal Law. Article 91 of MEPL only describes that 'significant pollution' and 'major damage' can cause criminal liability. Those standards are very subjective and not easy to enforce in practice when the competent administrative departments want to bring criminal cases to the court. It is not very clear to differentiate administrative illegal acts from crimes by current legislation. This problem has been noted. In the recent draft of the new RPPSAS calling for public opinion, Article 39 tries to lay down regulation on significant vessel-source pollution more clearly.

Third, the question that arises is who has the right to bring a criminal case to a court on behalf of the country? In the United States, the DOJ is entitled to prosecute. Unfortunately, we still have no idea which department is competent to do the same task in China. The MSA seems to be the leading department for the prevention of vessel-source pollution. However, it is not authorized to prosecute directly. In addition, there are special Maritime Courts besides local courts in main coastal cities. Even the judicial interpretation about their jurisdiction does not include criminal cases on vessel-source pollution. Reasonably, it may be better if those cases can be dealt with in a specialized maritime court.[102]

5 CONCLUSION

Although serious sanctions can be applied against an individual, criminal liability is increasingly used by both the EU and the United States to create deterrence in order to deal with vessel-source pollution. Based on port state jurisdiction, the EU and the United States are following the same pattern, seeking to criminalize accidental pollution as well as operational pollution. However, in relation to the EU, a worrisome point persists that the term 'serious negligence' in the Directive is still not clearly interpreted. This may result in a heavier burden compared with MARPOL for shipping in European waters.

In China, although civil, administrative and criminal liabilities are all set by MEPL to cope with vessel-source pollution, administrative sanctions are almost the only measures taken in practice. The most urgent and realistic tasks for China now are to revise its legislation as well as enhance enforcement, so that criminal liability can be truly and effectively imposed. Experiences of both the EU and the United States can be helpful for Chinese authorities to make its legislations more clear as well as better coordinate its inter-departmental enforcement.

102. On the Jurisdiction of Maritime Court, Supreme Court of P. R. China (.........) effective since 18 Sep. 2001, <www.law-lib.com/law/law_view.asp?id=16275>, 8 Mar. 2009.

References

Anthony, O.G. 'Criminalization of Seafarers for Accidental Discharge of Oil: Is there justification in International Law for Criminal Sanction for Negligent or Accidental Pollution of the Sea?' *Journal of Maritime Law and Commerce* 37 (2006): 230-244.

Bederman, D.J. 'The Old Isolationism and the New Law of the Sea: Reflections on Advice and Consent for UNCLOS'. *Harvard International Journal Online* 49 (2008): 21.

Birnie, P.W. & A. Boyle. *International Law and the Environment,* 2nd edn. Oxford: Oxford University Press, 1992.

Boyle, A. 'EU Unilateralism and the Law of the Sea'. *International Journal of Marine and Coastal Law* 21 (2006): 15 et seq.

Dempsey, P.S. 'Compliance and Enforcement in International Law: Oil Pollution of the Marine Environment by Ocean Vessels'. Nw. *Journal of International Law and Business* 6 (1984): 459, 541.

Denza, E. 'A note on Intertanko'. *European Law Review* 33 (2008): 870–879.

Dickman D.G. 'Recent Development in the Criminal Enforcement of Maritime Environmental Law'. *Tulane Maritime Law Journal* 24 (1999) 27–32.

Duff, J.A. 'The United States and the Law of the Sea Convention: Sliding back from Accession and Ratification'. *Ocean and Coastal Law Journal* 11 (2005–2006): 6–33.

Embiricos, E. The Extra-Territorial Jurisdiction in Criminalization Cases: Sovereign Rights in Legislation and New Risks for the Shipping Industry, 6 <www.london-shipping-law.com/pdfs/Presentation-Mr Embiricos.pdf>, 25 December 2008.

Faure M.G. & J. Hu, *Prevention and Compensation of Marine Pollution Damage: Recent Development in Europe, China and the US.* Alphen aan den Rijn: Kluwer Law International, 2006.

Franckx, E. (ed.), *Vessel-Source Pollution and Coastal State Jurisdiction: The Work of the ILA Committee on Coastal State Jurisdiction Relating to Marine Pollution (1991-2000).* The Hague: Kluwer Law International, 2001.

Frank, V. 'Consequences of the *Prestige* Sinking for European and International Law'. *Journal of Marine and Coastal Law* 20 (2005): 1 et seq.

Foley, V.J. & C.R. Nolan, 'The Erika Judgment-Environmental liability and Places of Refuge: A Sea Change in Civil and Criminal Responsibility that the Maritime Community Must heed'. *Tulane Maritime Law Journal* 33 (2008): 53–78.

Gauci, G.M. 'Protection of the Marine Environment through the International Ship-Source Oil Pollution Compensation Regimes'. *Review of European Community & International Environmental Law* 8 (1999): 29–36.

Grasso, J.M. & A. Fennell, 'Criminal Prosecutions and the Maritime Industry: A World-Wide Trend'. <www.iosc.org/papers/IOSC%202005%20a159.pdf>, 15 November 2008.

Homer, A.W. 'Red Sky at Morning: The Horizon for Corporations, Crew Members, and Corporate Officers as the United States Continues Aggressive Criminal Prosecution of Intentional Pollution from Ships'. *Tulane Maritime Law Journal* (2007): 152–180.

Ketkar, K.W. 'The Oil Pollution Act of 1990: A Decade Later'. *Spill Science and Technology Bulletin* 7 (2002): 45–52.

Khee-jin Tan, A. *Vessel Source Marine Pollution: The Law and Politics of International Regulation.* Cambridge: Cambridge University Press, 2006.

Mensah, T. <www.london-shiing-law.com/pdfs/Eighth%20Cadwallader%20 Transcriptm.pdf>, 13 July 2009.

Mitchell, R.B. *International Oil Pollution at Sea – Environmental Policy and Treaty Compliance.* Cambridge, MIT Press, 1994.

Molenaar, E.J. *Coastal State Jurisdiction over Vessel-Source Pollution.* The Hague: Kluwer Law International, 1998.

Molenaar, E.J. 'Port State Jurisdiction: Toward Comprehensive, Mandatory and Global Coverage'. *Ocean Development & International Law* 38 (2007): 227–257.

Nordquist, M.H. & J.N. Moore (eds). *Current Marine Environmental Issue and the International Tribunal for the Law of Sea.* The Hague: Kluwer Law International, 2001.

Osante, J.-M.M. 'Competition and the European Union Directive on Criminal Penalties for Ship-Source Pollution'. *Journal of International Maritime Law* 14 (2008): 419, 420.

Pereira, R. 'On the Legality of the Ship-Source Pollution 2005/35/EC Directive – The Intertanko Case and Selected Others'. *European Energy and Environmental Law Review* 17 (2008): 374–383.

Sands, P. *Principles of International Environmental Law.* Cambridge: Cambridge University Press, 2003.

Stevenson, R. & B.H. Oxman. 'The Future of the United Nations Convention on the Law of the Sea'. *American Journal of International Law* 88 (1994): 496–499.

Varotsi, I.C. 'Recent Developments in the EC Legal Framework on Ship-Source Pollution: The Ambivalence of the EC's Penal Approach'. *Transportation Law Journal* 4 (2007): 374–386.

Wang, H. 'The EU Marine Oil Pollution Prevention Regime – Recent Development'. *European Environmental Law Review* (2004): 293.

Chapter 12

Criminal Liability of Master and Crew in Oil Pollution Cases: A Possible Conflict between the Law of the Sea Convention (UNCLOS), MARPOL and the European Directive 2005/35/EC

*Marc A. Huybrechts**

1 INTRODUCTION

The purpose of this article is not to proffer a detailed analysis of the present civil/criminal liability regimes that deal with ship-source oil pollution incidents; others will be very much involved in that theme on the occasion of the centennial celebration of the Dalian Maritime University.[1] This chapter will concentrate on whether there is a conflict between Public International Law, that is, specific provisions from the United Nations Convention on the Law of the Sea (UNCLOS)

* Emeritus Professor at the Universities of Leuven and Antwerp (Belgium), visiting Professor at the Dalian Maritime University.

1. For a good general introduction into the legal issues relating to marine pollution we refer to R.R. Churchill & A.V. Lowe, *The Law of the Sea*, 2nd edn (Manchester, 1998–1991). In Ch. 15, Marine Pollution, 241–287, the authors discuss the relation between de various international treaties such as Marpol (dealing mainly with prevention) and the CLC convention (dealing with liability and compensation schemes) starting from the general principles as contained in the Law of the Sea Convention.

Michael G. Faure, Han Lixin & Shan Hongjun, *Maritime Pollution Liability and Policy: China, Europe and the US*, 215–230.

and specific provisions from the International Convention for the Prevention of Pollution from Ships (MARPOL) 73/78 treaty and the European Directive 2005/35/EC.

2 GENERAL FEATURES OF CIVIL LIABILITY IN OIL POLLUTION CASES

Civil liability consequences are dealt with by a number of international conventions such as MARPOL73/78 and the Fund Convention. The purpose of this article is not to analyse local or national regulations such as the OPA's oil spill liability compensation scheme.[2]

The International MARPOL Treaty 73/1978 covers oil pollution discharges, whether accidental or operational. The liability aspects of a ship source oil pollution discharge (and the compensation thereof) are basically covered by the convention on Civil Liability for Oil Pollution Damage (CLC) Convention of 1969 as replaced and amended by the CLC 1992. The basic features of that Convention are that the tanker owner is strictly liable and has to pay compensation for pollution damage resulting from a spill of 'persistent' oil. On the other hand, the tanker owner is able to limit its liability to pay compensation in accordance with the liability limits set out in the convention.

It was the second convention that introduced compulsory insurance. (The first convention to introduce compulsory insurance was the not very well known and rarely applied international convention on nuclear ships signed in Brussels 1962.)

Indeed, the owners of tankers carrying more than 2,000 tonnes of persistent oil as cargo are required to maintain insurance to cover their operational liabilities. A special feature is that a claimant of an oil pollution incident can bring a direct action against the insurer to recover the pollution damage.

As is the case with most treaties dealing with limitation of liability, the liability limits are expressed by reference to the size of the vessel. One of the major compromises reached on the occasion of the drafting of the CLC convention was the principle of 'channelling of liability'. The liability for an oil pollution incident would be channelled through to the 'registered owner', who in turn has to carry compulsory insurance.

The channelling of liability to the registered owner has many advantages. Indeed, a claimant is offered an easy and rapid solution for determining to whom he has to submit his claim and he has the assurance that the registered owner carries an insurance policy that covers the incident. However, the principle of 'channelling of liability' also has a substantial draw back. It tends to create immunity for all other parties concerned in the case of a ship-source oil pollution incident, such as the charterer, the classification society or an operator who is not the registered

2. Oil Pollution Act (OPA), 33 USC.

owner. It should not be forgotten, however, that recourse actions by the registered owner against liable parties remain possible.[3]

The French *Erika* decision of 16 January 2008 has placed a clear end to the dream of immunity for all 'others', except the registered owner.[4] The MARPOL treaties provide for detailed rules in view of preventing oil pollution incidents.

For the purpose of this chapter, it is important to draw attention to the fact that MARPOL provides for defences that a tanker owner can invoke in his favour, *inter alia*, Regulation 11b of Annex 1 of MARPOL. Essentially, Regulation 11b provides for a 'due diligence defence' in respect to accidental discharges resulting from damage to the ship or its equipment, which defence is unavailable only where it is proven that the owner or the master acted recklessly or with intent and with the knowledge that the damage would probably result.[5] In many instances the compensation provided or obtained under the CLC Convention will not suffice to cover the entire pollution damage. In such an instance, the Fund Convention of 1992 and the supplementary Fund Convention of 2003 will come into play. The Fund Convention established the International Oil Pollution Compensation Fund 1992 (the 1992 Fund) to provide compensation for victims who do not obtain full compensation under the Civil Liability Convention. The fund is financed by contributions of oil importers. The total combined amount of compensation payable for one incident under the Civil Liability Convention and the Fund Convention can vary. The limits set by the 1992 Fund increased considerably in 2003 by an additional Protocol to the Fund Convention, known as the Supplementary Fund Protocol. This Supplementary Fund will apply where the total amount of compensation available from the ship's insurer and the 1992 Fund are insufficient to cover the total compensation costs. The Supplementary Fund Protocol established a supplementary fund to provide compensation for victims who do not obtain full compensation under the Civil Liability and Fund Conventions. The Supplementary Fund is financed in a manner similar to that of the 1992 Fund. Table 12.1 gives an indication of the amounts available under the four applicable sources (CLC 1969, CLC 1992, Fund 1992 and Supplementary Fund 2003). Maximum amounts of compensation available under the conventions (expressed in United States Dollar [USD] millions [rates as per April 2008]).

3. LCL 1969, Art. III, 5: Nothing in this Convention shall prejudice any right of recourse of the owner against third parties.
4. Tribunal de Grande Instance de Paris, 11ème Chambre,-4ème section, 16 Jan. 2008, C-9934895010, with a summary at page 81–99 of the procedural steps. The Paris Court of Appeals has mainly confirmed that decision on 30 Mar. 2010. The 'Total' Company lodged an appeal before the French Court of Cassation on 6 Apr. 2010 which is still pending. For a clear and solid analysis of this leading continental, case we refer to V.J. Foley & C.R. Nolan, 'The Erika Judgment-Environmental Liability and Places of Refuge: A Sea Change in Civil and Criminal Responsibility that the Maritime Community Must Heed', *Tulane Maritime Law Journal* 33 (2008): 42–78.
5. P.K. Mukherjee, 'Criminalisation and Unfair Treatment: The Seafarer's Perspective', *Journal of International Maritime Law* (2006): 327.

Table 12.1 Tankers

GT	*CLC1969*	*CLC1992*	*FUND1992*	*Supplementary Fund*
5,000	7.3	327.6	327.6	1210.5
25,000	5.4	27.7	327.6	1210.5
50,000	10.7	53.1	327.6	1210.5
100,000	21.5	104	327.6	1210.5
140,000	22.6	144.9	327.6	1210.5

3 SOME OF THE SPECIAL FEATURES OF THE MARPOL TREATIES 73/78

The MARPOL treaty distinguishes between operational and accidental discharges.

Operational discharges are prohibited, except under special conditions in a very limited way, in special zones and under special circumstances. With reference to accidental discharges, no sanctions are imposed provided that these accidental discharges result from damage to a ship or its equipment and provided all reasonable precautions have been taken by the tanker owner to prevent or minimize the discharge. This is more specifically dealt with in regulation 11b of Annex 1 of the MARPOL Convention.[6] However, this defence will disappear if the owner/master acted with intent or recklessly.

4 RATIFICATIONS OF MARPOL 73/78

Some 136 countries are party to the MARPOL agreement, and all of the European countries have ratified that convention. Although all Member States of the

6. Marpol Regulation 11 Exceptions:

Regulations 9 and 10 of this Annex shall not apply to:

(a) The discharge into the sea of oil or oily mixture necessary for the purpose of securing the safety of a ship or saving life at sea; or
(b) The discharge into the sea of oil or oily mixture resulting from damage to a ship or its equipment:
   (i) Provided that all reasonable precautions have been taken after the occurrence of the damage or discovery of the discharge for the purpose of preventing or minimizing the discharge; and
   (ii) Except if the owner or the Master acted either with intent to cause damage, or recklessly and with knowledge that damage would probably result; or
(c) The discharge into the sea of substances containing oil, approved by the Administration, when being used for the purpose of combating specific pollution incidents in order to minimize the damage from pollution. Any such discharge shall be subject to the approval of any Government in whose jurisdiction it is contemplated the discharge will occur.

European Union (EU) have signed the MARPOL Convention, the Community (according to the opinion of the European Court of Justice [ECJ])[7] does not feel bound by that treaty, in contrast with what was said by the European Court in relation to another International Treaty namely the General Agreement on Trade and Tariffs Treaty, which 'the Community' tacitly assumed according to its own words.

The European Community is not by itself a party to the MARPOL Convention, nor has the European Community assumed the powers previously exercised by the various Member States[8] (thus said the ECJ in n° 47, 48, 49 and 50) of the Decision C-308/06.

That is a somewhat strange conclusion because the objective of European Directive 2005/35 is, as claimed in its preamble, to implement 'certain' rules of the MARPOL Convention into Community Law, but that fact in itself is not enough for the European Community to be bound by the MARPOL Convention as a whole, thus said the Court. That reasoning of the Court in itself is quite strange, because we found in the section 'summaries of legislation' quite a different message on the website of the EU. Indeed, on the European website, one can read the following statement:

> The sinking of The Prestige in November 2002 and of the Erika in December 1999 highlighted the need to tighten the net in relation to ship-source pollution.
>
> However, accidents are not the main source of pollution: most of it is the result of deliberate discharges (tank/cleaning operations and waste oil disposals).
>
> These rules incorporate into Community Law the 1973 International Convention on the Prevention of Pollution from Ships and its 1978 Protocol (MARPOL Convention 73/78). This will make it possible to harmonize application of the provisions of this Convention.[9]

5 THE INTRODUCTION OF THE EU DIRECTIVE 2005/35

After a number of very serious maritime incidents and catastrophic oil spills, such as that of the *Erika* in 1999 and *Prestige* in 2002, the EU decided to act vigorously to combat the threat of oil pollution. An impressive number of measures were taken, such as the Erika 1 and Erika 2 package, but also the Directive 2005/35 of the European Parliament and Council of 17 September 2005.[10] By this European Directive, the EU created a legal framework for the various Member States to

7. European Court of Justice ECJ C-308/06.
8. European Court of Justice ECJ C-308/06,47,48,49 and 50,
9. See <http://europa.eu/scadpalus/leg/en/lvb/124123.htm>.
10. Official Journal of the European Union, L255/11 30 Sep. 2005.In the meantime the Directive has been amended :Directive 2009/123/EC of the European Parliament and of the Council of the 21 Oct. 2009 amending directive 2005/35/EEC on ship source of pollution and on the introduction of penalties for infringements, OJ L280, 27 Oct. 2009.

impose penalties, particularly criminal penalties, in the event of discharges of oil and other noxious substances from ships far beyond Community waters, that is, internal waters, the territorial sea of Member States, straits used for international navigation to the extent that the Member State exercises jurisdiction over such straits, the exclusive economic zone or equivalent zone of a Member State and the high seas.[11] The message of the European Directive is clearly spelled out in Article 1 – to incorporate international standards for ship-source pollution into Community Law and to ensure that persons responsible for discharges are subject to adequate penalties as referred to in Article 8, in order to improve maritime safety and to enhance protection of the marine environment from pollution by ships.

Article 4 of the Directive imposes an obligation on the Member States who have to ensure that ship-source discharges of polluting substances into any of the areas referred to in Article 3(1) are regarded as infringements if committed with intent, recklessly or by serious negligence. The European Directive continues by stating that these infringements are regarded as criminal offences by, and in the circumstances provided for in, framework decision 2005/667/JHA, supplementing this Directive.[12] The penalties must be effective, proportionate and dissuasive and must be applied to anyone deemed responsible (the shipowner, the owner of the cargo or any other implicated person).[13] For the most serious cases, that is, instances that cause significant and widespread damage to water quality or animal or vegetable species or parts of them or the death or serious injury of person, each Member State must include imprisonment among possible penalties. The European Directive, together with its supplementary decision, makes it very clear what the Community has in mind.

However, on 23 October 2007, the ECJ nullified this supplementary decision because it was held to be in violation of Article 82 of the basic treaty.[14] The Commissions is clearly not competent according to the Court to 'determine the type and the level of criminal penalties to be applied'.[15] The Court came to the conclusion that framework decision 2005/667 encroached on the competence spelled out in Article 80(2) EC, which is attributed to the Community, as well as it infringed on Article 47 EU, so that it had to be annulled all together.[16]

6 BUT WHERE IS THE CONFLICT BETWEEN UNCLOS/ MARPOL AND THE EU DIRECTIVE 2005/35?

Directive 2005/35 makes no distinction between operational and accidental discharges with reference to the imposition of criminal sanctions. Indeed, the

11. Article 3,1a–e of the Directive 2005/35.
12. See No. 8 of the preamble of the Directive 2005/35.
13. Article 8, 1 and 2 of the Directive 2005/35.
14. See European Court of Justice; case C-440/05.
15. M. Faure, 'The Continuing Story of Environmental Criminal Law in Europe after 23 of October 2007', *European Energy and Environmental Law Review* (2008): 68–75.
16. See N° 74 of the Decision.

Directive imposes criminal liability for infringements if they are committed with intent, recklessness or by serious negligence. The sanctions have to be applied irrespective of the flag, which the vessel would fly, and the Directive is applicable starting from internal waters, to territorial seas and into the high seas. The sanctions have to apply to the owner, the master, the crew, the salvor, the charterer and the classification society, in other words, everyone who is connected with the incident and shares some responsibility for the incident. The conflict lies in the fact that the Directive spells out that a specific defence that is provided for in the MARPOL treaty, to which all Member States of the Community are a party, cannot be invoked when the incident occurs in the territorial seas of the Member States. Outside the territorial seas, the owner, the master and the crew can rely on the MARPOL Defences. As a consequence, in some cases the defendants or the accused are deprived of the defence that is spelled out in Regulation 11b of Annex 1 of MARPOL, which is a binding International Treaty. We will comment on another possible conflict with UNCLOS later on. With reference to MARPOL, the result of the Directive is that the master of a vessel and crew members can be severely punished with criminal sanctions, even for crimes or wrongdoings that are not committed with any form of 'intent', in other words without a guilty mind or with a *mens rea* but for wrongdoings that came about by mere negligence if the court finds this particular form of negligence to be 'serious'. But what is to be considered as 'serious negligence'? The fear is that 'serious negligence' could also be applied to a minor mistake by a crew member that results in a major incident with catastrophic consequences. The EU Member States were obliged, according to the instruction of the European Directive, to incorporate the message contained in this directive into their national laws, by 31 March 2007.[17] For serious cases, the Member States are even obliged to introduce imprisonment in the inventory of possible sanctions to be applied. According to decision 2005/667/JHA, Article 6, the Member State shall take the measures necessary to ensure that the offence referred to in Article 2, when committed with serious negligence, is punishable by a maximum of at least between 2 and 5 years of imprisonment when the offence caused significant and widespread damage to water quality, to animals or vegetable species or to parts of them and the death or serious injury of persons. The conflict with MARPOL lies in the fact that the MARPOL treaty lays down a uniform set of rules, from which contracting states cannot depart. However, the European Directive puts Member States in conflict with their existing obligations under MARPOL in that within territorial waters the directive imposes criminal liability for all discharges caused by serious negligence and precludes any defendant from relying on the MARPOL defence under Regulation 11b of Annex 1. Within a Member State's economic zone, or on the high seas, the Directive imposes liability for serious negligence, for persons, other than the owner, the master or the crew.

---

17. Article 16 of the European Directive 2005/35.

7 POSSIBLE VIOLATIONS WITH THE UNITED NATIONS CONVENTION ON THE LAW OF THE SEA 1982

Article 230 of the United Nations Convention on the law of the sea, imposes only monetary penalties, with respect to violations of national laws and regulations or applicable international rules and standards for the prevention, reduction and control of pollution of the marine environment, committed by foreign vessels, beyond the territorial seas.[18]

The text of the treaty says monetary penalties only may be imposed. Of course, the argument could be raised that the EU is violating section 230 by obliging Member States to impose imprisonment for violations of rules on pollution of the marine environment.

A further argument is that the European Directive in fact hampers the right of innocent passage in the territorial seas, which is guaranteed under UNCLOS, by lowering the threshold of liability to one of 'serious negligence'.

8 THE ATTACK ON THE EUROPEAN DIRECTIVE 2005

A number of interested parties and national interest groups (INTERTANKO, INTERCARGO, the Greek shipowners, Lloyds Register Group and the International Salvage Union) objected to the Directive. A few arguments were advanced. The test of 'serious negligence' is vague and subjective and therefore fails to satisfy a fundamental principle of European Law, namely, that of 'legal certainty'.

Furthermore, it was claimed that the European Directive requires Member States to punish, not only intentional, but also 'serious negligence' ship-source pollution. Several concerned parties objected to the Directive and started proceedings in England. The United Kingdom implemented the Directive, but then the higher parties brought an action in the High Court of Justice (administrative section) against the Secretary of State competent for Transport. This resulted in a decision by which the High Court (United Kingdom) submitted a number of relevant issues on the legality of the European Directive to the ECJ in Luxembourg. It was Justice Hodge who rendered the interlocutory English Decision, 30 June 2006/4 July 2006, in which he raised the following questions:

> Whether it is lawful for the EU to impose criminal liability in respect of discharges from foreign flag ships on the high seas or in the Exclusive Economic Zone and to limit/exclude MARPOL defences in such cases.
>
> Whether it is lawful for the EU to exclude MARPOL defences for discharges in the territorial sea.
>
> Whether the imposition of criminal liability for discharges caused by 'serious negligence' hampers the right of innocent passage.

18. See the paras 1, 2 and 3 of Art. 230, UNCLOS.

Whether the imposition of criminal liability for discharges caused by 'serious negligence' satisfies the requirement of legal certainty.

9 THE DECISION OF THE EUROPEAN COURT ON THE LEGALITY OF EUROPEAN DIRECTIVE 2005/35 – CASE C-308/06 – GRAND CHAMBER OF 30 JUNE 2009

The decision by the European Court in this matter was certainly a big disappointment for the claimants. Personally, we think it is a very evasive and rather technical decision by which the Court is more or less turning a blind eye on very relevant issues for owners, operators and seafarers, namely their exposure to criminal sanctions in case of accidental discharges on the basis of what is labelled as 'serious negligence', contrary to what is said in the Law of the Sea Convention and in the MARPOL treaty. The following question was submitted to the Court: Is European Directive 2005/35, which is a secondary European legal source, in violation of higher legal norms such as the Law of the Sea Convention and the MARPOL Treaty?

A further question submitted to the Court was whether the concept of 'serious negligence' is clear enough to satisfy the requirement of legal certainty, which is a well-established legal requirement within the European legal system? As was said above, the outcome of the Court proceedings was somewhat of a disappointment. Indeed, the Court came to the conclusion that it was not in a position to 'assess the legality of the European Directive 2005/35/EC of the European Parliament and of the Council of September 7th, 2005' in the light of MARPOL or the International Convention for the Prevention of Oil Pollution from Ships 1973, as supplemented by the Protocol of 1978 (MARPOL 73/78), nor in the light of UNCLOS, signed in Montero Bay, December 1982. With reference to the issue raised in connection with the legal certainty that, according to the claimants, cannot be offered by a broad concept, such as 'serious negligence', the Court found that there was no problem and that the concept was sufficiently clear.

10 HOW DID THE COURT COME TO THESE CONCLUSIONS?

With reference to the question of compatibility of the Directive with MARPOL, the Court quickly came to the conclusion that they could not examine the Directive against MARPOL because the Community was never a party to MARPOL. The Court had to overcome a few arguments. Indeed, all of the Member States of the European Union are a party to MARPOL, as are most civilized nations of the world. However, the Court stated that the Community as such has never signed the MARPOL Treaty, nor has the Community assumed the powers previously exercised by the Member States in the field in which MARPOL applies.[19]

19. See European Court of Justice, court case C-308/06, n° 48.

They had to overcome a precedent by which an international treaty, to which a Community never had been a formal party, nonetheless had become binding on the Community.

This was the case with the General Agreement on Trade and Tariffs (GATT 1947).

By exercising the powers that previously were vested with the Member States, the European Community had assumed the obligations derived from GATT 1947. The Court referred to the leading case, *International Fruit Company*, of 12 December 1972, 21/72-24/72.[20] The Court said that the situation was different and cannot be applied to MARPOL. Another nasty argument had to be answered.

Indeed, the purpose of the Directive was to incorporate MARPOL as is clearly said on the website of the European Community (see above).

But the Court continued by saying:

> The mere fact that the Directive purports to incorporate into Community Law 'certain' rules of MARPOL is not sufficient for the Court to be entitled to review the Directive's legality in the light of the MARPOL Convention.[21]

Nor are regulations 9 and 11b of Annex 1 and regulation 5 and 6 of Annex 2 of MARPOL the expressions of customary rules of international law (see again § 50).

As already mentioned, the reasoning of the Court is rather strange, because on the website of the European Commission it is proudly stated that the Directive 'incorporates MARPOL 73/78 into Community Law'. But all of a sudden, the Court is of the opinion that only a 'certain' number of rules of MARPOL are incorporated and not MARPOL as such.

11 THE COMPATIBILITY OF THE DIRECTIVE WITH UNCLOS 1982

Unlike MARPOL, about which the ECJ was of the opinion that the European Community was not bound by it, the EU is definitely bound by the United Nations Convention on the Law of the Sea by its approval decision 98/392.[22]

However, the Court was of the opinion that UNCLOS does not establish rules intended to apply directly and immediately to individuals, and does not confer rights capable of being relied upon against States.[23] The consequence of this was that these individual parties, such as INTERTANKO and others had no standing in Court, and the Court could not assess the legality of the Directive against UNCLOS. UNCLOS is a convention dealing with rights between sovereign states and was not intended to be relied upon by individuals against states.

20. See § 48 of the decision and *International Fruit Company and Others* [1972] ECR 1219, para. 7.
21. See N° 50.
22. Decision 98/392, see N° 65.
23. C-308/06, N° 64.

The argument is troubling because the Directive spells out its desired and claimed compatibility with UNCLOS.[24] Europe is also criticized for its arrogant unilateralism.[25]

12 THE REQUIREMENT OF LEGAL CERTAINTY

Is the Directive by its vague/broad concept, 'serious negligence', in violation of the principle of legal certainty, which is a fundamental requirement of Community law?

The Court answered in the negative – the concept was clear enough! True, no definition is given of this concept, which is impossible to define, but, according to the Court, we all can understand what the concept deals with – it is an unintentional act or omission by one who breaks a 'patent' duty of care! The question could be raised, however, as to whether by simply switching the word 'serious' into 'patent' anything has been solved at all,

Meng Fang was very critical of the decision of the European Court:

> The negative impact of directive 2005/35/EC on the human rights of seafarers and others could be enormous, as the Master and seafarers of a tanker that runs aground in force majeure circumstances and spills oil would be prejudged as a criminal. It should be noted that the imposition of criminal liability is a serious matter that must be guided by certain fundamental principles. Firstly, sanctions must be proportionate to the degree of personal fault on the part of the accused. Secondly, criminal law must be drafted in a clear, concise and unambiguous way and must be of restrictive interpretation so that it does not result in unintentional or unfair penalties against the accused. Thirdly, it must respect the standards set by international conventions and customary international law, including the law on human rights.[26]

---

24. Article 9 Directive 2005/35/EC of 7 Sep. 2005:

    > Compliance with international law: Member States shall apply the provisions of this Directive without any discrimination in form or in fact against foreign ships and in accordance with applicable international law, including Section 7 of Part XII of the 1982 United Nations Convention on the Law of the Sea, and they shall promptly notify the flag State of the vessel and any other State concerned of measures taken in accordance with this Directive.

25. L. Nenguye & F. Maes, ' "The European Unions" Role in the Prevention of Vessel – Source Pollution and Its Internal Influence', *Journal of Maritime Law* (2009): 422. the authors said: 'However the EU's practice tends toward unilateralism, which has been noted by the rest of the world. In regulating for a global industry, such as shipping, the EU will need to be careful about the balance between strict legal regimes an international obligations.'
26. M. Fang, The analysis of criminalization of marine pollution from the adoption by the European Union of the Directive on ship-source pollution and on the introduction of sanctions for infringements, in MPLP 2009, International Conference on Marine Pollution Liability and Policy, Papers Collection, Dalian 2009, 429.

13 THE CONSEQUENCES OF THIS DECISION

As a consequence of this Court decision, the Directive still stands and the Member States are obliged to imply the directive in their national legislation. This means that the poor fate of the seafarer is not any better after this Decision. The question could be raised, however, whether the Directive will not have some counter-productive results. Indeed, masters in command of vessels in distress could be very reluctant to enter into a port or into a place of refuge for fear of being arrested. Frank Wiswall already drew our attention to this possibility.[27]

The *Erika* and *Prestige* cases illustrate the point. At worst, such criminalization may deter the master of a distressed vessel from requesting refuge and thus aggravate or increase the chances for hull breach, thereby defeating the entire purpose of the effort to guarantee places of refuge. At best, the result is a manifest injustice to the master, who in the jaws of crisis is forced to weigh his personal exposure to criminal liability against the safety of his ship, crew and cargo.[28] Also, litigating parties, amongst which was INTERTANKO, were very disappointed with the decision. The chairman of INTERTANKO, Mr Nick Fistes stated:

> These proceedings were brought to uphold the principle of the shipping industry being regulated on a global basis with regulations that are legally certain and capable of uniform application. Furthermore, they were brought to insure the fair treatment of seafarers and others engaged in the shipping industry.

There is another drawback that may have been created by that decision. During the deliberations of the legal committee of IMO, representatives of the Industry voiced their concern that the imposition of criminal sanctions to seafarers may have an adverse impact on recruitment. A number of delegations supported the Secretary General statement in his opening address to the meeting that there was an urgent need for the IMO guidelines.[29]

Concern was also expressed about confusion that arises from the failure to distinguish between investigation of an incident to prevent its re-occurrence and a criminal investigation.[30]

---

27. F.L. Wiswall, Jr, Penal Liability, Document 12 in C.M.I. Yearbook 2003, 468:

    Criminalization most often means that if the hull of a vessel is breached and the result is damage to the coastline, environment or related interests of a costal state, the ships master (and perhaps other officers as well) will be subject to arrest and imprisonment upon entry into the territorial jurisdiction of that state, even when such entry is made involuntarily in the course of a rescue.', O.G. Anthony, 'Criminalization of Seafarers for Accidental Discharge of Oil: Is there justification in international law for criminal sanctions for negligent or accidental pollution of the sea?, *Journal of Maritime Law and Commerce* 37, no. 2 (2006): 219–243, see 236–239.

28. *Ibid.*
29. Guidelines for the fair treatment of seafarers to be implemented, not only for the benefit of the seafarers involved, but also to ensure that a future generation of seafarers may not be deterred from choosing the sea as a career.
30. P. Griggs & R. Shaw in C.M.I. Newsletter, September/December 2008 at 16.

14 THE IMO GUIDELINES ON THE FAIR TREATMENT OF SEAFARERS

In coming to the conclusion of this chapter, we believe that attention must be drawn to the International Maritime Organization (IMO) Guidelines on the Fair Treatment of Seafarers. The guidelines on fair treatment of seafarers in the event of a maritime accident were adopted by the legal committee of IMO at its last session and by the governing body of the International Labour Organization (ILO) at its 296th session on 12 June 2006.

The Baltic and International Maritime Council (BIMCO), a body representing vessel owners, has engaged in a study concerning the situation of seafarers. The BIMCO study comes to the conclusion that most laws that are applied are in fact fair and that negligent behaviour resulting in pollution deserves punishment, but there is an unfortunate pattern of laws being applied unfairly, with seafarers being detained after an incident on the presumption of criminal negligence, and in many cases seafarers are held as financial security for whatever claims authorities may introduce.[31]

15 THE FRENCH *ERIKA* DECISION

The massive oil spill by the oil tanker *Erika*, flying a Maltese flag, resulted in a decision by the French Criminal Court of Paris on 16 January 2008. It is a very lengthy detailed decision numbering 358 pages. But what does the French *Erika* decision teach us?

The Indian master of the *Erika*, Mr Mathur Karun Sunder, was incarcerated for a number of weeks by the Paris examining magistrate, which sparked indignation in the Shipping Community. In fact, two arrest orders were issued – one on 17 December 2001 and one on 27 April 2005. By Court decision on 16 January 2008, some nine years after the incident, the Paris Court[32] found the owner and the operator of the *Erika* guilty, as well as the classification society RINA and the charterer TOTAL, despite the 'channelling of liability' principle as contained in the CLC Convention.[33] The decision opens the possibility for the various claimants

31. E. Mallach, *The Fair Treatment of Seafarers*, on the website of Dabelstein & Passehl, Hamburg, and report of BIMCO study at the BIMCO website.
32. Tribunal de Grande Instance de Paris, 11ième Chambre, 4ième section, 16 Jan. 2008 C-9934895010.
33. Foley & Nolan, *supra* n. 4, 42–78: 'But make no mistake, this narrow view of classification liability has been shattered by the Erika court's ruling. This is also the first significant extension of liability to a charterer, an entity that is not involved with any maintenance or repair of the vessel. Charterers are "on notice" that by simply hiring a problem ship, it could be exposed to liability for the wrongful and criminal conduct of others. In particular the Erika court's holding should be considered as a stern warning to ship owners, managers, classification societies, and oil companies alike: adhere to safe shipping practices or face criminal charges and potentially limitless civil liability for endangering seafarers and causing harm to the environment.'

to go after the operator, classification society and the charterer in satisfaction of their claims. If the principle of 'channelling of liability' of the CLC were to be applied, the only liable party would have been the registered owner.[34] The master was found not guilty and obtained a small compensation of Euro (EUR) 16,666 against the owner, the operator and the classification society.[35]

It is good to notice that the master in fact saved the life of his twenty-three crew members by requesting helicopter rescue in a moment when it was still possible.

16 LET US HAVE A LOOK AT THE FRENCH CRIMINAL CODE

In the meantime the French Republic adapted its penal law. The then Minister of Justice, Perben, introduced several penal sanctions in the case of marine pollution resulting from voluntary or involuntary discharges. The applicable sanctions are categorized according to their nature (i.e., voluntary and involuntary discharges) and according to their 'gravity' in the case of involuntary discharges. Involuntary discharges are sanctioned according to their gravity, that is

- two years imprisonment and a fine of EUR 200,000 in the case of carelessness or negligence
- five years imprisonment and a fine of EUR 500,000 in the case of either intentional breach of safety duties/legally determined cautious obligations or irreversible environmental damage. This fine, however, can be increased up to the equivalent of the value of the ship or up to two times the value of the cargo on board, seven years imprisonment and a fine of EUR 700,000 in the case of both intentional breach of safety duties/legally determined cautious obligations and irreversible environmental damage. This fine, however, can be increased up to the equivalent of the value of the ship or up to three times the value of the cargo onboard.

The applicable sanctions may be imposed on masters as well as on the legal representatives of the company operating the polluting vessel. They apply to all flags, irrespective of the sea area where the pollution occurs, taking into account, however, the relevant provisions of UNCLOS.

34. CLC Art. III, 4. 'No claim for compensation for pollution damage shall be made against the owner otherwise than in accordance with this Convention. No claim for pollution damage under this Convention or otherwise may be made against the servants or agents of the owner.'
35. For a critical comment on the recent decision by the French Appeals Court (30 Mar. 2010) confirming for the major part the first decision by French Criminal Court see: Mans Jacobbsson, L'Indemnisation des dommages resultant des atteintes à l'environment dans le cadre du regime international CLC/ FIPOL, DMF 715, at 475–477.

17 AN ACTUAL ASSESSMENT WITH REFERENCE TO THE TREATMENT OF SEAFARERS

A recent updated BIMCO study shows the developments in past cases that have taken place since *Prestige* and *Erika*. BIMCO comes to the conclusion that it is encouraging that the number of incidents remains relatively low. There are a total of forty-seven cases since 1996 in which masters or crew members have received a lot of attention by the authorities.

There are fourteen cases in which sanctions were taken against seafarers involved before any deliberate act or negligence had been admitted or proven in Court.

In thirty-three cases, sanctions were implemented against involved seafarers after deliberate acts or negligence had been admitted or proven in Court.

18 CONCLUSION

On account of the consequences of the European Directive 2005/35, crew members are now facing liability consequences for criminal prosecution, even for unintentional acts, but with extreme penalties. The French Statute imposes very severe criminal sanctions, with imprisonment sentences of 2 to 7 years. Crew members have become vulnerable, but also shore personnel, including high-ranking executives are potential targets for criminal prosecution, even for accidents resulting from negligence, which is qualified as failure to use such care as a reasonable person would use under the circumstances.

A 'patent' breach of the duty of care imposed on the crew member is considered to be 'serious negligence', but once again we can raise the question as to what under the circumstances is 'patent' and what is 'serious'. Is it also a 'serious' breach when the consequences of a minor mistake result in a major or dramatic oil spill? We are of the opinion that unavoidable and draconian exposure to criminal liability could have a negative impact in the case in which the master has to sail to a place or a port of refuge. Imposing unavoidable criminal liability could undermine and frustrate the efforts by the Comité Maritime International (CMI), which prepared the draft on 'Guidelines on Places or Refuge', Athens 2008. Being tough on seafarers may also have a very negative impact on the recruitment of new seafarers because it will act as a deterrent.[36]

Crew members should not be treated as scapegoats or as a 'security' held against vessel owners but should be given the protection, which is the minimum

36. M. Jacobsson, Former director International Oil Pollution Compensation Funds, 'The International Maritime Education', in *International Maritime Education Forum*, Dalian Maritime University, 6 Jun. 2009, 7: 'In many countries a career at sea seems not to be sufficiently attractive to talented young people today. It is submitted that the present tendency of increased criminalization of seafarers does not help in this regard, and the risk of seafarers becoming the victims of piracy is another deterrent factor.'

protection offered to them in MARPOL and also in the Law of the Sea Convention:[37]

> We would like to refer to the third limb of article 230 of UNCLOS, which says:
>
> In the conduct of proceedings in respect of such violations committed by a foreign vessel, which may result in the imposition of penalties, recognized rights of the accused, shall be observed. . . .

References

Anthony, O.G. 'Criminalization Seafarers for Accidental Discharge of Oil: Is There Justification in International Law for Criminal Sanctions for Negligent or Accidental Pollution of the Sea?'. *Journal of Maritime Law and Commerce* 37, no. 2 (2006): 219–243, see 236–239.

Churchill, R.R. & A.V. Lowe. *The Law of the Sea*. 2nd edn. Manchester: Manchester University Press, 1998–1991.

Fang, M. *The Analysis of Criminalization of Marine Pollution from the Adoption by the European Union of the Directive on Ship-Source Pollution and on the Introduction of Sanctions for Infringements, in MPLP 2009*. International Conference on Marine Pollution Liability and Policy, Papers Collection, Dalian 2009.

Faure, M. 'The Continuing Story of Environmental Criminal Law in Europe after 23 of October 2007'. *European Energy and Environmental Law Review* 17 (2008): 68–75.

Foley. V.J. & C.R. Nolan. 'The Erika Judgment – Environmental Liability and Places of Refuge – A Sea Change in Civil and Criminal Responsibility that the Maritime Community Must Heed'. *Tulane Maritime Law Journal* 33 (2008): 42–78.

Jacobsson, M. 'The International Maritime Education'. International Maritime Education Forum, Dalian Maritime University, 6 June 2009.

Mallach, E. *The Fair Treatment of Seafarers*, on the website of Dabelstein & Passehl. <www.da-pa.com/en/publications.php> 15 July 2010.

Mukherjee, P.K. 'Criminalization and Unfair Treatment: The Seafarer's Perspective'. *Journal of International Maritime Law* 12, no. 5 (2006): 327.

Wiswall, F.L. Jr. *Penal Liability*. Document 12, VancouverConference Documents C.M.I. Yearbook, 2003, <www.comitemaritime.org>.

37. See No. 16 of the preamble of the European Directive 2005/35.

Chapter 13

Criminal Liabilities for Ship-Source Pollution Based on the Laws of China

*Jiang Yuechuan**

1 PREFACE

In recent years, criminal penalties against crew members, shipowners and related personnel and enterprises resulting from ship-source pollution accidents have drawn increasing attention from the international community.

After the oil pollution accident of the Prestige, the European Union (EU) passed Directive 2005/35/EC in September 2005, stipulating that EU Member States should ensure that ship-source discharges of polluting substances of oil or noxious liquid substances in bulk be regarded as infringements if committed with intent, recklessly or by serious negligence and are subject to effective, proportionate and dissuasive penalties, which include both administrative penalties and criminal penalties.[1] Such penalties should be applicable not only to the shipowner or the master of the ship but also the owner of the cargo, the classification society and any other person involved.[2]

* Lecturer in Law School of Dalian Maritime University, No. 1 Linghai Road, Ganjingzi District, Dalian 116026, P.R.C. Tel.: 86-411-84733176, E-mail: lawyer_jiang@hotmail.com.

1. Refer to Art. 8 of Directive 2005/35/EC.
2. Refer to para. (7) of the Preface of Directive 2005/35/EC.

Michael G. Faure, Han Lixin & Shan Hongjun, *Maritime Pollution Liability and Policy: China, Europe and the US*, 231–246.

The Korean Court convicted the captain and the chief officer of the *Hebei Spirit*, causing a big stir in the international society. On 10 December 2008, the Korean Court convicted Captain Jasprit Chawla and Chief Officer Syam Chetan of the oil ship the *Hebei Spirit* of an oil leakage accident caused by the ship's collision with a runaway tugboat that belonged to Samsung Heavy Industries and sentenced them to imprisonment for eighteen months and eight months, respectively. The International Transport Workers' Federation (ITF), some shipowners' associations, and the Indian Government protested against this sentence and considered it unjust. They held that in this accident the *Hebei Spirit*, which was in anchorage at the time, was totally innocent and the collision was completely caused by the Samsung Heavy Industries' tugboat; the captain and the chief officer of the *Hebei Spirit* should have been praised instead of being scolded because the level of professionalism they exhibited in that accident proved them to be excellent mariners. The international community is also concerned that this sentence by the Korean Court might frustrate young people who want to enter the trade as mariners and worsen the present shortage of mariners.

The transportation of oil and chemical products along the Chinese coast has been quite busy. According to data published by the National Bureau of Statistics of China, People's Republic of China ('PRC') imported 178,880,000 tonnes of crude oil in the year 2008, with a 9.64% increase compared to that of the previous year, and 38,850,000 tonnes of refined oil, with a 15% increase.[3] Of this merchandise, 95% is dependant on marine transportation. Besides oil and chemical products for foreign trades, oil and chemical products for domestic trades combine for a huge quantity of transportation. Consider the example of crude oil transportation in 2007. In 2007, the oil-handling capacity of major ports in the country was approximately 280,000,000 tonnes, of which the oil-handling capacity for domestic trade was about 130,000,000 tonnes. Fleets engaged in the transportation of crude oil, refined oil, and chemical products along the Chinese coast are expanding with each passing day.[4] According to a report published by the Ministry of Transport of the PRC, by the end of 2007 there were 1,348 Chinese ships/8,186,000 deadweight tonnes (DWT) engaged in cross-province transportation of hazardous liquid cargo along the Chinese coast, which was 103.5% and 109.3% of those by the end of 2006. Among them there were 1,085 oil ships/7,670,000 DWT, which were six ships/590,000 more than those by the end of 2006; 206 chemical cargo ships/431,000 DWT, which were thirty-three ships/95,000 more than those by the end of 2006.[5]

As a result of the huge transportation of oil and chemical products along the Chinese coast, the large number of ships engaged in the transportation of oil and chemical products and the large number of fuel oil pollution accidents caused by other kinds of ships, ship-source pollution accidents are already taking place

3. <www.stats.gov.cn/tjgb/ndtjgb/qgndtjgb/t20090226_402540710.htm>.
4. According to Art. 4 of China Maritime Code, only ships flying the national flag of China can be engaged in cargo transportation and towage services between Chinese ports.
5. <www.moc.gov.cn/zhuzhan/tongjixinxi/qita/200803/t20080324_475006.html>.

frequently along the Chinese coast. As well, threats from major accidents such as those caused by the *Prestige* and *Hebei Spirit* are always present.

However, unlike in the United States and other countries in which those responsible for oil pollution accidents are subject to criminal penalties, in China at present administrative penalties are usually adopted to punish ship-source pollution acts in addition to civil compensations to protect the rights and benefits of victims. It seems that no person or company has ever been convicted of criminal liabilities in any Chinese court as a result of ship-source pollution accidents. However, because of the growing concern in all circles in China about environmental protection, we gradually recognize that criminal penalties against environmental pollution acts are important approaches to the reduction of pollution accidents. Therefore, it can be predicted that in the near future, cases will occur in which people or companies are investigated for criminal liabilities resulting from ship-source pollution accidents. Before the advent of that historical moment, it is obviously necessary to study the issue of criminal liabilities related to ship-source pollution under the present PRC laws.

China is also one of the important sailor-supplying countries in the world and has more than 500,000 mariners, of which about 170,000 hold qualification certificates for oceangoing merchant ships. Therefore, how to achieve a balance between environmental protection and the activity of mariners in their profession is also a problem that must be considered when criminal penalties are used to punish ship-source pollution.

Based on the discussion to this point, this chapter will examine the following problems:

1. Under what circumstances will ship-source pollution incur criminal liabilities according to the present laws of PRC?
2. What kind of people will be investigated for criminal liabilities under the laws of PRC?
3. What kind of criterion of liability is adopted in the Criminal Law of PRC for criminal liabilities in ship-source pollution accidents? Is it necessary to establish a strict responsibility system for criminal liabilities for ship-source pollution on a legislative basis?

2 WHAT KIND OF SHIP-SOURCE POLLUTION ACCIDENTS WILL INVOLVE CRIMINAL LIABILITIES UNDER THE PRESENT PRC LEGISLATION?

2.1 Legal Basis for Investigations into Criminal Liabilities for Ship-Source Pollution Accidents under the Present Laws of the PRC

In China, some laws or regulations related to environmental protection have specified that certain severe breaches of the law will be investigated for criminal liabilities, such as Article 43 of the Environmental Protection Law of the People's Republic of China, which specifies:

> In cases where violations of this Law result in pollution damage to the environment and cause heavy losses to public or private property or deaths or injuries to persons, those who are directly responsible may be prosecuted for criminal responsibility by judicial organs in accordance with the law,

Clause 3 of Article 91 of the Marine Environmental Protection Law of the PRC, which specifies:

> Whoever causes major marine environment pollution accident resulting in grave consequences of heavy losses of public and private property, human injuries and deaths of persons, shall be investigated for criminal responsibility according to law,

and Article 90 of the Law on Prevention and Control of Water Pollution, which specifies:

> Any one that commits any violation against the specifications in this law that constitutes a violation against security administration shall be penalized by security administration according to the law; if such a violation constitutes a crime, it shall be investigated for criminal liabilities according to the law.

However, only the code of criminal laws of China, that is the Criminal Law of the People's Republic of China (the Criminal Law for short hereinafter) has the authority to make such specifications as to how serious such violations have to be to constitute crimes, how to convict and how to sentence.

In the Criminal Law of China, the following specifications in the 'Specific Provisions' may be used to determine if a certain ship-source pollution accident constitutes a crime and to determine what kind of criminal penalty shall be given against that crime.

> 1. Article 133: 'A person who violates the laws and regulations on administration of communications or transportation, thus causing a serious accident and resulting in serious bodily injury or death or serious losses to public or private property, shall be sentenced to fixed-term imprisonment of not more than three years or criminal detention; if he flees after causing a communication or transportation accident or has any other especially flagrant circumstance, shall be sentenced to fixed-term imprisonment of not less than three years and not more than seven years and; if his fleeing causes another person's death, to fixed-term imprisonment of not less than seven years.'

The crime described in this Article is commonly known as the 'traffic accident crime'. Since Article 131 and Article 132 of the Criminal Law have respectively specified the 'major flight accident crime' and 'railway operation accident crime', the 'traffic accident crime' specified herein is applicable only to road traffic accidents and waterborne traffic accidents.

In China, there have been judicial precedents in which people were convicted of traffic accident crimes and thus assumed criminal liability as a result of waterborne traffic accidents. 30 January 2008, the *Jin Yuan You 9* collided with the *Jin Tai Shun* anchored in the anchorage station at the Yangtze Estuary shortly after raising anchor from the same anchorage, causing water seepage in the *Jin Tai Shun*, which sank later. All seventeen crew members on board fell overboard; fifteen died, one was reported missing and only one survived. This was a major waterborne traffic accident, after which the *Jin Yuan You 9* was found responsible for most of the accident by Shanghai Maritime Safety Administration through investigations. The shipowner of the *Jin Yuan You 9* (a Chinese company) signed a settlement agreement with the family members of the deceased and missing sailors and paid civil compensations. The Yangpu District People's Court of Shanghai reached a verdict after the first trial that the captain of the *Jin Yuan You 9* failed to keep watch and thus did not discern the *Jin Tai Shun* anchored nearby and, in the dodging process that followed, failed to take the correct dodging measures. Therefore, he committed a traffic accident crime and was sentenced to fixed-term imprisonment of five years. The captain lodged an appeal arguing that the sentence was too heavy. But the appeal was rejected by the court of second instance and the original sentence was sustained.

However, the laws and legal practices of China have not established a well-defined standard as to what kind of acts constitute traffic accident crimes by 'causing serious losses to public or private property' as specified in Article 133 of the Criminal Law when no bodily injury or death is caused and only ship-source pollution occurs.

In November 2000, the Supreme Court of the PRC published an Explanation of Several Problems by the Supreme Court Concerning the Practical Application of Relevant Laws in the Hearing of Criminal Cases of Traffic Accidents, which specifies the prosecution criteria for the 'traffic accident crime'. The prosecution criterion for traffic accident crimes that involve only property losses is:

Causing direct losses to public or private property, assuming all or main liabilities for the accident, and the compensation amount being more than 300,000 Yuan above the capability of the person.

However, this Explanation by the Supreme Court does not specify whether it is also applicable to waterborne traffic accidents, and judging by its contents, there is reason to believe that this Explanation is meant for road traffic accidents and not applicable to waterborne traffic accidents.

Therefore, in China, if we investigate a person responsible for ship-source pollution for his traffic accident crime according to Article 133 of the Criminal Law, problems may arise as to the prosecution criteria or criminal registration criteria, including:

(a) Should the prosecution criteria be determined by the actual amount of losses caused by pollution? Or should other consequences, such as influences on the society, be taken into consideration at the same time?

(b) If the amount of loss caused by the pollution is taken as the sole criterion, how much should it be?

(c) According to the Explanation by the Supreme Court, only when there is just property loss and no bodily injury or death and the compensation amount is more than 300,000 Yuan above the capability of the liable person will the liable person have to bear criminal liabilities in the case of a road traffic accident; if the liable person has already satisfied civil compensation liabilities and the compensation amount is less than 300,000 Yuan above the capability of the liable person, the liable person will not be investigated for criminal liabilities. Therefore, a question arises as to the situation that the relevant persons may not be investigated for criminal liabilities if the owner of the ship causing the accident has already assumed the civil compensation liabilities according to related international conventions (such as CLC1992) or specifications of the Chinese law.

Because the prosecution or criminal registration criteria for traffic accident crimes constituted by ship-source pollution accidents are still unclear for the time being, Chinese procuratorial organs and courts are bound to treat such cases with considerable uncertainty, which should be noticed by ship owners and related parties.

It should be also noted that according to the aforementioned Article 133, the premise for constituting a 'traffic accident crime' is 'violating the laws and regulations on administration of communications or transportation'. The laws and regulations on administration of communications or transportation herein refer to laws and regulations on the sailing and manipulation of ships when it comes to waterborne transportation, such as International Regulations for Preventing Collision at Sea 1972 and Regulations for Preventing Collision at Inland Waters of the People's Republic of China. Therefore, committers of this crime are mostly ship navigators, such as ship captains and deck officers on duty, and other personnel including, the company management staff ashore, will obviously not commit this crime. Of course they can still commit other crimes, such as the crime of major liability accident specified in Article 134 of the Criminal Law as a result of ship-source pollution accidents. Furthermore, according to this premise to the traffic accident crime, if any ship-source pollution accident happens but the related personnel do not violate any laws or regulations on administration of communications or transportation, they will not constitute traffic accident crimes.

2. Article 134: 'Anyone who violates the safety management rules and regulations, thus causing a serious accident involving injury or death or any other serious result, shall be sentenced to fixed-term imprisonment of not more than three years or criminal detention and, if the circumstance is especially flagrant, to fixed-term imprisonment of not less than three years and not more than seven years.

Anyone who forces workers to work in a hazardous way in violation of the rules, thus causing a serious accident involving injury or death or any other serious result, shall be sentenced to fixed-term imprisonment of not more than

> five years or criminal detention and; if the circumstance is especially flagrant, to fixed-term imprisonment of not less than five years.'

The crimes specified in this Article are known as the 'crime of major liability accidents' (Provision 1) and 'crime of forcing workers to work in a hazardous way in violation of the rules' (Provision 2).

It should be noted that the previous edition of this Article was:

> An employee of a factory, mine, forestry centre, construction enterprise or any other enterprise or institution who disobeys management and violates the rules and regulations, or forces workers to work in a hazardous way in violation of the rules, thus causing a serious accident involving injury or death or any other serious result, shall be sentenced to fixed-term imprisonment of not more than three years or criminal detention and; if the circumstance is especially flagrant, to fixed-term imprisonment of not less than three years and not more than seven years.

The Amendment of the Criminal Law of the People's Republic of China (VI) passed in June 2006 changes it to the current content. It is clear that the amended Article has a wider scope of application compared with the original Article. Moreover, according to the current Article, the subjects of major liability accidents are no longer limited to 'employees' of enterprises and public institutions. Anyone, including managers of enterprises and public institutions, may become subjects of this crime.

Ship-source pollution accidents can obviously constitute the crime of major liability accidents specified in this Article. For example, when a crew member violates relevant regulations in the loading and unloading of cargo of an oil ship and causes cargo oil leakage that leads to a major pollution accident, this may constitute this crime.

The 'crime of forcing workers to work in a hazardous way in violation of the rules' specified in Provision 2 of this Article may also occur during a ship-source pollution accident, for example, a shipowner forcing a captain to carry out an operation against relevant safety or pollution prevention management regulations, which leads to a ship-source pollution accident. It is certain that the criminal subjects of this crime usually include personnel who hold commanding and managing responsibilities for the operation of the ship, such as the captain, the management staff ashore and the ship owner, instead of low-rank sailors.

Like the traffic accident crime, the 'crime of major liability accidents' and 'crime of forcing workers to work in a hazardous way in violation of the rules' specified in this chapter in ship-source pollution accidents also have unclear prosecution criteria, which leads to great uncertainty.

> 3. Article 135: 'If safety facilities or safety conditions for production do not conform to the state's provisions, thus causing a serious accident involving injury or death or any other serious result, the management staff directly responsible and other responsible personnel shall be sentenced to fixed-term imprisonment of not more than three years or criminal detention and; if the

> circumstance is especially flagrant, to fixed-term imprisonment of not less than three years and not more than seven years.'

This Article has also been amended by The Amendment of the Criminal Law of the People's Republic of China (VI). The original Article stated:

> If a factory, mine, forestry centre, construction enterprise or any other enterprise or institution, whose labour safety facilities do not conform to the state's provisions, fails to take measures to hidden peril of the accident after the relevant department or its employee puts forward a demand, thus causing a serious accident involving injury or death or any other serious result, the persons directly responsible shall be sentenced to fixed-term imprisonment of not more than three years or criminal detention and; if the circumstance is especially flagrant, to fixed-term imprisonment of not less than three years and not more than seven years.

Before the amendment, the crime specified in this Article was usually referred to as the 'crime of major labour safety accident', because it was limited to accidents resulting from labour safety facilities that failed to conform to the related regulations and most of them were personnel injuries or deaths. After being amended, this Article is no longer limited to labour safety facilities, but extended to all accidents resulting from any 'safety facilities or safety conditions for production' that do not conform to the state's provisions. Therefore, the crime specified in this Article is now commonly referred to as the 'crime of major production safety accidents'.

Ship-source pollution accidents may also involve this crime. For example, if it is ordered that a single hull oil tanker built before a certain year is to be dismantled, and the ship owner still uses it for operation, resulting in a major pollution accident, this crime may be constituted.

> 4. Article 338: 'A person who, in violation of the state's regulations, discharges, dumps or disposes radioactive wastes, wastes carrying infectious pathogens, poisonous substances or any other dangerous substances to land, water or air, thus causing a serious accident of environmental pollution, shall be sentenced to fixed-term imprisonment of not more than three years or criminal detention and concurrently or independently, to a fine, if a serious result or a great loss of public or private property or bodily injury or death of another person is caused; and if the result is especially serious, to fixed-term imprisonment of not less than three years and not more than seven years and concurrently to a fine.'

This Article is included in the Section 6 'Crime of Undermining Protection of Environment or Resources', Chapter 6 of the Specific Provisions of the Criminal Law. The crime specified therein is referred to as the 'crime of major environmental pollution accidents'. According to this Article, anyone who, in violation of the related regulations (such as MARPOL73/78), discharges, dumps or disposes hazardous wastes, such as oil-containing waste water, cabin-cleaning water and ship rubbish into the sea, thus causing a serious accident of environmental pollution, may be investigated for this crime.

Although discharging, dumping or disposing of hazardous wastes into land, waters or the atmosphere are intentional acts in this Article, this crime includes both intentional crimes and unintentional crimes. The former means that a person is clearly aware that the discharge of hazardous waste will cause serious consequences of environmental pollution but still commits the discharge act and lets the pollution consequences go unchecked. The latter means that although the actor discharges the hazardous waste intentionally, but does not expect the serious consequences as specified in this Article, that is,

> Knows that his act would produce a dangerous result to the society but fails to know due to his negligence or, has already known it but readily believes that it can be avoided, thus causing the result.

In Article 15 of the Criminal Law.

2.2 Conditions for Investigations into Ship-Source Pollution Accidents for Criminal Liabilities

All aforementioned provisions summed up, we can see that in China, ship-source pollution accidents meeting the following two basic conditions may be investigated for criminal liabilities:

2.2.1 The Ship-Source Pollution Accident Causes Serious Consequences of Marine Environmental Pollution

From the articles reviewed earlier we can see that only when a ship-source pollution accident 'causes personal injuries or deaths', 'causes major losses to the public or private properties' and 'leads to serious consequences' will the related person be investigated for criminal liabilities. If the pollution consequences are slight, they should not be investigated for criminal liabilities but are usually treated with administrative punishments.

2.2.2 The Ship-Source Pollution Accident or the Pollution Consequence is Caused by a Violation of the Law Committed by the Actor

From the aforementioned articles we can also see that the pollution consequence alone is not adequate; only when the act causing the pollution accident is against related laws and regulations will the actor be investigated for criminal liabilities. In other words, if the ship has caused a pollution accident but there is no violation of the law, for example, if the pollution accident is caused by force majeure or an emergency discharge in the case of an emergency to pull the ship out of danger, causing the pollution accident is not against the law the actor should not be investigated for criminal liability.

2.3 Examples of Ship-Source Pollution Accidents That May Be Investigated for Criminal Liabilities

According to the earlier discussion, the situations in which ship-source pollution accidents may be investigated for criminal liabilities include:

(a) The actor clearly knows that pollution may be caused, but still intentionally discharges, dumps or disposes hazardous waste from the ship to the sea or the inland water, thus causing serious pollution, such as pollution caused by secretly discharging unprocessed oily water from the oil sludge tank of the ship.
(b) The actor, in violation of the related regulations (such as MARPOL73/78), discharges, dumps or disposes hazardous waste into the sea or inland water, not intending to cause pollution but still causing serious pollution in the objective point of view.
(c) The actor, in violation of sailing rules, causes a water-borne traffic accident thus leading to serious pollution. For example, the ship violates International Regulations for Preventing Collision at Sea 1972 and thus has a collision with another ship which leads to oil leakage and major pollution consequences.
(d) The actor performs operation against related rules during the operation of the ship, thus causing an accident which results in major pollution consequences. For example, supervision is not carried out according to relevant requirements during the loading and unloading of cargo oil of the ship and thus cargo oil leaks and results in a major pollution accident.
(e) The actor forces others to operate against the rules, thus causing a major ship-source pollution accident. For example, the ship owner forces the captain to set sail in a bad weather that does not meet the required conditions, thus causing a ship wreck and a major pollution accident.
(f) The safety facilities of the ship do not meet related requirements, thus causing an accident of the ship and serious pollution. For example, the actor is clear aware that one ship is an overage vessel but still puts it into operation, thus causing a major ship-source pollution accident.

3 WHAT KIND OF PERSON WILL BE INVESTIGATED FOR CRIMINAL LIABILITIES AS A RESULT OF SHIP-SOURCE POLLUTION ACCIDENTS UNDER THE LAWS OF THE PRC?

According to the Criminal Law and other relevant laws as well as regulations of China, I take the view that the subjects that may be investigated for criminal liabilities resulting from ship-source pollution accidents include:

3.1 Masters and Mariners

The crimes that may be committed by masters and other mariners include traffic accident crime, crime of major liability accident, crime of major environmental pollution accidents and so forth. In addition, masters and officers may also commit the crime of forcing workers to work in a hazardous way in violation of the rules.

3.2 Shipowners

Today, most shipowners are companies incorporated under the law of a country or other types of legal persons. As recognized by criminal law, subjects such as legal persons and so forth. (under the criminal law, they are collectively called 'units', in contrast to 'natural persons') can become criminal subjects. As specified in Article 30 of the Criminal Law, a company, enterprise, institution, organization or group that commits an act endangering society that is considered a crime under the law shall bear criminal liabilities. As specified in Article 31 of the Criminal Law, where a unit commits a crime, it shall be sentenced to a fine and concurrently, the persons who are directly in charge and the other persons who are directly responsible for the crime shall be given criminal punishment.

The aforementioned 'crime of major production safety accidents' as specified in Article 135 of the Criminal Law is a type of unit crime committed by a unit instead of a crime committed by a natural person, despite the fact that just the 'directly liable person in charge and other persons directly liable' are subjected to the criminal punishment according to this Article.[6]

In accordance with Article 346 of the Criminal Law,

> If a unit commits the crimes specified in Article 338 to 345, the unit will be sentenced to a fine, while the leading person with direct responsibility and other personnel directly responsible for such violations are to be punished in accordance with the stipulations of related articles.

Obviously, as for the crime of major environmental pollution accident as specified in Article 338, the criminal subjects may be natural persons or units. Therefore, the shipowners may be subjected to criminal punishment in violation of the said crime.

3.3 Ashore Executive Personnel of Shipowners

In some cases, although ship owners do not constitute the crime committed by units, the acts of executive personnel of shipowners may constitute the crime resulting from ship-source pollution accident. Where shipowners' acts constitute the crime, as the persons directly liable, the ashore executive personnel may also be subjected to criminal punishment. The crimes that ashore executive personnel bear the criminal liabilities include as follows:

---

6. Because Art. 135 of the *Criminal Law* does not stipulates the punishment for a unit, many experts in China are of opinion that it is a flaw of the Criminal Law; the relevant punishment imposed on the unit by fine is supposed to be supplemented therein.

(a) Crime of major liability accident and crime of forcing workers to work in a hazardous way in violation of rules as specified in Article 134.
(b) Crime of major production safety accidents as specified in Article 135 and
(c) Crime of major environmental pollution accident as specified in Article 338.

3.4 Functionaries of State Organs Responsible for Environmental Supervision and Control

The functionaries of state organs responsible for environmental supervision and control may be investigated for criminal liabilities as a result of a ship-source pollution accident. In view of the said case, the Article 94 of Marine Environment Protection Law of PRC specified that:

> Any person in charge of marine environment supervision and control who abuses his power, neglects his duty or engaged in malpractice for personal gains resulting in pollution damage to the marine environment, shall be given administrative sanctions pursuant to the law; if his act constitutes a crime, he shall be investigated for criminal liabilities under the law.

The Article 408 of the Criminal Law also specifies that:

> Any functionary of a state organ who is responsible for environmental protection, supervision and control, through his gross neglect of duty, causes a serious environmental pollution accident, which results in heavy losses of public or private property or the grave consequences of injuries or deaths of persons, shall be sentenced to fixed-term imprisonment of not more than three years or criminal detention.

3.5 Inspectors of Classification Society and Other Vessel Inspection Agencies

In China, the agencies engaged in ship survey are classified as two kinds: the Ship Inspection Bureau of the People's Republic of China in the nature of a state organ and its local affiliates all over China and the ship classification societies in the nature of a social organization, such as China Classification Society (CCS).

According to the regulations as specified in Article 28 of Regulations of the Peoples Republic of China Governing Survey of Ships and Shore Installations,

> Where surveyors of ship survey agency abuse powers, play favouritism and commit irregularities, neglect duties and conduct gross neglect of duties, they shall be given disciplinary sanctions or disqualified for inspection by their work units or higher authorities; in serious cases, where the acts constitute a crime – the criminal liabilities shall be investigated by the judicial department.

Therefore, in the case in which the surveyors of a ship survey agency abuse their powers, and so on, such as to issue an inspection certificate to a sub-standard ship, thus causing the ship-source pollution accident, the surveyors may bear the criminal liability.

Although the Ship Inspection Bureau and classification societies in themselves can become the criminal subjects according to the Criminal Law,[7] I take the view that in China, it is scarcely possible for them to bear the criminal liabilities resulting from a ship-source pollution accident. As a matter of fact, although, as specified in Article 30 of the Criminal Law, the state organs such as the Ship Inspection Bureau can also become the subject of crime committed by a unit, many experts in the theoretical circle of China have raised an objection to it, holding the opinion that the state organs are not qualified criminal subjects. As for the classification societies, in the Criminal Law, there is not any related clause according to which the classification society in itself can be found guilty in one case of ship-source pollution accident. Article 3 of the Criminal Law specifies that

> For acts that are explicitly defined as criminal acts in law, the offenders shall be convicted and punished in accordance with law; otherwise, they shall not be convicted or punished,

requiring a classification society to bear the criminal liabilities relating to the ship-source pollution accident has no legal ground.

4 CRITERION OF LIABILITY FOR CRIMINAL LIABILITIES IN SHIP-SOURCE POLLUTION ACCIDENT

According to the criminal theory of most countries, including China, to constitute a crime several conditions must be required,[8] one of which is 'subjective elements of a crime' meaning that a criminal subject holds the mental attitude toward his act and consequence of endangering society. Only when the subjective mental attitude that the criminal subject holds toward his act and the consequence of endangering society is intention or negligence can such act constitute a crime. On a theory basis, the intention and negligence of constituting a crime is referred to as 'culpability'. 'No culpability, no crime' is a basic principle followed by the criminal law of China.

Under the criminal law of China, an intentional crime is categorized into directly intentional crime and indirectly intentional crime. The former means a person is clearly aware that his own act will cause a dangerous consequence to society and expects the consequence, thus constituting a crime; the latter means although the actor knows his act will produce a dangerous consequence to the society but still lets the consequence go unchecked, thus constituting a crime.[9]

---

7. See Art. 30 of the *Criminal Law* of China.
8. These several conditions include (1) criminal subject; (2) subjective element of crime; (3) objective element of crime; (4) criminal object.
9. See Art. 14 of *the Criminal Law*.

Several crimes relating to the ship-source pollution accident can hardly involve the directly intentional crime, while the indirectly intentional crime may occur. For example, a ship discharges oily water on the sly, so as to pay fewer expenses for oily water processing, thus causing pollution and constituting the crime of major environmental pollution accident as specified in Article 338 of the Criminal Law.

A negligent crime means a criminal subject should know his act may produce a dangerous result to the society but fails to know it due to his negligence or, has already known it but readily believes that it can be avoided, thus causing the result. Among the previously mentioned several crimes that may relate to the ship-source pollution accident as specified in the Criminal Law, except some part of the crime of major environmental pollution crime falling under the intentional crime, others fall under the negligent crime.

But in global society, there exits a great dispute as to whether masters and other mariners should be investigated for criminal liabilities for the ship-source pollution accident caused by their negligence in sailing or management of ship. As a matter of fact, once the 2005/35/EC Directive was passed by EU, it came under fire by maritime communicates; five organizations led by INTERTANKO for the benefit of shipowners lodged a lawsuit in the English High Court, claiming that this Directive shall be submitted to the European Court of Justice (ECJ) for examination. One of the issues about the Directive they requested ECJ to review is whether the standard of criminal liabilities for discharges resulting from 'serious negligence' (as specified in Article 4 of the Directive) breaches the right of innocent passage and whether the standard of liability in the Directive of 'serious negligence' satisfies the requirement of legal certainty. International Chamber of Shipping (ICS)–International Shipping Federation (ISF) said in a report:

> The industry is not opposed to appropriate punishment of deliberate violations of environmental rules, but the principle of criminalizing accidents is neither just nor reasonable given the hazards of the sea.

Under the Criminal Law of China, the ship-source pollution accident caused by mariners' negligence in sailing or management of ship, if a crime is constituted, shall separately fall under the traffic accident crime as specified in Article 133 thereof, the crime of major liability accident as specified in Article 134 thereof, and the crime of major environmental pollution accident as specified in Article 338 thereof. However, to date, no master or mariner has ever been investigated for criminal liabilities for a ship-source pollution accident caused by his negligence in sailing or management of ship, which may be attributable to the doubt of judicial and administrative organs of China about whether this kind of negligence of master or mariner constitutes a crime. However, based on the current Criminal Law of China, in consideration of aforementioned judicial precedents in which the captain was convicted of traffic accident crime and sentenced to a fixed-term imprisonment of five years because of the death and missing status of fifteen people, it is increasingly more likely that the masters or mariners who cause major ship-source pollution as a result of their negligence in sailing or management of a ship will be convicted of crime in the court of China.

As far as I am concerned, it is immoral to give mariners criminal punishment for their negligence in sailing and management of ship. Even if the negligence of this kind has caused great property loss and serious consequences of environmental pollution, mariners shall be censured and punished only by way of civil compensation and administrative penalty rather than imposed criminal liabilities. Additionally, as viewed from the policy of China, China is one of the main sailor-supplying countries in the world, and thus it is important to ensure China's position in the international mariner market, assure Chinese young men to select the occupation of mariner and relieve them from worries about imprisonment resulting from negligence in the work. For the reasons mentioned earlier, I take the view that, on the one hand, China should make clear that the relevant accidents caused by mariners' negligence in sailing and management of ship usually will not involve criminal liability by promulgating judicial interpretation and even modifying the laws and so forth; on the other hand, the courts of China, before legal modification, should impose criminal punishment on mariners with great caution in processing the ship-source pollution accident resulting from negligence of mariners in sailing and management of ship.

Additionally, in the common law countries, for some specific environmental pollution acts, the pollution actors are usually investigated for criminal liabilities subject to the liability without fault, that is as long as the actors commit the act of discharging a particular pollutant and result in environmental pollution, they shall bear the criminal liabilities, not conditioned by the criminal intention or negligence of them. Some Chinese experts also make a point that no-fault-liability environmental crime should be introduced into the laws of China to inhibit the increasingly more serious issue of environmental pollution. However, the prevailing opinion that Chinese experts hold are: first, the present criminal law of China does not recognize the crime without any fault, the subjective mental attitude of intention or negligence still constitutes the prerequisite of constituting a crime. Second, China should not introduce the concept of no-fault crime in the future, because the implementation of no-fault crime, in most cases, is to transform the administrative illegality to criminal illegality, and we should not treat the act that can only bring civil liability or administrative liability as a crime merely because of the needs of public interest.[10]

In my opinion, no-fault liability should not be implemented in terms of ship pollution at a minimum, for the following reasons. First, 'the negligence of actors are hard to testify' is the prerequisite of the implementing of no-fault liability, but in the ship-source pollution case involving mariner negligence, it is generally easy to testify the negligence of mariners. Second, in the United Kingdom, the United States and other countries, even if actors commit no-fault crime, they are generally sentenced to a fine as criminal punishment, whereas for the ship-source pollution accident caused by mariner negligence, the same effect can be reached by fining them as administrative punishment; therefore, it is not necessary for such an

10. C. Xingliang, *Philosophy of Criminal Law*, 3rd edn (Beijing: China University of Political Science and Law Press, 2004), 207–208.

accident to be deemed as a crime, and imprisonment and other criminal punishments shall not be readily imposed on the masters and mariners. Third, the implementation of no-fault liability can dampen the occupational confidence of mariners; therefore, it is unwise in regard to ensuring the supply of mariners.

Reference

Xingliang, C. *Philosophy of Criminal Law*. 3rd edn. Beijing: China University of Political Science and Law Press, 2004.

Chapter 14

Civil and Criminal Liabilities Concerning Marine Pollution in Hong Kong

Li Lianjun & Qin Mu***

1 INTRODUCTION

1.1 SHIP-SOURCE MARINE POLLUTION

Pollution damage[1] to the sea and countries with coastlines caused by oil tankers and other marine vessels remains a remarkable threat to the environment and

* Li Lianjun (B.Sc., M.Sc., LLM, FCIArb), a solicitor admitted in Hong Kong, England and Wales and a partner at Richards Butler in association with Reed Smith LLP. Mr Li is a fellow of Chartered Institute of Arbitrators and a Panel Arbitrator of China Maritime Arbitration Commission and China International Economic and Trade Arbitration Commission. He is also a Council Member of the China Maritime Law Association and a visiting professor to Dalian Maritime University.

** Qin Mu graduated with an LLB in Maritime Law from the School of Law, Shanghai Maritime University in 2007. He then obtained an LLM in Commercial Law from the School of Laws, University College London in 2008. He is now a legal assistant in Richards Butler in association with Reed Smith LLP.

1. See a detailed discussion of the definition of 'Pollution Damage' at W. Oosterveen, 'Some Recent Developments Regarding Liability for Damage Resulting from Oil Pollution – From the Perspective of an EU Member State', *Environmental Law Review* 6, no. 4 (2004): 229 et esq.

Michael G. Faure, Han Lixin & Shan Hongjun, *Maritime Pollution Liability and Policy: China, Europe and the US*, 247–260.

various economic activities related to the polluted coasts. With larger tankers having been built and larger scales of oil trade developed, the potential threat of substantial marine pollution keeps growing.

As a result of the increase in risk, the recent decade has witnessed serious accidents causing ship-source marine pollution, for example, the *Erika*, the *Prestige*, the *Hebei Spirit* and the recent explosion at Dalian Port, China. In the case of the *Erika*, cleanup operations took place along some 400 kilometres of polluted coastline and over 250,000 tonnes of oily waste was collected from the shoreline.[2] The *Prestige* case was initiated in Spain, but the persistence of the spilled cargo and the wind and current eventually caused heavy losses to both Spain and France.[3] The *Hebei Sprit* case involves a Hong Kong–flagged vessel,[4] and the estimated total loss to South Korea amounted to United States Dollar (USD) 411 to 437 million, save that significant costs incurred during the cleanup and compensation.[5] The recent explosion of oil pipes at the port of Dalian, China, has caused a massive pollution to the waters of Dalian. The explosion happened during MV 'Cosmic Jewel' was discharging oil at the largest crude oil terminal in China. As a result of the blast about 1,500 tonnes of crude spilled into the Yellow Sea.[6] Consequently the port had been closed for a week and substantial loss was caused.[7] Substantial legal implications are expected to take place following this serious accident.

In order to regulate the liability of responsible parties involved in marine pollution accidents on an international and collective level, a series of major international conventions regarding ship-source marine pollution have been promulgated under the auspices of the International Maritime Organization (IMO) in the aftermath of the *Torrey Canyon* incident in 1967,[8] among which many have been amended to catch up with the development of the shipping industry. Generally, these conventions include the 1969 International Convention on Civil Liability for Oil Pollution Damage as replaced by its 1992 Protocol and amended in 2000 (CLC 1992), 1992 International Convention on the Establishment of an International Fund for Compensation for Oil Pollution Damage as amended in 2000 (IOPC Fund 1992) and the 1973 International Convention for the Prevention of Pollution from Ships and its amending Protocol of 1978 (MARPOL73/78).

2. IOPC Fund Report, 'Erika, France, 12 December 1999' <www.iopcfund.org/erika.htm>, 9 Jan. 2009.
3. IOPC Fund Report, 'Prestige, Spain, 13 November 2002' <www.iopcfund.org/prestige.htm>, 9 Jan. 2009.
4. K. Wallis, 'Hong Kong Blames Korea Tugs for Hebei Spirit Spill', *Lloyd's List News*, 24 Mar. 2009.
5. Press Release of IOPC Fund, 'Hebei Spirit Incident: Recourse Action against Samsung Heavy Industries (SHI)' <www.iopcfund.org/pr-pdf/Hebei%20Spirit%20recourse%20action_e.pdf>, 6 Feb. 2009.
6. Gary Dixion, 'VLCCs Divert from Dalian', *Tradewinds*, 19 Jul. 2010.
7. Aaron Kelley, 'Dalian Reopens Terminals', *Tradewinds*, 20 Jul. 2010.
8. Oosterveen, *supra* n. 1, 223.

1.2 Relevant Legislations in Hong Kong

As early as in 1976, a report for the Intergovernmental Oceanographic Commission of United Nations Educational, Scientific and Cultural Organization (UNESCO)[9] already pointed out some major waste spills, especially oil spills, in Hong Kong waters from 1968 to 1974. The report also suggested that the harbour of Hong Kong was reaching its capacity for effective dispersal of marine pollutants.

Today, the threat of potential marine pollution is especially critical in East Asia, particularly areas around China. According to the statistics by the International Energy Agency, China's imports of crude oil was expected to reach a stunning 5.1 million barrels per day in order to meet its anticipated oil demand of 9.6 million barrels per day in 2014.[10] As a major shipping centre of East Asia and China, Hong Kong is expected to experience a greater amount of tankers and container vessels arriving, departing, passing by and registering. As a result, the risk for marine pollution in Hong Kong is particularly higher than most of other areas in Asia. Thus, although efforts have been made to prevent ship-source marine pollution through strict management and training, it is crucial for the authorities of Hong Kong to establish an effective legal system to deal with the civil and criminal liabilities of the parties involved in future marine pollution.

There are two major pieces of ship-source marine pollution legislation in Hong Kong, namely, Chapter 413 of Merchant Shipping (Prevention and Control of Pollution) Ordinance, in which civil liabilities regarding causing oil, garbage, sewage and other hazardous emissions to the sea are regulated (Cap. 413), and Chapter 414 of Merchant Shipping (Liability and Compensation for Oil Pollution) Ordinance, in which criminal liabilities regarding pollution are regulated (Cap. 414). By these two Ordinances, Hong Kong has adopted the three major international conventions concerning civil and criminal liabilities in ship-source marine pollution described earlier.

This article will first introduce the civil liability and then will focus on the criminal liability regarding ship-source marine pollution under the legislation of Hong Kong.

2 MARINE POLLUTION LEGISLATIONS OF HONG KONG

2.1 Shipowners' Strict Liability

In the case of the *Sea Empress*,[11] David Steel J held that even though the pollution was caused by a normal fact of life rather than an extraordinary incident, the

9. G.B. Thompson, 'Report on Marine Pollution Problems in Hong Kong', paper given at International Workshop on Marine Pollution in East Asian Waters (Paris, 1 Feb. 1976).
10. International Energy Agency, 'Medium-Term Oil Market Report' (June 2009), 67, 110.
11. *Environment Agency v. Milford Haven Port Authority and Andrews*, [1999] 1 Lloyd's Rep. 673.

authority's plea of guilty was accepted on the basis of strict liability and without admission of fault. Subsequently, in the *Erika* case,[12] the European Court of Justice imposed a strict liability on the shipowner, although the fuel was being transported by a third party when the incident took place. The Court suggested that the shipowner was deemed as possessing the polluting fuel immediately before it become waste and might thus be regarded as having produced the waste. Therefore, even though the pollution has been caused by the act of a third party, it is likely that the shipowner will be held strictly liable. Following this principle, Hong Kong law imposed a strict liability on the shipowner, which complied with CLC 1992. However, Cap. 414 also has some minor adjustments compared with the provisions in CLC 1992.

2.1.1 Provisions Regarding Strict Liability under the CLC 1992

The CLC 1992 lays down the principle of strict liability for ship owners.[13] Article III, paragraph 1 of the CLC 1992 provides:

> Except as provided in paragraphs 2 and 3 of this Article, the owner of a ship at the time of an incident, or, where the incident consists of a series of occurrences, at the time of the first such occurrence, shall be liable for any pollution damage caused by the ship as a result of the incident.

However, it should be noted that CLC 1992 imposes liability only on the 'registered owner' of the vessel that caused the pollution and excludes liability, whether under the CLC 1992 or otherwise, of various parties, including managers, operators, charterers, pilots, salvers and the owner's servants or agents. This is the so-called 'channelling' of liability to the registered owner and is a feature of CLC 1992, which targets simplified and efficient use of insurance capacity.[14]

However, under CLC 1992, the liability of the shipowner may be avoided if he can prove that the damage resulted from the three exonerating circumstances, namely, an act of war or a natural phenomenon of an exceptional character, or action or inaction of a third party in the intent to cause damage, or negligence of governmental authorities in maintaining lights or other navigational aids.[15]

12. *Commune de Mesquer v. Total France SA and Total International Ltd* (C-188/07), [2008] 3 CMLR 16.
13. E. Keeble & L. Coates, 'Does the Polluter Pay on the High Seas?', *Environmental Law Review* 5, no. 3 (2003): 193.
14. C. de la Rue & P. Murray of Ince & Co, 'Oil Pollution from Ships – Current Legal Issues', paper given at Shanghai International Maritime Forum 2005 (Shanghai, 5–7 Jul. 2005). The authors also mentioned that the liability regime in the new International Convention on Civil Liability for Bunker Oil Pollution Damage is a single-tier regime and measures have been taken to preserve rights of recovery from other parties in addition to the registered owner where each party may be held jointly or severally liable.
15. See CLC 1992, Art. III, para. 2.

2.1.2 Hong Kong Legislation

Hong Kong has enacted the same scheme described earlier regarding the three exonerating circumstances[16] and similar regime provided by the CLC 1992 in Cap. 414. However, the strict liability of shipowners has been extended to causing of a relevant threat of potential contamination, as section 6 of Cap. 414 provides:

> (1) Where, as a result of any incident any oil is discharged or escapes from a ship, the owner of the ship shall be liable, except as otherwise provided by this Ordinance, for any pollution damage caused in Hong Kong.
> (1A) Where as a result of any incident a relevant threat of contamination occurs, the owner of the ship that caused such threat shall be liable, except as otherwise provided by this Ordinance, for-
> (a) the cost of any reasonable measures taken for the purpose of preventing such contamination or minimizing any damage in Hong Kong arising from a discharge or escape of oil from the ship; and
> (b) any damage caused outside the ship in Hong Kong by any measures so taken.

The aforementioned regime of 'channelling' of liability to the registered owner in CLC 1992 has been preserved where the operator will be liable in the absence of registered owner by section 2 of Cap. 414, which states:

> 'owner' (. . .), in relation to a ship, means the person or persons registered as the owner of the ship or, in the absence of registration, the person or persons owning the ship, except that in relation to a ship owned by a State which is operated by a person registered as the ship's operator, it means the person registered as its operator.

Therefore, shipowners are strictly liable for pollution spilled from their vessels as well as creating a potential threat of contamination in the waters of Hong Kong. The claimants, on the other hand, will face fewer obstacles in identifying the party who is responsible for the damage.

2.2 Limitation of Liability and Cover by the IOPC Fund

Although shipowners may be confronted with the risk for strict liability when causing marine pollution or threat of such pollution in Hong Kong, the local legislation has provided two layers of protection for shipowners, namely the limitation of liability and payment to cover the compensation from the IOPC Fund 1992. To comply with the greater risk for marine pollution spilled from ships, the amount of the limitation has been increased compared to the original sum set out in the earliest version of CLC 1992, and the cap of the coverage of the IOPC Fund 1992 has also been increased to cope with larger accidents.

16. See Cap. 414, s. 7.

2.2.1 Provisions Regarding Limitation of Liability in the CLC 1992

Most marine risks are subject to the maritime law principle of limitation of liability that allows a shipowner to assess the total risk exposure in relation to the tonnage[17] or size of the vessel and not to the amount of damage caused by the incident.[18] Such limitation could be disallowed only if the accident in question would be found by a court to have occurred as a result of certain facts regarding the accident. Similar principles apply also to ship-source marine pollution, which is codified by Article V of the CLC 1992.

The amounts recoverable under the scheme set out in the CLC 1992 are calculated as Special Drawing Rights (SDRs).[19] According to CLC 1992, a shipowner will be entitled to limit his liability regarding any single incident to an aggregate amount that depends on the tonnage of the vessel involved. To be protected by such limitation of liability, a shipowner shall constitute a fund for the total sum of the limit of his liability with the Court or other competent parties subject to the articles under the CLC 1992.[20] The calculation of the amount is provided as follows:

(a) 4,500,000 units of account for a ship not exceeding 5,000 units of tonnage;
(b) for a ship with a tonnage in excess thereof, for each additional unit of tonnage, 631 units of account in addition to the amount mentioned in subparagraph (a);

provided, however, that this aggregate amount shall not in any event exceed 89,770,000 million units of account.[21]

Nevertheless, CLC 1992 also provided circumstances in which the shipowner is not entitled to avail himself of the benefit of the limitation set out under Article V. Once it could be proved that the pollution is caused by the shipowner's personal act or omission, provided such act or omission is committed intentionally or recklessly but with knowledge of the consequence of such act or omission,[22] he will lose the right to benefit from the above limitation.[23]

---

17. E. Gold, 'Liability and Compensation for Ship-Source Marine Pollution: The International System', paper given at the 3rd International Conference on Maritime Law (Piraeus, May 1998).
18. Department of Infrastructure, Transport, Regional Development and Local Government of Australia, 'Principles of Liability Limitation', <www.infrastructure.gov.au/maritime/liability/limitation.aspx>, 26 May 2008.
19. See CLC 1992, Art. V, para. 9(a).
20. See *ibid.*, Art. V, para. 3.
21. This is the latest amount updated by the annex of the Amendment to the Limitation Amounts in the Protocol of 1992 to Amend the International Convention on Civil Liability for Oil Pollution Damage 1992.
22. See CLC 1992, Art. V, para. 2.
23. Similar to the principle of limitation of liability in the shipping industry, for example, in the case of *Richard Irvin & Sons Ltd v. Aberdeen Near Water Trawlers Ltd* [1983] SLT 26, Lord Grieve held that the owners of the vessel that collided with the other vessel and caused the other vessel to sink cannot limit their liability because the collision was due to the inability to follow the

2.2.2 Local Legislation Regarding Limitation of Liability in Hong Kong

Relevant sections of Cap. 414 of Hong Kong is also similar to the above Articles of the CLC 1992, with a slight increase in the amount of the limitation for ships not exceeding 5,000 tonnes in order to fit the situation in Hong Kong. The calculation of the limitation amount could be found in the following paragraphs under Cap. 414:

Section 9 Limitation of liability under section 6

...

(2) ... 'the relevant amount' (有關數額) means-

(a) for a ship not exceeding 5,000 tons, 4,510,000 special drawing rights;

(b) for a ship exceeding 5,000 tons, 4,510,000 special drawing rights together with an additional 631 special drawing rights for each ton of its tonnage in excess of 5,000 tons, but not exceeding an aggregate amount of 89,770,000 special drawing rights.

2.2.3 Limitation Actions in Hong Kong

Once liability is incurred or allegedly incurred by the shipowner, the shipowner may apply to the court in Hong Kong for the limitation of liability to an amount based on section 9 of Cap. 414.[24] On such an application, the court shall decide whether the applicant has incurred the liability and whether he is entitled to limit it. If the liability falls into the category of section 9(3) of Cap. 414, which is similar to CLC 1992 Article V, the shipowner may not be able to limit his liability.

If the court finds that the applicant is entitled to limit his liability according to section 9 of Cap. 414, it shall then determine the limit of such liability and the applicant will be directed to pay into the court the amount of such limit.[25] Such payment should be made in Hong Kong Dollars (HKDs) and the conversion from SDRs to HKDs may be certified by the Monetary Authority.[26]

Subsequently, the court shall determine the amounts due regarding the liability to the claimants apart from the limit and distribute the payment to the court among those claimants on a prorated basis.[27] The court shall also order release of the assets arrested in respect to the liability after receiving the amount to be distributed.[28]

It should be noted that under section 10 of Cap. 414, if the shipowner has voluntarily made any reasonable sacrifice or taken any other reasonable measures

bridge order of a drunk officer and lack of a necessary on-board system. The European Court of Justice (Grand Chamber) held that the competent courts must decide whether that exclusion from liability is applicable to the liable party in the *Erica* case.

24. See Cap. 414, s. 10(1).
25. See *ibid.*, s. 10(2).
26. See *ibid.*, s. 10(3).
27. See *ibid.*, s. 10(2)(a)–(b).
28. See *ibid.*, s. 11(a).

to prevent or reduce the pollution damage to which the liability extends or might have extended, he shall be entitled to the distribution as if he had established a claim regarding the liability for an amount equal to the cost of such sacrifice or other measures.[29]

In addition, the court may postpone the distribution of a certain part of the amount to be distributed if such part is related to any claims that later may be established before a court outside Hong Kong.[30]

2.2.4 IOPC Fund 1992 Coverage of the Liability of Shipowners

In cases in which coverage by CLC 1992 is insufficient, another coverage provided by IOPC Fund 1992 will be in operation.[31] Although the burden of payment of compensation under CLC 1992 falls on the ship owners, the IOPC Fund provides for contributions by persons who have received oil in excess of a certain quantity in contracting states.[32] The amounts to be levied are decided each year by the Assembly of IOPC Fund.[33]

The calculation and the cap of coverage set out in IOPC Fund 1992 is as follows:

> (a) Except as otherwise provided in subparagraphs (b) and (c) of this paragraph, the aggregate amount of compensation payable by the Fund under this Article shall in respect of any one incident be limited, so that the total sum of that amount and the amount of compensation actually paid under the 1992 Liability Convention for pollution damage within the scope of application of this Convention as defined in Article 3 shall to exceed 203,000,000 units of account.
>
> (b) Except as otherwise provided in subparagraph (c), the aggregate amount of compensation payable by the Fund under this Article for pollution damage resulting from a natural phenomenon of an exceptional inevitable and irresistible character shall not exceed 203,000,000 units of account.
>
> (c) The maximum amount of compensation referred to in subparagraphs (a) and (b) shall be 300,740,000 units of account with respect to any incident occurring during any period when there are three Parties to this Convention in respect of which the combined relevant quantity of contributing oil received by persons in the territories of such Parties, during the preceding calendar year, equalled or exceeded 600 million tons.
>
> (d) Interest accrued on a fund constituted in accordance with Article V, paragraph 3, of the 1992 Liability Convention, if any, shall not be taken into

29. See *ibid.*, s. 10(6).
30. See *ibid.*, s. 10(7).
31. See IOPC Fund 1992, Art. 4. The 'secondary liability' of IOPC Fund was confirmed per Lord Gill in *Landcatch Ltd v. IOPC Fund*, [1998] 2 Lloyd's Rep. 552.
32. N. Singh, *International Maritime Law Conventions*, vol. 3 (London: Stevens & Sons, 1983), 2495.
33. *Brochure of the IOPC Fund 1992* (London, IOPC Fund 1992, November 2002), 3.

account for the computation of the maximum compensation payable by the Fund under this Article.[34]

2.2.5 Local Legislation Regarding IOPC Fund 1992 in Hong Kong

Part III of Cap. 414 generally adopted the terms of IOPC Fund 1992, with the entry cap of the distributors at a sum of 150,000 tonnes of annul imported or received oil.[35] The calculation of the coverage provided by IOPC Fund is incorporated in section 25(9) of Cap. 414.

Special proceedings regarding the claims involving the compensation by IOPC Fund are provided in section 27 of Cap. 414 and section 2A of the Hong Kong Admiralty Proceedings (Order 75), in which special provisions regarding the intervention of the IOPC Fund, the service of writ and notices are set out.

2.3 SUMMARY

It is clear that the legislation in Hong Kong adopts and follows CLC 1992 and IOPC Fund 1992 with some minor tailoring on the sum of the amount and its unique procedures to fit into the local practice.[36]

3 CRIMINAL LIABILITY

Apart from the civil liability, the liable party of the pollution may find itself criminally liable in Hong Kong. By section 3 of Cap. 413, Hong Kong adopted

---

34. See IOPC Fund 1992, Art.4, para. 4.
35. See Cap. 414, s. 22(5). According to a report dated 31 Dec. 2007 attached to the 2007 Annual Report of the IOPC Fund, the contributing oil in Hong Kong is 6,471,650 tonnes, which is 0.44% of the total contribution from all Member States.
36. The United Kingdom is also a party to both CLC 1992 and IOPC Fund 1992. The UK's international obligations under the above two conventions were given effect by the Merchant Shipping (Oil Pollution) Act 1971 and the Merchant Shipping Act 1974, which were consolidated into the Merchant Shipping Act 1995 (MSA 1995). The MSA 1995 was amended in 2006 with an implementation of the Supplementary Fund Protocol and any new instruments that modify or replace the existing oil pollution compensation regime.

    Section 157 of MSA 1995 caps owner liability to an amount based on the gross tonnage of the tanker involved in the incident. The maximum liability of the owner is approximately GBP 78 million. In order to meet potential financial obligations, owners of oil tankers are obliged to maintain insurance or other financial security. Once the owner's cap has been exceeded, the IOPC Fund 1992 becomes responsible for further claims up to a maximum of around GBP 176 million if the pollution damage is suffered in the territory of a state that is a party to the 1992 Fund convention.

    In practice, the Court of Appeal in the case of the *Sea Empress* suggested that the MSA 1995 could cover claims for economic loss (e.g., from fishermen) as long as they are sufficiently directly connected to the oil pollution. Thus, the geographical proximity to the incident will remain a key factor concerned by the English court when dealing with marine pollution cases in the future.

MARPOL73/78 with various subsequent amendments and the Protocols in relation to MARPOL73/78.

3.1 The Regime under MARPOL73/78

Under the MARPOL73/78 regime, deliberate discharges where contravention of the restrictions under which discharge is permitted, and accidental discharges where the pollution is caused by unexpected circumstances, for example, collision or shipwreck, are distinguished as criminal liability only arises if the shipowner or master acted with intent, or recklessly with knowledge that damage would probably result.[37]

According to Article 3 of MARPOL73/78, ships of any type operating in the marine environment, including hydrofoil boats, air cushion vehicles, submersibles, floating craft and fixed or floating platforms entitled to fly the flag of a Party to MARPOL73/78 or operating under the authority of such Party, fall within the purview of the provisions of the MARPOL73/78.[38]

The detailed requirements to prevent relevant marine pollution are contained in the five annexes of MARPOL73/78, which set out the criteria for oil, noxious liquid substances carried in bulk, harmful substances carried in packages, tanks, wagons or containers, sewage and garbage from ships.[39]

Article 4 of MARPOL73/78 provides the proceedings when violation of the requirements is detected, which are as follows:

> 1. Any violation of the requirements of the present Convention shall be prohibited and sanctions shall be established therefore under the law of the Administration of the ship concerned wherever the violation occurs. If the Administration is informed of such a violation and is satisfied that sufficient evidence is available to enable proceedings to be brought in respect of the alleged violation, it shall cause such proceedings to be taken as soon as possible, in accordance with its law.
> 2. Any violation of the requirements of the present Convention within the jurisdiction of any Party to the Convention shall be prohibited and sanctions shall be established therefore under the law of that Party. Whenever and violation occurs, the Party shall either:
> a) cause proceedings to be taken in accordance with its law; or
> b) furnish to the Administration of the ship such information and evidence as may be in its possession that a violation has occurred.
> 3. Where information or evidence with respect to any violation of the present Convention by a ship is furnished to the Administration of that ship, the Administration shall promptly inform the Party which has furnished the information or evidence and the Organization, of the action taken.

37. E. Denza, 'Case Comment: A Note on Intertanko', *European Law Review* 33, no. 6 (2008): 871.
38. Singh, *supra* n. 32, 2272.
39. See *ibid.*, 2273–2274 for a more detailed description of the five annexes.

> 4. The penalties specified under the law of a Party pursuant to the present Article shall be adequate in severity to discharge violations of the present Convention and shall be equally severe irrespective of where the violations occur.

Apparently, the regime of MARPOL73/78 is to let the authorities of subscribed countries make their own criminal sanctions when marine pollution takes place in their waters under the general requirements of MARPOL73/78. Thus, it is critical to look at the law in Hong Kong, particularly Cap. 413 and outline the criminal liabilities a shipowner may face for violation of the requirements set out in the annexes of MARPOL73/78.

3.2 Local Legislation Regarding Criminal Liability in Hong Kong

The general regulations are provided under section 3 of Cap. 413, which are as follows:

> (g) that contraventions of the regulations[40] shall be offences punishable by penalties not exceeding-
> (i) on conviction upon indictment,[41] a fine of $5,000,000 and, in the case of an individual, imprisonment for 2 years;
> (ii) on summary conviction,[42] a fine of $500,000;
> (h) that in the case of any such contravention an offence is committed by-
> (i) the master and the owner of the ship concerned; and
> (ii) where the contravention is due to the act or omission of another person, that person;
> (i) for denying entry of any ship to the waters of Hong Kong in respect of which any such contravention is believed to have occurred;
> (j) for detaining any ship in respect of which any such contravention is believed to have occurred and for notifying the relevant consular officer (if any) of the detention and of any proceedings against the ship;'[43]

In Hong Kong, even when the pollution is directly caused by the act of an individual, his supervisory party, for example, his employer, may be charged for a criminal offence with strict liability in order to effectively promote the objectives of the relevant statute.[44] Thus, once the ship-source marine pollution has occurred,

40. Detailed regulations regarding various types of pollutions, preventions and report of accidents could be found in Cap. 413A–413M.
41. See Cap. 227 Magistrates Ordinance, s. 2.
42. See Cap. 228 Summary Offences Ordinance.
43. The sanctions here are the maximum extent of which the regulations can include subject to the specific offences in the regulations.
44. According to Lord Scarman in *Gammon v. A-G of Hong Kong* [1984] 3 WLR 437 PC, once the statute is concerned with an issue of public safety or social concern, the presumption of *mens rea* can be displaced and the supervising party of the person who directly committed the offence will

the shipowner,[45] the company in relation to the ship,[46] the operator, the demise charterer, the master of the ship[47] or any other person who is deemed to have contravened the regulations[48] could each be criminally liable. If the crime is one where the magistrate is authorized or empowered or required to commit the accused person to prison for trial before the court (i.e., an 'indictable offence'), an individual may also be held criminally liable for the pollution and hence be imprisoned for a certain period.[49] However, regarding offences of particular regulations,[50] it shall be a defence for a charged person to prove that he took all reasonable steps to comply with the regulations.

Further, the vessel involved in the pollution may be detained by the authorities and it shall not proceed to sea before released by the authority. If a party takes a ship that was detained because of violation of relevant regulations of Cap. 413 and MARPOL73/78 to sea, it will also be criminally liable under the law of Hong Kong. Section 4 of Cap. 413 provides:

> (1) Where a ship is authorized or ordered to be detained under regulations made under this Part, and after such detention or after service on the master of any notice of or order for such detention, the ship proceeds or attempts to proceed to sea before having been released by a competent authority, the master of the ship commits an offence and is liable to a fine of $500,000 and to imprisonment for 2 years; and if the owner or agent or any person who sends the ship to sea is party or privy to the offence he also commits an offence and is liable to the same punishment.

---

have a strict liability. In *Alphacell v. Woodward* [1972] 1 QB 127, the appellants had been convicted under s. 2(1) of the Rivers (Prevention of Pollution) Act 1951 because the House of Lords decided the offence was one of strict liability because it concerns issues of public safety. This principle was followed by the court in Hong Kong in the case of *HKSAR v. Paul Y – ITC Construction Ltd* [1998] 2 HKLRD 35 CA, in which two employees of the defendant company created noise pollution while operating the machines but the company was found strictly liable by the court for 'causing' the criminal offence as if it had done the work itself. More recently in the case of *Secretary for Justice v. Flame Construction & Others* [2001] HKEC 1544, the court found that the company was criminally liable although the pollution was initially caused by its workman.

45. For example, Cap. 413A Merchant Shipping (Prevention of Oil Pollution) Regulations, reg. 37.
46. 'Company' in relation to a ship means (a) the owner of the ship; or (b) any other person who has assumed responsibility for the operation of the ship, and on assuming such responsibility has agreed to take over all of the duties and responsibilities imposed in respect to the ship by the International Convention for the Prevention of Pollution from Ships, 1973, as modified by the Protocol of 1978. See Cap. 413K Merchant Shipping (Prevention of Pollution by Sewage) Regulation, s. 2.
47. For example, Cap. 413J Merchant Shipping (Prevention of Garbage) Regulation, s. 7.
48. For example, Cap. 413C Merchant Shipping (Reporting of Incidents) Regulations, reg. 8.
49. See Cap. 413H Merchant Shipping (Safety) (Dangerous Goods and Marine Pollutants) Regulation, Part V, s. 21(a).
50. See Cap. 413A, reg. 37; Cap. 413B, reg. 29; Cap. 413C, reg. 8; Cap. 413D, reg. 10; Cap. 413E, reg. 10 and Cap. 413K, s. 30.

(2) Where a ship proceeds to sea in contravention of subsection (1) whilst having on board a public officer who is acting in the execution of his duty, the master and owner of the ship-

(a) in addition to any punishment to which they may be liable under subsection (1), both commit an offence against this subsection and are each liable to imprisonment for 6 months and to a fine of $20,000 and an additional fine of $1,000 for each day during the period from the date on which the ship proceeded to sea until the date on which the officer returns to Hong Kong or, if he does not return directly to Hong Kong, would have returned if he had travelled by the quickest practicable route; and

(b) are jointly and severally liable to pay to the Government all expenses incidental to the taking of the officer to sea and to securing his return to Hong Kong.

In addition, once an accident has occurred and hence created the potential risk for oil or other kinds of pollution in Hong Kong waters, the Director of the Marine Department of Hong Kong may give directions to the shipowner, master, salver or any other person in possession or control of the vessel in order to avoid or reduce pollution or the risk for pollution.[51] Failure to comply with such directions is also a criminal offence,[52] in which the accused may be fined HKD 5,000,000 upon conviction upon indictment and HKD 500,000 upon summary conviction.[53] Nevertheless, in the proceedings for such offence, it shall be a defence for the accused party if it can show that it used due diligence to secure compliance with the above directions, or it can reasonably believe that such compliance would have involved a serious risk to human life.[54]

4 CONCLUSION

It is a fundamental principle that the sea and its environment represent a universal resource, whose preservation calls for international responsibility. Thus, the solution to the ship-source marine pollution shall be a collective one.[55] As a shipping centre with a high level of risk for ship-source marine pollution, Hong Kong has subscribed to the major international conventions in respect to the civil and criminal liabilities of the parties involved in pollution accidents.

Under the legislation of Hong Kong, a claimant does not need to commence complicated litigation to identify the liable party of the pollution, nor to prove fault or negligence of that party, because the registered owner of the ship will be under strict liability. On the other hand, a shipowner could seek coverage of his liability from the limitation provided by the CLC 1992 and further support from the IOPC

51. See Cap. 413, s. 6.
52. See *ibid.*, s. 8(1).
53. See *ibid.*, s. 8(4).
54. See *ibid.*, s. 8(3).
55. V. Terrier, 'Are "Black Tides" Inevitable?', *Constitutional Law Journal* 6, no. 2 (2001): 37.

Fund where the liability of the pollution has been distributed among the oil importers and receivers.

A shipowner, an operator, a salver or a crew member could be subject to criminal liability if he does not comply with regulations in respect to pollution, prevention and report of accidents. However, it may be a defence for a person charged under some regulations to show that he took all reasonable steps to ensure compliance with the regulations.

References

de la Rue, C. & P. Murray of Ince & Co. 'Oil Pollution from Ships: Current Legal Issues'. Paper presented at the Shanghai International Maritime Forum 2005, Shanghai, 5–7 July 2005.

Denza, E. 'Case Comment: A Note on Intertanko'. *European Law Review* 33, no. 6 (2008): 871.

Gold, E. 'Liability and Compensation for Ship-Source Marine Pollution: The International System'. Paper presented at the 3rd International Conference on Maritime Law. Piraeus, 28–30 May 1998.

Dixon, Gary. 'VLCCs Divert from Dalian'. *Tradewinds* 19 July 2010.

International Energy Agency. 'Medium-Term Oil Market Report' (June 2009), 67, 110.

Keeble, E. & L. Coates. 'Does the Polluter Pay on the High Seas?'. *Environmental Law Review* 5, no. 3 (2003): 193–197.

Kelley, Aaron. 'Dalian Reopens Terminals'. *Tradewinds* 20 July 2010.

Oosterveen, W. 'Some Recent Developments Regarding Liability for Damage Resulting from Oil Pollution – From the Perspective of an EU Member State'. *Environmental Law Review* 6, no. 4 (2004): 229 et seq.

Singh, N. *International Maritime Law Conventions*, vol. 3. London: Stevens & Sons, 1983.

Terrier, V. 'Are "Black Tides" Inevitable?'. *Constitutional Law Journal* 6, no. 2 (2001): 37.

Thompson, G.B. 'Report on Marine Pollution Problems in Hong Kong'. Paper presented at the International Workshop on Marine Pollution in East Asian Waters, Paris, 1 February 1976.

Wallis, K. 'Hong Kong Blames Korea Tugs for Hebei Spirit Spill'. *Lloyd's List News*, 24 March 2009.

Part IV
United States

Chapter 15

A Comparison of the Recovery of Compensation for Injury to Natural Resources under the 92 CLC and Fund Conventions with the US Oil Pollution Act of 1990

*Robert Force**

1 INTRODUCTION

As the title states, this chapter will compare the approach of compensation for injury to natural resources caused by the discharge of oil under several international Conventions to which the United States is not a party with compensation under US law. The Conventions are the International Convention on Civil Liability for Oil Pollution Damage (CLC) 1992,[1] the International Convention on Civil Liability for Oil Pollution Damage 1969,[2] the International Convention on the Establishment of an International Fund for Compensation for Oil Pollution

* Niels F. Johnsen Professor of Maritime Law and Director Emeritus, Tulane Maritime Law Center. Visiting Professor of Law, Dalian Maritime University.

1. Protocol of 1992 to amend the International Convention on Civil Liability for Oil Pollution Damage, 27 Nov. 1992, 1956 U.N.T.S. 255 (the CLC).
2. International Convention on Civil Liability for Oil Pollution Damage, 29 Nov. 1969, 973 U.N.T.S. 3.

Michael G. Faure, Han Lixin & Shan Hongjun, *Maritime Pollution Liability and Policy: China, Europe and the US*, 263–283.

Damage 1992[3] and the International Convention on the Establishment of an International Fund for Compensation for Oil Pollution Damage 1971.[4] For simplicity, the chapter will focus on the current 1992 Conventions – the CLC and Fund Convention (collectively the Conventions). The Conventions will be compared with the compensation regimes in the Comprehensive Environmental Response Compensation and Liability Act (CERCLA),[5] the Oil Pollution Act of 1990 (OPA),[6] and regulations promulgated by the Department of the Interior (DOI)[7] and the Nation Oceanic and Atmospheric Administration (NOAA).[8]

2 OVERVIEW OF MAJOR DIFFERENCES RELATING TO COMPENSATION

There are some dramatic differences in the way the Conventions approach the subject of compensation for injury to, or destruction of, natural resources as a result of marine oil pollution compared to the way compensation is provided in the United States. The first of these differences, to some extent, may stem from the role of tort law and the elements of compensable damages recoverable by tort victims in the United States. In brief, as will be demonstrated, under the Conventions' approach,[9] compensation is payable only when the damage can be quantified in monetary terms. In other words, how much will it cost to restore the damaged resource to baseline? Other factors, such as loss of use, loss or enjoyment or loss of existence, are not compensable, unless the 'owner' (or the person having control or management) of the resource, usually a governmental entity, has suffered an economic loss, such as might occur where user fees are not collected because the resource is not available for use because of its polluted state. An example of this might occur where a government is unable to charge fishermen a fee to fish in a river during the time that oil is being removed. The policy behind the CLC and Fund Convention approach is that you cannot give a monetary award for something that does not have a monetary value.

In my opinion, assessing damages for injury to, or destruction of, natural resources often is the most difficult problem in the area of environmental liability. The reason for this is easy to understand. It is simply that the assessment of damages to compensate the public fully may require that a monetary value be

3. Protocol of 1992 to amend the International Convention on the Establishment of an International Fund for Compensation for Oil Pollution Damage, 17 Nov. 1992, 1953 U.N.T.S. 373 (the Fund Convention).
4. International Convention on the Establishment of an International Fund for Compensation for Oil Pollution Damage, 18 Dec. 1971, 1110 U.N.T.S. 57.
5. 42 USC §§ 9601 et seq.
6. 33 USC §§ 2701 et seq.
7. 43 CFR Part 11.
8. 15 CFR §§ 990.10–990.66.
9. See the Section on the Conventions, *infra*.

placed on an injury that, in some respects, has no monetary value. There is an imperfect analogy in certain forms of tort litigation, most prominently personal injury litigation. For example, when a person has been injured in an automobile accident, most of the items of recovery are pecuniary losses, such as repairs to a car, replacement of damaged clothing, reimbursement of medical expenses, compensation for loss of wages and so forth. This, however, is not true of all personal injury damages. In the United States, under the prevailing state and federal tort[10] regimes, money damages may be awarded for injuries to non-economic interests under appropriate circumstances.[11] We call these damages 'non-pecuniary' damages. In the United States, one who is injured by the negligence of another is entitled to recover for 'pain and suffering'. The United States simply has a different approach to making tort victims 'whole' than some other countries.

In the end, it depends on how we conceive of and hence define the term 'injury'. Some aspects of damage to natural resources can be quantified using market techniques. If a strand of mature trees has been destroyed and if that area is used only for commercial logging, it is possible to put a value on those lost trees by looking at the market. This type of assessment has commercial integrity. This is the case regardless of whether the trees are owned by the government, which has no plans to cut the trees down and sell them. But paying the market price does not replace the trees. It may take ten or fifteen years for replacement saplings to mature.

Another problem arises from the fact that the same type of trees may have other value. The trees may serve as a nesting area for an endangered species of bird. The trees may be 200 years old. The trees may be part of a beautiful landscape utilized by artists and frequented by nature lovers. Should those values be included in the calculation of natural resource damages? If so, how does one put a monetary value on these non-commercial values? How does one put a value on something society or some part of society values but not in a monetary sense?

There can be at least two responses to these questions. The easiest response is that you do not calculate in monetary terms values that are non-monetary. If something does not have monetary value, then we should not try to put a monetary value on it because it is too arbitrary. This is the approach taken by the CLC and Fund Conventions. The other approach is that we should do the best that we can to put a monetary value on these non-commercial values. That is the approach used in the United States.

In an actual case, how does one put an economic value on a non-economic injury to natural resources held in trust for the public? When compensating for values that transcend market values or that cannot be fully measured by true market

10. Marine pollution is a species of tort under US law at least from the perspective of awarding damages to persons whose interests have been adversely affected by pollution, such as damage to property, etc.
11. A. Kanner & T. Nagy, 'Measuring Loss of Use Damages in Natural Resource Damage Actions', *Columbia Journal of Environmental Law* 30 (2005): 417.

values, on what basis are those values calculated? Admission to a natural resource such as a national park or aquatic park may be free or may charge a very small fee to encourage and enable people to appreciate the beauty of the site or to learn about nature. The market price in the case of free admission is zero or a very low price; hence the admission fees are grossly insufficient to sustain and maintain the site. These are not fair indictors of the use or service value of the resource.

Some people who purport to be knowledgeable assert that the answer is provided by using a technique known as 'contingent valuation' (CV), which is designed to capture non-economic values and translate them into a dollar figure. Although controversial,[12] it has been strongly argued that CV methodology is an appropriate technique for determining non-use values, such as option and existence values,[13] but there are scientists who take a different view.[14] I am not aware of a reported decision that has based damages on this technique. On the other hand, some have advocated a position that would forgo reliance on CV evidence because it is controversial and subject to claims that it is grossly unreliable. Instead, these CV critics would simply permit the fact finder, jury or judge, to assess compensation for injuries to natural resources the same as is done in all tort cases that allow plaintiffs to be compensated for non-economic injuries.[15]

As will be discussed later, the NOAA regulations applicable to compensation for injury to natural resources caused by oil spills and the amended DOI regulations applicable to releases of hazardous substances, both of which are discussed later in this paper, have adopted an approach called 'compensatory restoration' that, for practical purposes, could render the debate over the use of CV moot.

In sum, the first major difference between the US approach and the Conventions is that, in the United States, injury to natural resources encompasses non-economic injuries, whereas under the Conventions only economic loss is compensable.

A second major difference between the Conventions and the US approach is that the Conventions focus solely on the goal of restoring the injured resource to the condition it would have been in if the injury had not occurred. Although the Conventions recognize the right to compensation for steps taken outside the immediate area of the injured resource, those steps must contribute to the restoration of the injured resource. The United States uses a more flexible approach. For example, the OPA provides that compensation for injury to natural resources may include 'the cost of restoring, rehabilitating, replacing, or acquiring the equivalent, of the damaged natural resources'.[16]

---

12. Report of the National Oceanic and Atmospheric Administration (NOAA) Panel on Contingent Valuation, 58 Fed. Reg. 4601 (1993).
13. A strong case supporting CV techniques is made in J.C. Dobbins, 'The Pain and Suffering of Environmental Loss: Using Contingent Valuation to Estimate Nonuse Damages', *Duke Law Journal* 43 (1994): 879. The meaning of 'option' and 'existence' values will be explained, *infra*.
14. J.A. Hausman (ed.), *Contingent Valuation: A Critical Assessment* (Amsterdam: North-Holland, 1993).
15. Kanner & Nagy, *supra* n. 11, 417.
16. 33 USC § 2706(d)(1)(A).

Even without statutory authority, courts take a flexible approach. In *Commonwealth of Puerto Rico v. SS Zoe Colocotroni*,[17] the court recognized the impracticality of restoration of the damaged resource and suggested acquisition of equivalent resources. In that case, oil ruined mangrove swamps, a condition that could not be remediated. The District Court had awarded natural resource damages based on the price of replacing 92 million invertebrate organisms that had been killed by the oil. The Court of Appeals reversed, finding this measure of damages to be unreasonable because the Commonwealth did not actually intend to replace those organisms. As a matter of fact, because the oil had remained in the mangrove forests, any replacement would have resulted in the death of any newly acquired creatures. In remanding the case, the appellate court suggested, without ordering the District Court to do so, that an appropriate measure of damages might be the acquisition of equivalent resources, that is, the creation of new mangrove forests in a different location in an area unaffected by the oil spill.

In contrast, under the Conventions, where a resource sustains injuries that make restoration technically infeasible or impractical in terms of cost, there would be no recovery of compensation for the lost natural resources. The public has simply suffered an irremediable loss.

A third difference, although somewhat related to compensation for non-economic injuries, is that OPA allows recovery of damages for the 'loss of use of' natural resources.[18] It may be feasible to restore a resource to baseline, but it may take years to do so. US law allows for compensation not only for economic losses that occur during the period of restoration but also for non-economic injuries as well.

3 THE APPROACH UNDER THE CONVENTIONS

Article I of the CLC states:

> 6. 'Pollution damage' means:
>
> (a) loss or damage caused outside the ship by contamination resulting from the escape or discharge of oil from the ship, wherever such escape or discharge may occur, provided that compensation for impairment of the environment other than loss of profit from such impairment shall be limited to costs of reasonable measures of reinstatement actually undertaken or to be undertaken.[19]

The Fund Convention adopts the CLC definition of pollution damage.[20] That is all these Conventions say about natural resource damages. The 1992 Conventions

17. 628 F.2d 652 (1st Cir. 1980).
18. 33 USC § 2702(b)(2)(A).
19. The italicized portion was added by the 92 CLC.
20. Fund Convention Art. 1(2).

incorporated a new definition of pollution damage, which codified the 1980 Resolution adopted by the 1971 Fund Assembly.[21] The new definition states:

> That compensation for impairment of the environment other than loss of profit from such impairment shall be limited to costs of reasonable measures of reinstatement actually undertaken or to be undertaken.[22]

The Funds Convention's position on compensation is straightforward. Simply stated, environmental injury is compensable if it results in an actual, quantifiable economic loss and reasonable expenditures for restoration. The position was stated and explained by the Director in his comments about the Fund's position on the *Haven* incident:[23]

> The IOPC Fund does not reject all types of claims for compensation resulting from damage to the marine environment. It accepts claims which, in accordance with the terminology used in the present document<1>, relate to 'quantifiable elements' of damage to the marine environment, for example:
> (a) Reasonable costs of reinstatement of the damaged environment; and
> (b) loss of profit (income, revenue) resulting from damage to the marine environment suffered by persons who depend directly on earnings from coastal or sea-related activities, e.g. loss of earnings suffered by fishermen or by hoteliers and restaurateurs at seaside resorts.

In this regard, the Fund Assembly in 1980 adopted a Resolution that provides:

> [T]he assessment of compensation (for environmental damage) to be paid by the International Oil Pollution Compensation Fund is not to be made on the basis of an abstract quantification of damage calculated in accordance with theoretical models.[24]

The Fund denied the claim in the Patmos incident on that basis.[25]

The first Working Group was established in 1980 by the 1971 Fund Assembly to consider the Fund's general policy in respect to the payment of claims, with particular regard to their expeditious settlement. This led to the publication of the 1971 Fund's first *Claims Manual*, which gave advice to claimants on the presentation of claims and the information that should be provided in support of different

21. *Ibid.*, at 114.
22. *Ibid.*, at 104.
23. Criteria for the Admissibility of Claims for Compensation, Fund 7/4, Seventh Intercessional Working Group, Agenda Item 2, 4 Jan. 1994, 4–6.
<1> For the purpose of the present document, the expression 'quantifiable elements' means damage to the environment in respect to which the value of the damage can be assessed in terms of market prices: the expression 'non-quantifiable elements' means damage in respect to which the quantum of the damage cannot be assessed according to market prices.
24. *Ibid.*
25. Criteria for the Admissibility of Claims for Compensation, Fund 7/4, Seventh Intercessional Working Group, Agenda Item 2, 4 Jan. 1994, 3. *Also see*, J. Nichols, 'Admissibility of claims: development of the IOPC Funds' policy, The IOPC's 25 years of compensating victims of oil pollution victims', 103, 114.

types of claims. The *Manual* has been revised several times in the light of developments, the most recent edition being published in December 2008.

Finally, the Report of the Seventh Intercessional Working Group contains the text of criteria used by the Fund in determining the admissibility of various claims. The following criteria apply to claims for environmental damage:

2.7 Environmental Damage[26]

2.7.1 The Working Group based its consideration of the admissibility of claims relating to environmental damage on the definitions of pollution damage laid down in the 1969 Civil Liability Convention, the 1971 Fund Convention and the 1992 Protocols thereto. It also took note of Resolution N°3 adopted by the Assembly and the conclusions of the 5th Intersessional Working Group which had been endorsed by the Assembly. The Working Group took the view that the IOPC Fund should maintain its position that claims relating to the impairment of the environment should be accepted only if the claimant had sustained a quantifiable economic loss, and that the loss must be such that it could be quantified in monetary terms. (Emphasis added).

2.7.2 The Working Group considered in particular the question as to whether the IOPC Fund should pay compensation for the cost of measures to reinstate the marine environment, and took the view that this matter would have to be decided on the basis of the definition of 'pollution damage' laid down in the 1992 Protocol to the Civil Liability Convention.

2.7.3 The Working Group agreed that measures for reinstatement of the environment would have to fulfil the following criteria in order to be admissible for compensation:

The cost of the measures should be reasonable;

The cost of the measures should not be disproportionate to the results achieved or the results which could reasonably be expected; and

The measures should be appropriate and offer a reasonable prospect of success.

2.7.4 The Working Group stated that the test of reasonableness laid down in Article 1.6 of the 1992 Protocol to the Civil Liability Convention should be an objective one. i.e., that the measures should be reasonable from an objective point of view in the light of the information available when the specific measures were taken,

2.7.5 The Working Group noted that compensation should be paid only in respect of measures actually undertaken or to be undertaken.

2.7.6 The Working Group considered that it would normally be necessary to carry out an in depth study before any measures of reinstatement were undertaken, and that the cost of such studies would qualify for compensation

26. Summary of the Conclusions of the Intercessional Working Group, Annex I, 3–4, in IOPC, Report of the Seventh Intercessional Working Group, Fund/WGR7/21, Assembly 17th Session Agenda Item 5, 20 Jun. 1994.

only if they fulfilled the requirements generally applied by the IOPC Fund in this regard.

Environmental compensation claims were made as a result of the *Nissos Amorgos* incident in Venezuela in 1997.[27] A tanker ran aground in the Maracaibo Channel in the Gulf of Venezuela. The Venezuelan Government made claims of alleged damage to beds of clams, the cost of restoring the quality of water along the affected coast and the cost of replacing damaged sand on beaches used as a tourist locale. The Fund rejected the claim relating to clams on the ground that it 'was based on a theoretical population model that relied on false assumptions regarding clam mortality'. The claim for the cost of restoring the seawater quality was also rejected on the grounds that it was based 'on abstract quantification'. Of interest, the Fund also rejected the claim for restoration of the beach on the ground that it had paid for the removal of the oil from the beach. Furthermore, the Fund maintained:

> That the claimed theoretical cost of replacing the sand was based on more than twice the quantity that had actually been removed as a result of the clean-up operations, and that, in any event, the natural and constant replenishment of sand in the Gulf of Venezuela made replacement unnecessary.[28]

In addition, the Fund asserted 'that the alleged damage to the beach as a tourist resort was based on a theoretical model and, in any case, duplicated claims for economic losses by private claimants in the tourism sector that had already been paid by the 1971 Fund.[29] Although the Fund's positions are being challenged in a Venezuelan court, the Fund reports that experts appointed by the court to study the alleged damage to the clam beds have agreed with the Fund's position.

The Working Group established in 2000 considered a proposal to introduce 'the concept of compensation for environmental damage as a violation of "collective property"'.[30] Under this approach, compensation would be paid 'on the basis of international rights under other Conventions'. The amount of compensation would 'be based on the conclusions of environmental impact studies conducted in accordance with procedures adopted by the 1992 Fund'. More important, the 'Working Group also examined a proposal to change the 1992 Fund's policy on environmental damage so that compensation would no longer be limited to cases in which the claimant had suffered economic loss and to allow compensation to be calculated using theoretical models'. These proposals were rejected because they deviated from the definition of 'pollution damage' in the 1992 Conventions.

> As regards reinstatement measures, the Working Group focused on the development of additional specific criteria, recognizing that most major oil spills do

27. The Fund's treatment of the various claims is described in Nichols, *supra* n. 25, 103, 115–116.
28. *Ibid.*
29. *Ibid.*
30. The deliberations of this Working Group are described by Nichols, *supra* n. 25, 116–117.

> not cause permanent damage to the marine environment due to its great potential for natural recovery. The aim of any reasonable measures of reinstatement should be to bring the damaged site back to the same ecological condition that would have existed had the oil spill not occurred, or at least as close to it as possible (i.e., to re-establish a biological community in which the organisms characteristic of that community at the time of the incident are present and functioning normally). Measures taken at some distance from, but still within the general vicinity of, the damaged area might be acceptable, so long as it could be demonstrated that they would enhance the recovery of the damaged components of the environment.[31] (Emphasis added).

Observe that measures may not be taken 'at some distance from . . . the general vicinity of the damaged area', unless 'they would enhance the recovery of damaged components of the environment'.

The following specific admissibility criteria were also developed in respect of reinstatement measures:

> the measures should be likely to accelerate significantly the natural process of recovery
> the measures should seek to prevent further damage as a result of the incident
> the measures should, as far as possible, not result in degradation of other habitats or in adverse consequences for other natural or economic resources
> the measures should be technically feasible
> the costs of the measures should not be out of proportion to the extent and duration of the damage and the benefits likely to be achieved.[32]

The 1992 Fund accepted these recommendations, and these criteria are included in the current 2008 Fund *Claims Manual*.[33] As will be demonstrated, these criteria are in stark contrast to the US approach that allows compensation for acquiring alternative resources to replace a resource that cannot feasibly be restored and that also allows the acquisition of alternative resources to compensate for the time that the damaged resources are unavailable to the public even though the alternative resources do not contribute at all to the restoration of the damaged resources.

It is notable that an identical approach towards injury to natural resources was adopted in the International Convention on Civil Liability for Bunker Oil Pollution Damage 2001[34] (Bunker Convention) and in the International Convention on Liability and Compensation for Damage to in Connection with the Carriage of Hazardous and Noxious Substances by Sea 1996 (HNS).[35] As in the CLC, both of the

31. *Ibid.*, at 117.
32. *Ibid.*
33. IOPC Claims Manual 36, 2008.
34. International Convention on Civil Liability for Bunker Oil Pollution Damage, 21 Mar. 2001.
35. International Convention on Liability and Compensation for Damage to in Connection with the Carriage of Hazardous and Noxious Substances by Sea, 3 May 1996.

conventions limit the compensation for natural resource damages to costs of reasonable measures of reinstatement. Therefore, neither compensation for non-economic losses nor acquisition of comparable resource or use of theoretical models for damage evaluation will be permitted under the conventions. This indicates that a narrow understanding of pollution damage is well established under the international regime.[36]

4 THE APPROACH IN THE UNITED STATES

Before examining statutes, regulations and cases, a reference to some basic terms may be helpful. In the United States, we think of natural resources as having uses or as providing 'services' to use the terminology of the regulations.[37] Some uses are 'consumptive', that is, they exploit a resource such as logging, fishing, desalinization to creative potable water and so on. Other uses are non-consumptive, such as bird watching and swimming in the ocean. In other words, the resource is used but remains intact.

Another way of looking at natural resources is by classifying them as having 'active-use' values or 'passive-use' values. 'Active-use values' include all of the consumptive and non-consumptive uses given as examples previously because they all involve people using or taking advantage of a resource. Yet there is still another category of uses of natural resources that we classify as 'passive-use' values. One example is where a person has no present intent to use a resource but wants to preserve the resource in case he changes his mind. In other words, he wants to preserve his options. We refer to this as a resource's 'option value'. There are other passive-use values, but the most extreme example of passive-use value is 'existence value'. When existence value is included in assessing damage to natural resources, this means that the government has determined that society is enhanced by the mere existence of natural resources, regardless of whether the resources can be exploited. Conversely, the recognition of existence value also reflects the view that society suffers a loss when, for example, a resource becomes rare or extinct. Existence-use values are recognized in the United States. For example, we have

36. The author wishes to express his gratitude to Zuzanna Pepowska, Assistant Professor of Law, Nicolaus Copernicus University, Torun, Poland, for providing this paragraph.

37. In *Kennecott Utah Copper Corp. v. US Dept. of the Interior*, 88 F.3d 1191 (D.C. Cir. 1996), the Court explained the concept of services:

> In the 1986 Regulations, [the Department of the] Interior developed a concept called 'services' for measuring the level of restoration of an injured resource. The resulting rules quantified the damages to the injured resource in terms of 'the extent to which natural resource services have been reduced as a result of the injuries'. Under the services approach, the restoration level of an injured resource is measured by looking at the level of services the resource provided rather than the physical and biological characteristics of the resource. ('Restoration or replacement measures are limited to those actions that restore or replace the resource services to no more than their baseline.')

enacted legislation to protect 'endangered species',[38] sometimes at great cost to private interests, such as by restricting the use of a certain area of privately owned land to protect the habitat for an endangered species.

This approach to natural resources has been conceptualized in the view that damage to natural resources reduces the services they provide not only in regard to use or active values but also with regard to passive values, such as option and existence values. Under US environmental regulations implementing CERCLA and OPA, natural resources are equated with the 'services'[39] they provide to people or groups of people. For example, the amended CERCLA

Regulations provide:

> (c) Compensable value. (1) Compensable value is the amount of money required to compensate the public for the loss in services provided by the injured resources between the time of the discharge or release and the time the resources are fully returned to their baseline conditions, or until the resources are replaced and/or equivalent natural resources are acquired. The compensable value can include the economic value of lost services provided by the injured resources, including both public use and non-use values such as existence and bequest values.[40]

The regulations promulgated by NOAA in regard to damages in oil pollution incidents provide: 'Services (or natural resource services) means the functions performed by a natural resource for the benefit of another resource and/or the public'.[41] The regulations also state:

> Restoration means any action (or alternative) or combination of actions (or alternatives) to restore, rehabilitate, replace, or acquire the equivalent of injured natural resources and services.[42]

Finally, by way of introduction, recovery of damages in pollution cases has been an evolutionary process in the United States. Initially the Clean Water Act merely allowed the US Government to recover its cleanup costs. Then states were given the right to recover their cleanup costs. Later recovery was permitted for damage to natural resources. The most comprehensive array of damage remedies for both governmental and private parties is contained in OPA. As will be seen, the measure of recovery has also been evolutionary.

---

38. The Endangered Species Act, 16 USC §§ 1531-44.
39. The services concept will be explained, *infra*. As will be explained, the CERCLA Regulations originally applied both to discharges of oil and to releases of hazardous substances. Beginning in 1996 separate regulations for oil were promulgated by NOAA.
40. 43 C.R.R. § 11.83 (c), 73 F.R. 57267 (2 Oct. 2008).
41. 15 CFR § 990.30.
42. *Ibid.*

5 THE STATUTES: OPA 90

The following provisions are contained in OPA 90 in Title 33 of the US Code:

§ 2701

(20) 'natural resources' includes land, fish, wildlife, biota, air, water, ground water, drinking water supplies, and other such resources belonging to, managed by, held in trust by, appertaining to, or otherwise controlled by the United States (including the resources of the exclusive economic zone), any State or local government or Indian tribe, or any foreign government;

§ 2702(b)

(2) Damages

The damages referred to in subsection (a) of this section are the following:

(A) Natural resources

Damages for injury to, destruction of, loss of, or loss of use of, natural resources, including the reasonable costs of assessing the damage, which shall be recoverable by a United States trustee, a State trustee, an Indian tribe trustee, or a foreign trustee.[43]

* * *

§ 2706

Natural resources

(a) Liability

In the case of natural resource damages under section 2702(b)(2)(A) of this title, liability shall be –

(1) To the United States Government for natural resources belonging to, managed by, controlled by, or appertaining to the United States;

(2) to any State for natural resources belonging to, managed by, controlled by, or appertaining to such State or political subdivision thereof;

(3) to any Indian tribe for natural resources belonging to, managed by, controlled by, or appertaining to such Indian tribe; and

(4) In any case in which section 2707 of this title applies, to the government of a foreign country for natural resources belonging to, managed by, controlled by, or appertaining to such country.[44]

(c) Functions of trustees

(1) Federal trustees

The Federal officials designated under subsection (b)(2) of this section –

(A) Shall assess natural resource damages under section 2702(b)(2)(A) of this title for the natural resources under their trusteeship;

(B) may, upon request of and reimbursement from a State or Indian tribe and at the Federal officials' discretion, assess damages for the natural resources under the State's or tribe's trusteeship; and

---

43. Observe that damages are recoverable not only for injury to natural resources but also for the loss of use of those resources, for example, during the period they are being restored to their pre-pollution condition.
44. Subsection (b) provides for the process of selection trustees.

(C) Shall develop and implement a plan for the restoration, rehabilitation, replacement, or acquisition of the equivalent, of the natural resources under their trusteeship.[45]

(d) Measure of damages

(1) In general

The measure of natural resource damages under section 2702(b)(2)(A) of this title is –

(A) The cost of restoring, rehabilitating, replacing, or acquiring the equivalent of, the damaged natural resources;

(B) The diminution in value of those natural resources pending restoration; plus

(C) The reasonable cost of assessing those damages.

(2) Determine costs with respect to plans

Costs shall be determined under paragraph (1) with respect to plans adopted under subsection (c) of this section.

(3) No double recovery

There shall be no double recovery under this Act for natural resource damages, including with respect to the costs of damage assessment or restoration, rehabilitation, replacement, or acquisition for the same incident and natural resource.

(e) Damage assessment regulations

(1) Regulations

The President, acting through the Under Secretary of Commerce for Oceans and Atmosphere and in consultation with the Administrator of the Environmental Protection Agency, the Director of the United States Fish and Wildlife Service, and the heads of other affected agencies, not later than 2 years after August 18, 1990, shall promulgate regulations for the assessment of natural resource damages under section 2702(b)(2)(A) of this title resulting from a discharge of oil for the purpose of this Act.

(2) Rebuttable presumption

Any determination or assessment of damages to natural resources for the purposes of this Act made under subsection (d) of this section by a Federal, State, or Indian trustee in accordance with the regulations promulgated under paragraph (1) shall have the force and effect of a rebuttable presumption on behalf of the trustee in any administrative or judicial proceeding under this Act.

(f) Use of recovered sums

Sums recovered under this Act by a Federal, State, Indian, or foreign trustee for natural resource damages under section 2702(b)(2)(A) of this title shall be retained by the trustee in a revolving trust account, without further appropriation, for use only to reimburse or pay costs incurred by the trustee under subsection (c) of this section with respect to the damaged natural

45. State, Indian and Foreign Trustees perform similar functions with the exception of subs. (B).

resources. Any amounts in excess of those required for these reimbursements and costs shall be deposited in the Fund.

(g) Compliance

Review of actions by any Federal official where there is alleged to be a failure of that official to perform a duty under this section that is not discretionary with that official may be had by any person in the district court in which the person resides or in which the alleged damage to natural resources occurred. The court may award costs of litigation (including reasonable attorney and expert witness fees) to any prevailing or substantially prevailing party. Nothing in this subsection shall restrict any right which any person may have to seek relief under any other provision of law.

6 REGULATIONS AND JUDICIAL DECISIONS

As stated above in 33 U.S.C. § 2706(e), Congress has delegated to administrative agencies the responsibility for formulating regulations prescribing the methods for calculating damages for injury to or destruction of natural resources. Before 5 February 1996, damages for injury to, or destruction of, natural resources caused by oil or hazardous substances were calculated according to regulations promulgated by the US Environmental Protection Agency, an agency within the DOI.[46] Thus, the regulations promulgated before 1996 and, more importantly, judicial decisions determining the validity of the regulations, are relevant to discharges of oil that injure natural resources.

The first regulations relating to oil and chemical pollution damages for injuries to natural resources were promulgated by an administration that was pro-business. Those regulations and what happened to them are discussed in the following section.

Among other matters, the regulations mandate a Natural Resource Damage Assessment. The regulations provide for two types of assessments. One type, type A, relies on computer models with the input of crucial data from the site. The other, type B, requires extensive examination of the site and laboratory analysis. The background legislation and regulations regarding the damage assessment process is explained in *National Association of Manufacturers v. United States Department of the Interior*,[47] a case involving challenges to the Type A natural resource damage regulations promulgated by the DOI. Pertinent sections of the court's opinion explain in detail the process for assessing natural resource damages.

Parts of these regulations were also challenged by some states and pro-environmental groups in the case of *State of Ohio v. US Department of the Interior*.[48] This case considered in particular, whether the regulations could properly

46. 43 CFR §§ 11.10–11.93 (Code of Federal Regulations).
47. 134 F.3d 1095 (D.C. Cir. 1998).
48. 880 F.2d 432 (D.C. Cir. 1989).

limit natural resource damages to the lost-use value if the cost of restoring or replacing the resource would exceed that value and whether lost-use values could be measured exclusively by market values when available. The holding of the court is clear:

> We hold that the regulation limiting damages recoverable by government trustees for harmed natural resources to 'the lesser of' (a) the cost of restoring or replacing the equivalent of an injured resource, or (b) the lost use value of the resource is directly contrary to the clearly expressed intent of Congress and is therefore invalid. We also hold that the regulation prescribing a hierarchy of methodologies by which the lost-use value of natural resources may be measured, which focuses exclusively on the market values for such resources when market values are available, is not a reasonable interpretation of the statute.[49]
>
> The second issue focused on the emphasis the regulations placed on market value in determining 'use' value. As stated by the court:
>
> The regulations establish a rigid hierarchy of permissible methods for determining 'use values,' limiting recovery to the price commanded by the resource on the open market, unless the trustee finds that 'the market for the resource is not reasonably competitive.' If the trustee makes such a finding, it may 'appraise' the market value in accordance with the relevant sections of the 'Uniform Appraisal Standards for Federal Land Acquisition'. Only when neither the market value nor the appraisal method is 'appropriate' can other methods of determining use value be employed.[50]
>
> In striking down the regulations the court stated:
>
> There are many resources whose components may be traded in 'reasonably competitive' markets, but whose total use values are not fully reflected in the prices they command in those markets. Interior itself provides ample proof of the inadequacy of the 'reasonably competitive market' caveat. For example, DOI has noted that 'the hierarchy established in the type B regulation' would dictate a use value for fur seals of $15 per seal, corresponding to the market price for the seal's pelt. Another example of DOI's erroneous equation of market price with use value is its insistence that the sum of the fees charged by the government for the use of a resource, say, for admission to a national park, constitutes 'the value to the public of recreational or other public uses of the resource.' because 'these fees are what the government has determined to represent the value of the natural resource and represent an offer by a willing seller'. This is quite obviously and totally fallacious; there is no necessary connection between the total value to the public of a park and the fees charged as admission, which typically are set not to maximize profits but rather to encourage the public to visit the park. In fact, the decision to set entrance fees far below what the traffic would bear is evidence of Congress's strong

49. *Ibid.*, at 438.
50. *Ibid.*, at 462.

> conviction that parks are priceless national treasures and that access to them ought to be as wide as possible, and not, as DOI would have it, a sign that parks are really not so valuable after all.

After 5 February 1996, natural resource damages resulting from marine oil pollution are assessed in conformity with regulations issued by NOAA.[51] These procedures are explained in the case of *General Electric Co. v. US Dept. of Commerce*.[52]

7 LOSS OF SERVICES PENDING RESTORATION AND THE CONCEPT OF 'COMPENSATORY RESTORATION'

7.1 Loss of Use

One of the characteristics of the US compensation scheme for injury to natural resources is that compensation may be recovered for the public's loss of services during the period of time while resources are being restored to baseline. The OPA specifically includes 'loss of use of [ ] natural resources' as a compensable item.[53] By contrast, neither the Clean Water Act (CWA)[54] nor CERCLA[55] expressly refers to loss of use. Nevertheless, the courts have interpreted those statutes as authorizing recovery for loss of use as illustrated in *Kennecott Copper Corporation v. US Department of the Interior*.[56]

7.2 Compensatory Restoration

In October 2008, the DOI published regulations amending the CERCLA regulations discussed earlier. Although these regulations apply to the discharge of hazardous substances and not to oil, they reflect the evolutionary process of formulating appropriate criteria for determining compensation for damage to natural resources. The new regulations also demonstrate the current thinking of a major US agency charged with implementing the congressional decision that polluters should pay for damage to natural resources. Furthermore, as will be seen, the CERCLA regulations and the OPA regulations are very similar in their approaches to this issue.

The regulations revise the Type B rules and are a direct response to the *State of Ohio* and *Kennecott Copper* cases mentioned earlier. In these regulations, the term 'restoration' is used 'as an umbrella term for all types of actions that the natural

51. 15 CFR 990.10 et seq.
52. 128 F.3d 767 (D.C. Cir. 1997).
53. 33 USC §§ 2702(b)(2)(A) and 33 USC § 2706(d)(1)(B).
54. 3 USC § 1321(f)(4).
55. 42 USC §§ 9607(a)(4)(C) and (f).
56. 88 F.3d 1191 (D.C. Cir. 1996).

resource damage provisions of the CERCLA and the CWA authorize to address injured natural resources, including restoration, rehabilitation, replacement or acquisition of equivalent resources'.[57] The regulation purports to shift from an approach that calculates 'loss-of-use damages' based on a determination of the value of lost services (including passive-use values) pending restoration by expanding the scope of the term restoration to include measures that provide alternative services, thereby compensating the public for the period that the public is denied use of the injured resource. The amended regulations attempt to avoid the need to quantify the cost of a loss of services, which may be difficult to do to in some cases, by substituting measures that will permit providing alternative services under the rubric of restoration.

Under this approach there are two critical questions: (1) Which services will the public lose during the period of time it takes to restore the injured resource? And (2) What steps can be taken to make the public whole by providing the public with equivalent, alternative services? Earlier, in the explanatory section of the amended regulation, the agency explains:

> Estimating option and existence value through the use of contingent valuation methodologies remains controversial. We note, however, that our revision's focus on compensating for public losses pending restoration with restoration actions rather than monetary damages for the economic values of the losses will provide options for comparing functional losses to functional gains expected from restoration actions, which will reduce the need for trustees to seek to recover the monetary value of passive economic losses, such as option and existence value.[58]

This approach is similar to the approach in the NOAA regulations promulgated to implement the natural resource damages provisions in OPA. The NOAA regulations contain the following definitions:

> Restoration means any action (or alternative) or combination of actions (or alternatives) to restore, rehabilitate, replace, or acquire the equivalent of injured natural resources. Restoration includes:
>
> (a) Primary restoration, which is any action, including natural recovery, that returns injured natural resources and services to baseline; and
>
> (b) Compensatory restoration, which is any action taken to compensate for interim losses of natural resources and services that occur from the date of the incident until recovery.
>
> Services (or natural resource services) means the functions performed by a natural resource for the benefit of another natural resource and/or the public.[59]

57. 73 F.R. 57259 (8 Oct. 2008) (emphasis added).
58. 73 F.R. 57261 (2 Oct. 2008).
59. 15 CFR § 990.30.

A recent oil spill incident reveals how NOAA applies its regulations in assessing damages for injury to natural resources caused by the discharge of oil into the navigable waters of the United States. In January 2009, NOAA and various other federal and state agencies issued a Draft Assessment and Restoration Plan and Environmental Assessment for the 26 November 2004, M/T Athos I Oil Spill on the Delaware River near the CITGO Refinery in Paulsboro, New Jersey.[60] The Athos sailed from Venezuela for the CITGO Refinery in Paulsboro, New Jersey. She was a single bottom, double-sided vessel, carrying a cargo of 13 million gallons of Bachaquero Venezuelan crude. As the vessel was being docked at the refinery, she struck several submerged objects and spilled approximately 263,371 gallons of oil into the Delaware River. Over the following months, the oil spread downriver, threatening natural resources over 115 river miles and 280 miles of shoreline and affecting three states – New Jersey, Pennsylvania and Delaware.

A determination was made that 'primary restoration' actions would be of little use, and natural recovery was the alternative selected for primary restoration. The trustees then identified four areas of natural resource injuries: (1) shorelines, (2) birds and wildlife, (3) aquatic resources, and (4) recreation use. The trustees assessed the magnitude of the injury and the time for full recovery.

> Injury assessments conducted by the Trustees and other experts identified the following injuries to natural resources and recreational services from the spill:
>
> - Shoreline – 1,729 acres were very lightly, lightly, moderately, or heavily oiled.
> - Tributaries – Six tributaries, with a total area of 1,899 acres, were exposed to very light to moderate oiling.
> - Aquatic – 412 acres were exposed to Athos oil.
> - Birds – 11,869 estimated dead (includes direct and indirect losses, a majority of which were swans and geese).
>
> Recreational services – An estimated 41,709 trips on the river were affected by the spill, with an estimated lost value of $1,313,239.[61]

In as much as the NOAA regulations allow for compensatory restoration pending full recovery of the injured resources, the trustees 'identified and evaluated a wide range of project alternatives capable of restoring ecological services comparable to those lost'. Basically, this approach accepts the reality that the public will be denied access to certain injured natural resource services until the resource is fully recovered and develops strategies to provide alternative natural resource services in the affected area. The projects selected include oyster reef enhancement, marsh restoration, grassland restoration, acquisition of a wet meadow, pond and pasture enhancement, shoreline restoration, habitat restoration, dam removal, boat ramp

60. <www.darrp.noaa.gov/pdf/DraftAthosDARP_Final_2pdf>. This Draft has been circulated for public comment and is not final. Based on input from the public, the agency may make revisions (hereinafter Draft).
61. Draft, ix.

installations, and trail and habitat improvements. None of these projects will lead to an earlier recovery of the area injured by the spill, but all are intended to provide the public with alternative services to compensate for the services that will not be available as a result of the damage caused to the resources.

This approach should be contrasted with the Convention approach, which would provide no compensation for any of these projects because they do not lead to the restoration of the injured resources.

8 THE FUTURE

Notwithstanding the Fund's approach, a growing awareness of the need for recognizing natural resource damage in a broader sense is apparent. After the *Erika* and *Prestige* incidents, the European Commission emphasized the view that compensating environmental damage under CLC should be extended.[62] Its position was that limiting liability to reasonable costs of reinstatement contradicts the well-established principle under EU law that the polluter pays.[63]

In light of the above, it is important to note the new European solutions towards compensating environmental damage adopted under the Directive on environmental liability of 2004.[64] The Directive has many similarities to the NOAA final rule of 1996 and Natural Resource Damage Assessment of 2008. Under its provisions the polluter is responsible for undertaking remedial measures that include actions to restore, rehabilitate or replace damaged natural resources and services impaired.[65] Therefore, liability is not limited to the cost of returning damaged resources to the baseline condition (so-called primary remediation). Where full restoration of natural resources or environmental services is not possible, the polluter is required to provide a similar level of natural resources and services at an alternative site (so-called complementary remediation). Replacement does not necessarily have to be in the vicinity of a damaged area or to have any enhancing effect on it. Besides primary and/or complementary remediation, the polluter is responsible for compensatory remediation, that is, compensating the public for loss of services provided by a natural resources pending restoration action. Similar to the US solutions, rather than quantifying the monetary value

62. Communication from the Commission to the European Parliament and the Council on safety of the seaborne oil trade, 21 Mar. 2000, COM (2000) 142 final; Communication from the Commission to the European Parliament and the Council on the second set of Community measures on maritime safety, 6 Dec. 2000, COM (2000) 802 final.
63. Article 174(2) of the Treaty Establishing European Union states that Community policy on the environment shall be based on the principal that environmental damage should as a priority be rectified at the source and that the polluter should bear the cost of that. [2006] OJ C321 E.
64. Directive 2004/35/EC of the European Parliament and of the Council of 21 Apr. 2004 on environmental liability with regard to the prevention and remedying of environmental damage, 21 Apr. 2004, [2004] OJ L143/56.
65. Articles 6 and 2(11) of the Directive. The author wishes to express his gratitude to Zuzanna Peplowska, Assistant Professor of Law, Nicolaus Copernicus University, Torun, Poland, for providing this paragraph and the following one.

of a lost resource and its services, providing new natural resources and services of the same kind as the damaged ones is preferred. Only when the latter is not possible should the monetary value of the injured resources/services be estimated. Therefore, the Directive does allow the use of economic valuation techniques. Nevertheless, the responsible party is not required to pay the monetary value of interim losses. Rather, the polluter will be obliged to undertake remedial measures that cost the same as the estimated monetary value of injured resource/service.[66] The Directive leaves the determination of valuation techniques to the national laws of Member States.

Presently, the Directive explicitly excludes from its application environmental damage arising from an incident for which liability or compensation falls within the scope of the CLC, FUND, HNS or Bunker Conventions.[67] The Directive itself provides for a revision of the oil pollution damage exclusion by 2014.[68]

Another sign of the future is illustrated by the decision of the European Court of Justice (Grand Chamber) in the case of *Commune de Mesquer v. Total France SA and Total Internat'l. Ltd.*[69] This case arose out of the *Erika* oil spill off the coast of France and the question involved the scope of Directive 75/442 that deals with 'waste'. The Court held that when oil is spilled into water as a result of a marine casualty and becomes mixed with water and sand it is 'waste', because that term is used in Directive 75/442, and, under the 'polluter pays' principle, liability may be imposed 'on the persons who cause the waste, whether they are holders or former holders of the waste or even producers from which the waste came', as long as they have contributed to the risk'. The Court points out that, in contrast to Directive 2004/35 EC, Directive 75/442 does not exempt from its scope incidents covered by the CLC or Fund Conventions. Although some may find the result comforting, the Court had to take a torturous route to reach it.

9 CONCLUSION

The approach to pollution damage to natural resources in the United States varies in three significant ways from the approach of the Conventions.

First, consistent with the US approach to damages recoverable in tort cases, compensation for non-economic losses may be recovered under OPA and CERCLA for injury to natural resources.

66. Annex II to the Directive.
67. Article 4(2) excludes from the Directive's application pollution damage covered by International Conventions listed in Annex IV, being CLC, FUND, Bunker Convention, HNS and Convention on Civil Liability for Damage Caused during Carriage of Dangerous Goods by Road, Rail and Inland Navigation Vessels (CRTD), 10 Oct. 1989. Neither CRTD not the HNS have come into force.
68. Article 18(3)(a) of the Directive.
69. Case C-188/07, <http:/eur-lex.europa.eu/LexUriServ.do?uri=CELEX:62007J0188:EN:HTML>.

Second, both OPA and CERCLA recognize that sometimes a resource may never recover, or the recovery period is so long that it is the equivalent of destruction, or the expense of full restoration might be so costly that it would be considered unreasonable. Therefore, both statutes provide that replacement of the damaged resource or the acquisition of comparable resources may be the appropriate remedy.[70]

Third, compensation is recoverable for interim losses of a resource, that is, loss of the use of a resource while it is being restored. Thus, under regulations promulgated by NOAA and the new regulations promulgated by the DOI, restoration includes not only primary restoration of the injured resource, but also compensatory restoration, whereby equivalent, alternative resources are made available to the public.

In contrast, none of these types of compensation appear compensable under the Conventions, which emphasize what the NOAA regulations call 'primary restoration'. The Conventions do not include so-called compensatory restoration unless those measures contribute to primary restoration, nor do they provide for replacement of an impaired resource or the acquisition of comparable resources as US law permits.

References

Dobbins, J.C. 'The Pain and Suffering of Environmental Loss: Using Contingent Valuation to Estimate Nonuse Damages'. *Duke Law Journal* 43 (1994): 879–946.

Hausman, J.A. (ed.). *Contingent Valuation: A Critical Assessment.* Amsterdam: North-Holland, 1993.

Kanner, A. & T. Nagy. 'Measuring Loss of Use Damages in Natural Resource Damage Actions'. *Columbia Journal of Environmental Law* 30 (2005): 417–448.

Nichols, J. 'Admissibility of Claims: Development of the IOPC Funds' Policy – The IOPC's 25 Years of Compensating Victims of Oil pollution Victims'. 103, 114, 2003 available on the web site of the IOPC at <www.iopcfund.org/npdf/jub_en.pdf>.

70. 33 USC § 2702(b)(2)(A), § 2706(d)(1)(A) (OPA), 42 USC § 9607(a)(4)(C), § 9607(f) (CERCLA).

Chapter 16

National Interests Influence in Compensation for Damages Resulting from Oil Pollution: The Experience of the United States and Its Reference to China

Li Tiansheng & Han Lixin***

1 INTRODUCTION

Two key issues are considered in the compensation for damage resulting from oil pollution: one is the scope of compensation, and the other is to ensure the realization of compensation. In respect to the scope of compensation for damage resulting form oil pollution, a controversial issue is whether a pure economic loss should be accounted for recovery. When it comes to implementation of compensation, according to Chinese law, many tankers and vessels are not requested to establish compulsory insurance and Chinese law makes no provisions for funds for compensation for damage resulting from oil pollution. This situation does not enable

* Ph.D. Candidate in Maritime Law, Law School of Dalian Maritime University-Dalian, 116026, China.

** Professor of Law School of Dalian Maritime University, Ph.D. in Law, Ph.D. Supervisor, the Director of Maritime Law Department, and the Director of Institute of International Maritime Law, Dalian, 116026, China. Arbitrator of CMAC.

Michael G. Faure, Han Lixin & Shan Hongjun, *Maritime Pollution Liability and Policy: China, Europe and the US*, 285–297.

victims of oil pollution to be fully protected. An examination of the US compensation for damage resulting from oil pollution may provide some references to the issues whether Chinese law should provide recovery for pure economic loss arising from oil pollution,[1] and how Chinese law should regulate the compulsory insurance and funding to allow compensation for damage resulting from oil pollution.

2 THE US DEVELOPMENT OF COMPENSATION FOR PURE ECONOMIC LOSS RESULTING FROM OIL POLLUTION AND ITS REFERENCE TO CHINA

2.1 The US Development of Compensation for Pure Economic Loss Resulting from Oil Pollution

The US is not a contracting party to any of the international conventions concerning the compensation for damages resulting from oil pollution. The US statutes and case law apply to all issues of compensation for damage resulting from oil pollution. At to the case law, the important case is *Robins Dry Dock v. Flint.*[2] In this case, the petitioner sought review of a judgment that allowed respondents to recover for the loss of use of the owner's steamer, in action alleging the petitioner was liable to respondents because petitioner's negligence damaged the steamer, causing a delay in the steamer's availability to respondents in violation of the terms of their contract with owner. But the United States Supreme Court reversed this judgment. Mr Justice Holmes delivered the opinion of the court that

> No authority or ownership need be cited to show that, as a general rule, at least, a tort to the person or property of one man does not make the tortfeasor liable to another merely because the injured person was under a contract with that other unknown to the doer of the wrong.

And '[t]he law does not spread its protection so far'. This case established a rule that recovery of economic loss was not available if that loss did not result from physical damage to the property in which the claimant had proprietary or ownership interest. This rule was popularly known as the 'bright line rule', or 'Robins Dry Dock rule'. Under this rule, pure economic loss, including medium and long-term loss of natural resource and loss of profit of oil pollution, was not allowed for recovery. This rule was affirmed and was extended in many cases, such as

1. See IOPC Fund *Claim Manual 2005* s. II: Under certain circumstance compensation is also payable for loss of earnings caused by oil pollution suffered by person whose property has not been polluted (Pure economic loss). For example, a fisherman whose nets have not been contaminated may nevertheless be prevented from fishing because the area of the sea where he normally fishes is polluted and he cannot fish elsewhere. Similarly, an owner of a hotel or a restaurant located at a contaminated public beach may suffer losses because the number of guests falls during the period of the pollution. Chinese law has no provisions about this issue.
2. *Robins Dry Dock v. Flint*, 275 THE US 303 (1927).

Akron Corp. v. M/T Cantigny,[3] in which the court noted, 'a party may not recover for economic losses not associated with physical damages'. In *State of Louisiana v. M/V Testbank* case,[4] the Robins Dry Dock rule was specifically applied to environmental damages. Though application of this rule to claims under the Trans-Alaska Pipeline Authorization Act (TAPAA) was questioned, for example, in *Slaven v. BP America*,[5] the district court held this rule and incorporated common law proximate cause principles. However, the Robins Dry Dock rule still is a 70-year-old dominant doctrine denying recovery for pure economic loss resulting from tortious conduct in maritime cases.[6]

Also, various US statutes apply to oil pollution damage. Historically, the following statutes were applicable to marine oil pollution claims. The Federal Water Pollution Control Act (FWPCA), commonly known as the Clean Water Act (CWA),[7] is a prevailing federal statute to establish a strict liability system and applies to claims for the federal government's costs for cleaning up water pollution. Other important relevant statutes include: (1) the Outer Continental Shelf Lands Act ('OCSLA'),[8] which applies to activities of facilities on the outer continental shelf and vessels operating off-shore transporting oil from such facilities; (2) the Deepwater Port Act,[9] which imposes strict liability on vessels and operators for cleanup expenses and damages resulting from discharge of oil from a vessel operating in a safety zone surrounding a designated deep water port facility; (3) Tran-Alaska Pipeline Authorization Act (TAPAA),[10] which provides for strict liability for oil pollution from vessels transporting oil through the Trans-Alaska Pipeline; and (4) the Refuse Act,[11] which as a part of the Rivers and Harbours Act of 1899, prohibits the dumping of refuse, including oil, into the navigable waters of the United States. All of the statutes mentioned here basically only cover cleanup costs incurred by state and federal authorities and do not abrogate the Robins Dry Dock rule. There are two main justifying reasons for the Robins Dry Dock rule: one is that most of economic loss without physical damages is based on contracts that are unknown or unforeseen to liable parties; hence it is a too heavy duty for liable parties to provide recovery for economic loss without physical damages. The other reason is that recovery for pure economic loss will cause too many claims for the courts to handle. However, an exception to this rule has been recognized with regard to claims of commercial fishermen who have been held to have a property interest in the fishing resource.[12]

3. *Akron Corp. v. M/T Cantigny*, 706 F.2d 151 (5th Cir. 1983).
4. *State of Louisiana v. M/V Testbank*, 752 F.2d 1019 (5th Cir. 1985).
5. *Slaven v. BP America*, No. CV90-0722, C.D. Cal. (11 Feb. 1992).
6. C.B. Kende, 'The United States Approach', in *Liability for Damage to the Marine Environmental*, ed. C.M. De La Rue (London: Lloyd's of London Press Ltd, 1993), 144.
7. 33 USC 1251 et seq.
8. 46 USC 1331–1356.
9. 33 USC 1501 et seq.
10. 43 USC 1651–1655.
11. 33 USC 407.
12. See *Union Oil Company v. Oppen*, 501 F.2d 558 (9th Cir. 1974).

On 24 March 1989, the supertanker *Exxon Valdez* ran aground on Alaska's Bligh Reef, tearing open its hull and spilling 11 million gallons of sticky crude oil into Prince William Sound. It was the largest oil spill pollution accident in US history, with 1,609 km of coastline polluted. Cleanup costs were United States Dollar (USD) 2 billion, and the spill resulted in a loss of USD 8 billion in total. After this disaster, more than 170 civil actions were brought, including as litigants fishermen, a canning factory, real estate owners, miners and administrative authorities. However, because most of the losses were not 'physical damage', the majority of the plaintiff's damage cannot be supported under the Robins Dry Dock rule and the other statutes. Based on this background, the Oil Pollution Act of 1990 (OPA 1990) was signed by the US president on 18 August 1990. OPA 1990 has totally changed previous laws in categories of damages in that under the Act, the responsible parties are liable for damages resulting from following:

> (a) Damages caused by the loss of natural resources, including the reasonable costs of assessing the recoverable damage.
> (b) Damages for injury to economic losses resulting from destruction of real or personal property.
> (c) Damages for loss of subsistence use of natural resources by any claimant who so uses natural resources.
> (d) Damages equal to the net loss of taxes, royalties, rents, fees, or net profit shares due to the injury, destruction, or loss of real property, personal property, or natural resources, which shall be recoverable by the Government of the United States, a State, or a political subdivision thereof.
> (e) Damages equal to the loss of profits or impairment of earning capacity due to the injury, destruction, or loss of real property, personal property, or natural resources, which shall be recoverable by any claimant and
> (f) Damages for net costs of providing increased or additional public services during or after removal activities recoverable by a State, or a political subdivision of a State.[13]

Clearly, under OPA 1990, a wide range of pure economic loss that claimants have incurred without physical damages can be compensated. Actually, almost all conceivable damage resulting from oil pollution can now be compensated.

2.2 Lessons from the US Development Concerning the Compensation for Pure Economic Loss Resulting from Oil Pollution for China

It is a basic legal principle that claimants may not recover for economic loss if that loss did not result from physical damage to property in which the victim had proprietary or ownership interest. This principle originated in English courts

13. 33. USC 2702(b)(2).

during the nineteenth century and is still applied in English courts today.[14] Also, the Robins Dry Dock rule and all of the previous laws did not provide recovery for pure economic loss. Why did the US law abrogate such an important principle and take such a different position from English courts on the issue of compensation for pure economic loss resulting from oil pollution? An analysis of this issue can provide useful reference for Chinese legislation.

In our opinion, different legislations just reflect different national interests. Theories are always based on certain reasons that are consistent with certain requirements of society, the country or the historical period and are not based on only theories themselves. However, with the different national circumstances and different economic and social developments of the two countries, it is inevitable to make different laws. Compensation for pure economic loss resulting from oil pollution is a typical example. Specifically, there are two important differences between the United States and the United Kingdom: one is that the United States is a dominant cargo owner's country, whereas the United Kingdom is a dominant shipping country; the other is that the United States is a country with the largest amount of oil importation, importing 669 million tons of oil in 2007.[15] Above all, the importance of environmental protection is increasingly more obvious for nations, societies and people. The United States is a country with the largest amount of oil import, but at the same time it is not a dominant shipping country, which means most of the oil imported from foreign countries is transported by foreign vessels. For oil pollution damage, the primary liable parties are foreign ships, but the main damage and losses are in the US environment and other victims in the United States. Oil pollution can cause major damage to the environment, natural resources and marine-related industries. The *Exxon Valdez* case made it clear that many claimants' benefits need to be protected, but the basic legal principle that denies recovery for pure economic loss without physical damage is no longer consistent with the interests of the United States. This is the main reason why the United States abrogated such an important principle and took a position so different from that of English courts on compensation for pure economic loss resulting from oil pollution. Juristic theories and legal principles change with the development of history, just like Holmes said, the life of the law has not been logic, it has been experience.[16]

China is a state that is a party to the International Convention on Civil Liability for Oil Pollution Damage 1992 (CLC 1992) and to the International Convention on Civil Liability for Bunker Oil Pollution Damage 2001 (Bunker Convention 2001). However, the scope of compensation for damage resulting from oil pollution is not clear. Also, according to the Article 2(3) of CLC 1992 and Article 1(9) of Bunker Convention 2001, the language 'provided that compensation for impairment of the environment other than loss of profit' does not indicate whether pure economic loss

14. *Landcatch Ltd v. IOPC Fund.* 1999 SLT 1208, 1999 SCLR 709.
15. See <www.lrn.cn/bookscollection/magazines/maginfo/2008maginfo/qingbao200804/200809/t20080904_272259.htm>.
16. O.W. Holmes, *The Common Law*, ed. M. Howe, vol. I (Boston: Little Brown, 1963 [1881]), 5.

can be recovered, and further it does not provide the meaning of pure economic loss.

There are many disputes on compensation for pure economic loss in China. With respect to legislation, the following legislations have provisions concerning the scope of compensation for damage resulting from oil pollution: (1) General Principles of Civil Law of the People's Republic of China (PRC). This law provides for liability for loss resulting from pollution or damage of environment;[17] (2) the Environmental Protection Law of the PRC, which provides for liability for loss directly resulting from pollution of environment;[18] (3) the Law on Prevention and Control of Water Pollution of the PRC, which has same provision as the Environmental Protection Code of the PRC;[19] and (4) the Marine Environmental Protection Law of the PRC, which provides for liability for damage to marine environment. In accordance with this law, whoever causes pollution damage to the marine environment shall remove the pollution and compensate the losses; in the case of pollution damage to the marine environment resulting entirely from the intentional act or fault of a third party, that third party shall remove the pollution and be liable for the compensation. For damages to marine ecosystems, marine fishery resources and marine protected areas that cause heavy losses to the State, the department is vested with power by the provisions of this law to conduct marine environment supervision and administration shall, on behalf of the State, put forward compensation demands to those held responsible for the damages.[20] However, each of the statutes mentioned do not indicate the specific scope of the 'damages', but just operate as the principle rules. As a civil law system country, Chinese civil law theories basically do not support recovery for pure economic loss in torts because the causal relationship between the cause and the damage is too distant. In the *Minrangong 2* case (1999), the Guangzhou Maritime Court noted that medium- and long-term damage to fishery resources cannot be recovered from liable parties of the oil pollution, but the Senior Court of the Guangdong Province reversed this judgment and held that it should be recovered. This case reflects that the compensation for pure economic loss is controversial in China. In Article 7(1) of 'Certain Rules about Trial of Case of Compensation for Damage Resulting from Oil Pollution by Vessel (discussion draft, 2005)',[21] the Supreme People's Court provided for no recovery for pure economic loss resulting from oil pollution. In Article 68(3) of the 'Regulations of the PRC for the Prevention of Pollution of Marine Environment by Vessels (final draft, 2008)',[22] the loss of profits as a result of environmental damage is not recoverable. The loss of profits is the main type of pure economic loss without physical damage to the property resulting from oil pollution, in which the claimant had no proprietary or ownership interest, such

17. Article 124, *General Rules of Civil Law of the P.R.C.*
18. Article 41, *the Environmental Protection Code of the P.R.C.*
19. Article 55, *the Water Pollution Control Code of the P.R.C.*
20. Article 90, *the Marine Environmental Protection Code of the P.R.C.*
21. Author's translation from the original Chinese version.
22. Author's translation from the original Chinese version.

as losses of beach tourism companies, losses of restaurant owners whose source of shellfish has been cut off by an oil spill, and so on. By definition, the pure economic loss should be a certain loss without physical damage to the properties used or owned by victims.[23]

Most of the statutes and theories in China seem not to support compensation for pure economic loss. However, taking into account the experience of the United States, China should legislate to provide for recovery for pure economic loss resulting from oil pollution. The supporting reasons are the following:

(1) The situation in China is comparable to the one in the United States. As a large oil-importing country, China imported 200.67 million tons of oil from foreign countries in 2008,[24] and more than 90% of imported oil was shipped by foreign tankers.[25] As a matter of fact, China is not a dominant shipping country, at least in shipping of oil. Based on the experience of the United States, for oil pollution damage in China, the primary liable parties in oil pollution damage cases in China are foreign vessels, and the main damage is to the environment of China and other victims in China. Some may point out that China needs to promote the capacity of ocean transport for oil, but if China legislates to provide for recovery for pure economic loss resulting from oil pollution by tankers, it is not beneficial to this promotion. These opinions are incorrect because the promotion of the capacity of ocean transport for oil depends on development of the Chinese tanker fleet and on the improvement of the competitiveness of the Chinese shipping industry, but does not depend on expenses for destroying the marine environment and harms to the interests of the victims of oil pollution.

(2) Marine environmental protection is increasingly more important for China. From 2005 to 2006, there was an average of one oil spill accident every three days in China; a total of 253 accidents of oil spill in Chinese coastal and inland waters, including 9 incidents involving more than 50 tons of oil spilled; and a total amount of 2,956 tons of oil spilled.[26] The situation in regard to marine environmental protection is very serious in China. Through legislation to provide for recovery for pure economic loss resulting from oil pollution, on the one hand, both the victims of oil pollution in Chinese territory and marine environmental resources can obtain compensation; on the other hand, legislation will encourage carriers to improve their equipment, vessels and diligence to reduce oil pollution. In fact, environmentalists directly promoted of the adoption of

23. See CMI Guidelines on Oil Pollution Damage part II, Art. 3(c) 'Pure economic loss' means financial loss sustained by a claimant otherwise than as a result of such physical loss of or damage to property; (d) 'Property' means anything in which the claimant has a legally recognized interest by virtue of a proprietary or possessory right.
24. See <www.cec.gov.cn/info/NewsDetail.jsp?news_id=56488>.
25. See <www.ce.cn/cysc/ny/shiyou/200608/11/t20060811_8091400.shtml>.
26. See <http://news.sohu.com/20071113/n253214107.shtml>.

OPA 1990 in the United States, and given the environmental situation in China, the importance of environmental protection cannot be overemphasized.

(3) Marine-related industries are developing rapidly in China. Along China's extensive coastline, marine-related industries are flourishing, for example, tourism, fisheries, marine processing and so forth. With the change in economic structure, in the last couple of years, tertiary industry in China has been developing very rapidly. Beach tourism and other marine-related industries are important parts of China's tertiary industry. As the economic benefits of marine-related industries grow and the majority of the economic benefits are pure economic losses for oil pollution, China should enact laws to provide for recovery for them. As mentioned earlier, similar to the practice in the United States, the benefit for parties in the marine-related industries are more important than the benefit for parties who are in the oil-shipping industry in China.

The conclusion is that all of the changes and legislation in the United States serve interests of the United States. China should also shake off shackles of theories and understand that providing recovery for pure economic loss is a more important interest for China.

3 THE US PRACTICE IN COMPULSORY INSURANCE AND FUND FOR DAMAGE RESULTING FROM OIL POLLUTION AND ITS REFERENCE TO CHINA

3.1 The US Practice in Compulsory Insurance and Fund for Damage Resulting from Oil Pollution

3.1.1 The US Practice in Compulsory Insurance for Damage Resulting from Oil Pollution

The purpose of compulsory insurance is to ensure the realization of compensation for damage arising from oil pollution. Both CLC 1992 and the Bunker Convention 2001 provide for a compulsory insurance system.[27] CLC 1992 requires ships registered in contracting states and carrying more than 2,000 tons of oil in bulk as cargo to maintain insurance or other financial security. The Bunker Convention 2001 applies to ships over 1,000 gross tons. The two conventions provide for the establishment of compulsory insurance by guarantee from a bank, a certificate delivered by an international compensation fund or similar financial institutions. The US CWA provides for the establishment financial responsibility for

27. See Art. VII, para. 1, of CLC1992; Art. VII, para. 1 of Bunker Convention 2001.

compulsory insurance, but the OPA 1990 revised the financial responsibility requirements of the CWA for:

> The responsible party for –
>
> (1) any vessel over 300 gross tons (except a non-self-propelled vessel that does not carry oil as cargo or fuel) using any place subject to the jurisdiction of the United States; or
>
> (2) any vessel using the waters of the exclusive economic zone to tranship or lighter oil destined for a place subject to the jurisdiction of the United States; shall establish and maintain, in accordance with regulations promulgated by the Secretary, evidence of financial responsibility sufficient to meet the maximum amount of liability to which the responsible party could be subjected under section 1004(a) or (d) of this Act, in a case where the responsible party would be entitled to limit liability under that section. If the responsible party owns or operates more than one vessel, evidence of financial responsibility need be established only to meet the amount of the maximum liability applicable to the vessel having the greatest maximum liability.[28]

Comparing CLC 1992 and Bunker Convention 2001, the provisions quoted above provide for a wider responsibility of compulsory insurance. First, OPA 1990 requires more vessels to establish and maintain compulsory insurance, with all vessels over 300 gross tons included, which is different from the requirements of 2,000 tons under CLC 1992 and 1,000 tons under the Bunker Convention 2001. Second, the requirements of financial responsibility of OPA1990 apply to a broader geographical scope in which the place is subject to the jurisdiction of the United States, including the exclusive economic zone. Finally, OPA 1990 provides for the financial responsibility sufficient to meet the maximum amount of liability.

On methods of financial responsibility, OPA 1990 provides:

> Financial responsibility under this section may be established by any one, or by any combination, of the following methods which the Secretary (in the case of a vessel) or the President (in the case of a facility) determines to be acceptable: evidence of insurance, surety bond, guarantee, letter of credit, qualification as a self-insurer, or other evidence of financial responsibility. Any bond filed shall be issued by a bonding company authorized to do business in the United States. In promulgating requirements under this section, the Secretary or the President, as appropriate, may specify policy or other contractual terms, conditions, or defences which are necessary, or which are unacceptable, in establishing evidence of financial responsibility to effectuate the purposes of this Act.[29]

Under this provision, the number of methods of financial responsibility is greatly increased compared to the methods of the CLC 1992 and the Bunker Convention 2001. Financial responsibility can be established by one method or combination of methods; besides insurance, surety bond, guarantee, letter of credit and

28. 33 USC 2716(a).
29. 33 USC 2716(e).

qualification as a self-insurer; almost all certifications of property and economic benefits can be accepted as evidence of financial responsibility. Having multiple methods is more convenient for liable parties and vessels to establish and maintain financial responsibility.

3.1.2 The US Practice in Establishing a Fund for Damage Resulting from Oil Pollution

As mentioned earlier, the purpose of compulsory insurance is to ensure the realization of compensation for damage arising from oil pollution. Without compulsory insurance, the shipowner can easily become bankrupt and therefore unable to provide compensation to victims of oil pollution. Compulsory insurance costs are paid by shipowners. However, to protect the development of shipping industry and set up marine commercial insurance, there is a limitation to the shipowners' liability in all relevant domestic and international laws. This means that in the case of a large oil spill some damage will not be sufficiently recovered. Hence a fund for oil pollution damage is another important device to ensure the realization of compensation for damage arising from oil pollution. The difference with insurance is that the main part of the funds is financed through a tax imposed on petroleum enterprises. OPA 1990 created the Oil Spill Liability Trust Fund and transferred the other funds to this fund.[30] The sources of tax imposition are: (1) crude oil received at a US refinery and (2) petroleum products entered into the United States for consumption, use or warehousing.[31] OPA 1990 provides a limit on the total amount available from the Oil Spill Liability Trust Fund for one incident of USD 1 billion, and within that overall limit restricts damage for injury to natural resources to USD 500 million per incident.[32] Following are the usages of the Oil Spill Liability Trust Fund:

> The Fund shall be available to the President for–
>
> (1) the payment of removal costs, including the costs of monitoring removal actions, determined by the President to be consistent with the National Contingency Plan, . . . by Federal authorities; or . . . by a Governor or designated State official . . . ;
>
> (2) The payment of costs incurred . . . for assessing natural resource damage and for developing and implementing plans for the restoration, rehabilitation, replacement, or acquisition of the equivalent of damaged resources . . . ;
>
> (3) the payment of removal costs determined by the President to be consistent with the National Contingency Plan as a result of, and damage resulting from, a discharge, or a substantial threat of a discharge, of oil from a foreign offshore unit;

30. 26 USC 9509(a)(b).
31. 26 USC 4611(a).
32. 26 USC 9509(c)(2)(A).

(4) The payment of claims in accordance with section 2713 . . . for uncompensated removal costs determined by the President to be consistent with the National Contingency Plan or uncompensated damage;

(5) The payment of Federal administrative, operational, and personnel costs and expenses.[33]

These provisions indicate that the main uses of this fund are cleanup costs, recovery for damage of natural resources and uncompensated damage. In sum, the US fund for oil pollution in the United States provides for a higher level compensation for damage arising from oil pollution, especially providing adequate protection to natural resources and marine environment when they are damaged by oil pollution incidents. The US did not ratify the CLC or the Fund Convention, instead, opting to create an indigenous and 'tough' statute regarding marine oil pollution.[34] The reason why the United States did not join the international convention regime is clearly stated in OPA 1990 as that

> It is the sense of the Congress that it is in the best interests of the United States to participate in an international oil pollution liability and compensation regime that is at least as effective as federal and state laws in preventing incidents and in guaranteeing full and prompt compensation for damage resulting from incidents.[35]

For the United States, the limitation to the total amount available from these conventions is too small to prevent oil incidents and cover the losses resulting from oil pollution.

3.2 Lessons from the US Practice Concerning Compulsory Insurance and the Fund for Damage Resulting from Oil Pollution for China

3.2.1 The Regime of Compulsory Insurance in China and Ways to Improve It

China ratified CLC 1992 and the Bunker Convention 2001. According to the two conventions, tankers over 2,000 tons of international transportation shall be required to maintain insurance or other financial security for cargo oil or bunker oil pollution in China, and all vessels over 1,000 gross tons shall be required to maintain insurance or other financial security for bunker oil pollution from 2009 on China. However, tankers less than 2,000 tons of international transportation, and all of tankers of inland transportation are not required to maintain compulsory insurance for cargo oil or bunker oil pollution. Vessels less than 1,000 gross tons

33. 33 USC 2712(a).
34. *Congressional Record-House*, 3 Aug. 1990, at H6920, per Mr Bonior.
35. Section 3001 of the OPA1990.

are not required to maintain compulsory insurance for bunker oil pollution, except requirements for compulsory insurance for tankers by the People's Congress of Shenzhen City and by the Shanghai Maritime Safety Administration. Based on the experience of the United States, all vessels over 300 gross tons at any place subject to the jurisdiction of the United States are required to establish and maintain financial responsibility. Establishment of compulsory insurance for more vessels and wider geographical scope, not only can ensure the realization of compensation for damage arising from oil pollution but can also strengthen the protection of the marine environment by full and prompt compensation. Especially, as mentioned earlier, as the interests of marine-related industries becomes larger and more important in China, marine oil pollution issues are becoming increasingly more serious; therefore, China should establish a regime of compulsory insurance for all vessels as soon as possible. For the methods of establishment of compulsory insurance, the experience of the United States is a valuable reference for China. China should consider insurance, surety bond, guarantee, letter of credit and all certifications of property and economic benefits as compulsory insurance with the appropriate specified procedure. Multiple methods of establishing compulsory insurance can make the responsible party capable of maintaining his guarantee for compensation and reduce his costs.

3.2.2 Establishing a Fund for Oil Pollution in China

China is a State party to the Fund Convention 1992, but the Fund Convention 1992 applies to Hong Kong only; it does not apply to the mainland of China. There is no regime of fund for oil pollution existing in mainland of China. As mentioned earlier, the United States did not ratify the oil pollution – related conventions because they think that these conventions do not provide full and prompt compensation for damage resulting from oil pollution and are not effective in preventing oil incidents. In fact, the degree of protection under CLC 1992 and the Fund Convention 1992 is too low for the United States and the interests of the US victims and the US marine environment needs to be protected at a higher level. This is the reason why the United States enacted OPA 1990. On the contrary, ten or more years ago China was not a country with a large amount of oil transportation. As a low-level developing country, it did not sufficiently understand the importance of marine environmental protection. China considered that it was not favourable to join the Fund Convention 1992/1971 because it would have the duty to contribute fees to the fund and could receive equal benefit. So it is also not a surprise that China did not establish a fund for oil pollution at that time. However, with the rapid development of the economy, the situation has totally changed in China: the marine-related industries are more and more important in the tertiary industry; China has become a large oil-importing country, and the government, the society and most people have now realized the importance of marine environmental protection. The situation and national interests of China are increasingly similar to those of the United States, and it is therefore time to establish a fund for oil pollution for China. Basically, it is cost-effective for China to join the Fund

Convention 1992 now or China should consider legislating a fund for oil pollution, with higher limitation and wider scope of compensation as well, with the reference to the United States in providing for a higher level protection to the victims, natural resources, marine environment and economy of China.

4 CONCLUSION

The US regime of compensation for damage resulting from oil pollution is different from the regimes of most other countries and international conventions. The compensation for pure economic loss arising from oil pollution and compulsory insurance and fund for oil pollution of the US are typical legal systems thereof. The reasons why the United States abrogated such a basic principle originating in English courts during the twentieth century and did not ratify CLC 1992 and the Fund Convention 1992/1971, and instead, decided to create an indigenous and tough statute, are: (1) the United States is the largest oil-importing country and at the same time is not a dominant shipping country; (2) the United States thinks that marine environmental protection is more important than the protection of the oil shipping industry and a higher level marine environmental protection is consistent with US national interests; and (3) the United States also thinks that a heavy responsibility of shipowners is effective in preventing oil incidents and guaranteeing full and prompt compensation for damages resulting from oil pollution. Because China is experiencing rapid development, the Chinese situation of a large amount of oil imports, oil transportation and marine environmental protection are increasingly more similar to those of the United States and the interests of the marine-related industries are growing as well. The experience of the United States becomes a valuable reference to China. China should provide for a larger scope of compensation for damages resulting from oil pollution, requiring more tankers and other vessels to maintain compulsory insurance by various methods and establishing the regime of a fund for oil pollution to protect the overall national interests of China.

References

Holmes, O.W. *The Common Law*, edited by M. Howe. Vol. I. Boston: Little Brown, [1881] 1963.

Kende, C.B. 'The United States Approach'. In *Liability for Damage to the Marine Environmental*, edited by C.M. De La Rue. London: Lloyd's of London Press Ltd, 1993.

Chapter 17

Study on Ocean Pollution Damage Legislation between the United States and China: A Comparative Perspective

*Guo Ping**

1 INTRODUCTION

Environmental pollution is much more than a domestic nuisance. It ignores boundaries and therefore poses problems that require world solutions. Oil pollution from ships was the first form of pollution to be recognized as an international problem, and the US Government convened in 1926 the Washington Conference in an effort to determine solutions. It was probably typical of the problems that were to be encountered in effective protection for the environment that the draft convention it produced was never opened for signature. A further attempt by the British Government for another conference in 1935 under the auspices of the League of Nations never came to fruition because of the likelihood of war. Until 1954, besides the Brussels Convention on the Prevention of pollution of the Sea by Oil (OILPOL), only some wildlife conservation treaties existed, and they were of questionable success.

* Professor of Maritime Law and Environment Law, Fulbright Scholar, Law School of Dalian Maritime University, member of CMAC and IIML.

Michael G. Faure, Han Lixin & Shan Hongjun, *Maritime Pollution Liability and Policy: China, Europe and the US*, 299–320.

The marine environment is polluted by a variety of sources – wastes are intentionally dumped at sea, including sewage sludge, chemicals, industrial wastes and so forth. Pollution is also deposited into the marine environment from the atmosphere. The most serious pollution, however, comes from land-based sources such as industrial and municipal waste that is discharged at the mouth of rivers, ocean out-falls and run-off from crop lands, urban areas, construction and mining sites. Ocean shipping accounts for approximately half of all marine pollution, primarily oil pollution. Most of this comes from standard ship operations such as deballasting and tank washing; only about 10% of marine oil pollution results from tanker accidents.[1]

2 THE LEGISLATION OF THE UNITED STATES ON MARINE POLLUTION DAMAGE

Environmental protection laws reflect a sense of community, wherein the health and safety of individuals are as important as the economic well-being of the nation as a whole. This philosophy was incorporated into one of modern federal environmental statutes, the National Environmental Policy Act of 1969 (NEPA), which established a national policy to ensure for all Americans safe, healthful, productive and aesthetically and culturally pleasing surroundings.[2]

The Environmental Protection Agency (EPA) was created in 1970 and administers nine major environmental statutes passed by Congress between 1969 and 1980: the Clean Air Act; Federal Water Pollution Control Act; Comprehensive Environmental Response; Compensation and Liability Act; Marine Protection, Research and Sanctuaries Act; Solid Waste Act (more commonly known as the Resource Conservation and Recovery Act); Federal Insecticide, Fungicide and Rodenticide Act; Toxic Substances Control Act; Noise Control Act; and Safe Drinking Water Act. The Departments of Interior, Commerce, Transportation, Labor, and the Council on Environmental Quality have primary responsibility for administering numerous other environmental laws, such as Coastal Zone Management Act, Deepwater Port Act, Endangered Species Act, Federal Land Policy and Management Act, Fishery Conservation and Management Act, Marine Mammal Protection Act, National Environmental Policy Act, National Forest Management Act, National Park System Mining Regulation Act, National Wildlife Refuge System Act, Occupational Safety and Health Act, Oil Pollution Act of 1990, Outer Continental Shelf Lands Act Amendments of 1978 (OCSLA), Shore Protection Act and Wild and Scenic Rivers Act.[3]

1. T.J. Schoenbaum, *Admiralty and Maritime Law* (Hornbook Series), 3rd edn (West Group Publishing, 2001), 876.
2. S. Ferrey, *Environmental Law: Examples and Explanations*, 2nd edn (New York: Aspen Publishers, 2004), 2.
3. *Ibid.*, 3–4.

There are various legal regimes that impose liability for maritime pollution. The main are based on federal statutes.[4] There are three federal statutes that play important roles for pollution damages resulting from maritime activities. They are Oil Pollution Act of 1990 (OPA), Clean Water Act (CWA) and Comprehensive Environmental Response, Compensation, and Liability Act (CERCLA).

The Federal Water Pollution Control Act (FWPCA) as amended by the CWA in section 311[5] specially provides for oil and hazardous substance liability. Because of the promulgation of OPA 1990, some provisions of the CWA have been superseded or modified by OPA. As to pollution from hazardous substances other than oil, CERCLA plays a role as the supplement of the CWA. The former did not expressly repeal or supersede the latter, so in fact both statutes co-exist.

Under pre-OPA federal law, at least five statutes dealt with the issue of oil spill liability and compensation. Each was different, and each was inadequate. The FWPCA (or CWA), which has jurisdiction over most large tanker and inland barge incidents, has historically provided only partial protection.

On 24 March 1989, the *Exxon Valdez* ran aground on Bligh Reef in Prince William Sound, creating was the largest oil spill more than 261,904.76 barrels from a vessel in US history. During the weekend of 23–24 June 1989, three other incidents involving oil tankers occurred.[6] Investigation of the possible causes of these incidents revealed that there appeared to be similarities among these incidents, and these incidents and the *Exxon Valdez* spill demonstrated that the system of regulation and control in place at the time could be improved to prevent future spills and to enhance the capability of oil tanker operators, the oil industry and others to respond effectively to those spills that do occur despite preventive measures.[7]

Congress thought that the goal of compensating those injured may be complicated by questions of jurisdiction of the various federal agencies. To buttress this point, Congress cited as an example the fact that when an oil spill occurs it may be necessary to track down the source of the spill before the applicable statute is applied and compensation claims presented.[8] It also noted that, as documented in a report prepared by the US General Accounting Office (GAO) in 1983, gross inadequacies exist in the FWPCA's provisions dealing with the spiller's responsibility for cleanup costs. In individual cases the percentage of cleanup costs expended by the government that was recovered in courts was even lower. Funds

4. R. Force & J. Force, *Marine Pollution* (vol. 1) materials used for students in fall 2007, law school of Tulane University, 4.
5. See 33 USCA § 1321.
6. These accidents resulted in the spilling of more than 34,000 barrels of oil in US waters. In Rhode Island Sound, a Greek-registered tanker grounded and spilled approximately 10,000 barrels of No. 2 heating oil. A barge collided with an oil tanker in the Houston Ship Channel in Texas, and released 4,000 barrels of crude oil. In the Delaware River, a Uruguayan tanker grounded and spilled 20,000 barrels of heavy oil.
7. O.O. Awoyemi, *The Anatomy of United States' Oil Pollution Act of 1990*, SJD dissertation, 2002, Tulane University School of Law, 7.
8. *Ibid.*, 8–10.

also were expended in the costly, time-consuming litigation required to recoup cleanup expenditures.

Congress also thought that another problem with the FWPCA was that its section 311(k) on revolving fund was inadequate to cover large oil spills such as in the *Exxon Valdez* incident.

The court had found that the FWPCA was the only statute to provide a remedy for recovering oil spill cleanup costs. The government could not use common law nuisance or maritime tort theories to recover its excess costs.

Each of these inadequacies of the FWPCA was adequately addressed in OPA 1990. On 18 August 1990, President George Bush signed OPA 1990 into law. It sets greater monetary liability limits for spills and assesses liability for a broader area of damages. At the same time it allows states to enact or maintain more stringent liability laws. OPA 1990 greatly improved the previously existing scheme, but it still fails to properly define its relationship with international and state law. By maintaining the state's rights to set different and higher liability limits, it fails to set a uniform standard on which oil carriers and producers can depend. Additionally, by failing to conform to international agreements, it isolates US law from the international legal community.[9] The United States has not approved any international convention on civil liability for pollution from ships that are mostly accepted by the international community to date, although it signed the 1969 International Convention relating to the Intervention on the High Sea in Cases of Oil Pollution Casualties and actively participated in the development of its failed 1984 Protocol. However, the Senate still refused to pass legislation that would preempt state oil pollution laws. Thus, the House conceded, and the Oil Pollution Act of 1990 passed, ending any real chance for a future US adoption of CLC.

In fact, OPA 1990 superseded three previous federal statutes – the Trans-Alaska Pipeline Authorization Act (TAPAA), which established the Trans-Alaska Pipeline Liability Fund[10]; the Deepwater Port Act (DPA), which established the Deepwater Port Liability Fund[11]; and the 1978 Amendments to the Outer Continental Shelf Lands Act (OCSLA) which created the Offshore Oil Pollution Compensation Fund,[12] which imposed strict liability for oil pollution in specified circumstances.[13] All of these previous statutes provided varying and uneven liability standards and scope of coverage for cleanup costs and damages associated with activities covered by each individual.

OPA 1990 contains liability limits that are significantly higher than the limits contained in CWA, FWPCA and other previous legislations. The OPA 1990 limit for vessels is a 3,900% increase on the old liability limits of the FWPCA.[14]

---

9. P. Peterson & D. Kopec, 'Crude Legislation: Liability and Compensation under the Oil Pollution Act of 1990', *Rutgers Law Journal* 23 (1992): 599.
10. See 43 USCA § 653.
11. See 33 USCA § 1517.
12. See 43 USCA § 1812 (a).
13. R. Force & J. Force, *supra* n. 4, 4.
14. Awoyemi, *supra* n. 7, 341.

The US Congress amended OPA 1990 on 20 November 1995, via the Edible Oil Regulatory Reform Act. This amendment resulted in the amendment of sections 2704(a)(1) and 2716(a) of OPA 1990.[15] In 1998 a further amendment added subsection (c)(4) to section 2704.[16]

Apart from those mentioned previously, two other sources should not be ignored – general maritime law and state law. An action may be brought to impose liability for damages resulting from an oil or hazardous substance discharge on the basis of nuisance theory either in a federal district court under the court's admiralty jurisdiction or in a state court under the 'saving to suitors' provision. Also, the action may be brought based on state law, under either common law principles or special legislation enacted in some states.

In general, from the aspect of federal statutes in United State, CWA applies to the liability for pollution damage resulting from both oil and harmful substance, OPA only provides for the liability for oil pollution damage; as for the liability of pollution damage resulting from hazardous substance other than oil, CERCLA shall apply.

There is very little legislative history on CERCLA.[17] This scarcity may be due to the rapidity with which the statute was enacted. Yet, two broad policies of CERCLA have been identified. First, Congress wanted to deal effectively with hazardous waste disposal problems not previously addressed by earlier environmental regulatory statutes, and second, Congress wanted to create an exhaustive system of liability reaching all parties responsible for the improper disposal of hazardous materials.

3 THE MAIN PROVISIONS OF AMERICAN LEGISLATION ON LIABILITY FOR MARINE POLLUTION

3.1 Scope of Application

3.1.1 Section 1321(b)(3) of the CWA

The discharge of oil or hazardous substances (i) into or upon the navigable waters of the United States, adjoining shorelines, or into or upon the waters of the contiguous zone; or (ii) in connection with activities under the OCSLA or the DPA, or which may affect natural resources belonging to, appertaining to or under the exclusive management authority of the United States (including resources under the Magnuson-Stevens Fishery Conservation and Management Act) is prohibited.

15. 33 USCA. § 2720, Pub. L. 104-55, s. 2(d)(1).
16. 33 USCA § 2704(c) (4), Pub. L. 105-383, s. 406(2) added para. 4.
17. F.A. Grad, 'A Legislative History of the Comprehensive Environmental Response, Compensation and Liability (Superfund) Act of 1980', *Columbia Journal of Environmental Law* 8 (1982): 1.

3.1.2 Section 2702 of OPA 1990

Notwithstanding any other provision or rule of law, and subject to the provisions of this Act, each responsible party for a vessel or a facility from which oil is discharged, or which poses the substantial threat of a discharge of oil into or upon the navigable waters or adjoining shorelines or the exclusive economic zone is liable for the removal costs and damages specified in subsection (b) of this section that result from such incident.[18]

It is noted that the vessel under OPA 1990 is not restricted to vessels that carry oil as cargo; it includes all vessels that have oil on board whether as cargo or bunkers.

The definition of 'oil' under OPA 1990 is similar to that of under the CWA. According to section 2701(23) of OPA 1990, the word 'oil' excludes any substance that is specifically listed or designated as a hazardous substance under subparagraphs (A) through (F) of section 101(14) of CERCLA.

3.1.3 Section 9601(8) of CERCLA

The term 'environment' means (A) the navigable waters, the waters of the contiguous zone and the ocean waters of which the natural resources are under the exclusive management authority of the United States, and (B) any other surface water, ground water, drinking water supply, land surface or subsurface strata or ambient air within the United States or under its jurisdiction.

3.2 Scope of Claims for Marine Pollution Damages

Although FWPCA provided for natural resources damages, it failed to address cleanup costs incurred by state governments or private interests and did not provide compensation for other damages suffered. Instead, these issues remained under the control of state regulation or other federal laws applicable only in specific locations. Maritime scholars had long criticized the FWPCA as a woefully inadequate attempt to address the grave problems caused by oil spills in the United States.[19]

Under CWA the owner or operator shall be liable for the costs of removal of oil or hazardous substances and costs or expenses incurred by the Federal Government or any State Government in the restoration or replacement of natural resources.[20]

Both CWA and OPA clarify the meaning of 'removal costs' not only to include the actual costs incurred after the oil or harmful substance is discharged, but also to include the expensed incurred to prevent, minimize or mitigate a substantial threat of a discharge of oil or substance.[21]

18. See 33 USC § 2702.
19. Awoyemi, *supra* n. 7, 12–13.
20. See subs. (f)(4) of CWA.
21. See § 1321 (a)(25) of CWA and § 2701 (31) of OPA.

According to section 2702(b)(1)(2) of OPA 1990, each responsible party for a vessel or a facility from which oil is discharged shall be liable for removal costs and damages (A) for injury to, destruction of, loss of or loss of use of natural resources, including the reasonable costs of assessing the damage, which shall be recoverable by a trustee; (B) for injury to, or economic losses resulting from destruction of, real or personal property, which shall be recoverable by a claimant who owns or leases that property; (C) for loss of subsistence use of natural resources, which shall be recoverable by any claimant who so uses natural resources that have been injured, destroyed, or lost, without regard to the ownership or management of the resources; (D) to the net loss of taxes, royalties, rents, fees or net profit shares due to the injury or loss of real property, personal property, or natural resources, which shall be recoverable by the US Government or a State; (E) to the loss of profits or impairment of earning capacity as a result of the injury, or loss of real property, personal property, or natural resources, which shall be recoverable by any claimant; (F) for net costs of providing increased or additional public services during or after removal activities, including protection from fire, safety or health hazards, which shall be recoverable by a State.

Section 2701(20) of OPA 1990 provides that natural resources includes land, fish, wildlife, biota, air, water, ground water, drinking water supplies and other such resources belonging to, managed by, held in trust by, appertaining to or otherwise controlled by the United States (including the resources of the exclusive economic zone), any State or local government or Indian tribe or any foreign government.

Under section 9607 of CERCLA, the following items shall be claimed against the responsible party: (A) all costs of removal or remedial action incurred by the US Government or a State or an Indian tribe not inconsistent with the national contingency plan; (B) any other necessary costs of response incurred by any other person consistent with the national contingency plan; (C) damages for injury to, destruction of or loss of natural resources, including the reasonable cost of assessing such injury, destruction or loss resulting from such a release; and (D) the costs of any health assessment or health effects study.

3.3 Persons Liable for Marine Pollution Damage

Under CWA, only the owner or operator of a vessel, onshore and offshore facility shall be liable for pollution damages.[22] However, the CWA provides that if it is proved that the discharge was caused solely by an act or omission of a third party, the third party shall be liable for pollution damage.[23]

Under OPA 1990, there are three groups of persons who may be liable: responsible parties, guarantors and third parties. Responsible parties include the owner and operator of a vessel, onshore and offshore facility, deepwater port and

22. See § 1321 (f) (1) (2) (3) of CWA.
23. See § 1321 (g) of CWA, which was superseded as to oil by OPA 1990.

pipeline. Also, it includes a demise charterer.[24] If it is proved that the discharge of oil or hazardous substance was caused by defences to exempt the liability of owner or operator, they may claim for actual removal costs against the US Government.[25]

Under CERCLA, (1) the owner and operator of a vessel or a facility; (2) any person who at the time of disposal of any hazardous substance owned or operated any facility at which such hazardous substances were disposed of; (3) any person who by contract, agreement or otherwise arranged for disposal or treatment, or arranged with a transporter for transport for disposal or treatment, of hazardous substances owned or possessed by such person, by any other party or entity, at any facility or incineration vessel owned or operated by another party or entity and containing such hazardous substances; and (4) any person who accepts or accepted any hazardous substances for transport to disposal; or treatment facilities, incineration vessels or sites selected by such person, from which there is a release or a threatened release that caused the incurrence of response costs of a hazardous substance shall be liable for the removal costs or damages that are covered by CERCLA.[26]

3.4 Defences for Marine Pollution Liability

All federal statutes, such as CWA, OPA 1990 and CERCLA impose strict liability on the owner or operator of a vessel from which oil or hazardous substances were discharged for the costs incurred by the United States in the removal of such oil or hazardous substance. However, there are few defences that the responsible party may invoke.

Subsection (f) of CWA provides that only the following defences are available when the discharge was caused by (A) act of god; (B) act of war; (C) negligence on the part of the US Government; (D) act or omission of a third party regardless of whether such act or omission was negligent; or any combination of these causes.

This provision has been superseded with respect to oil spills by section 2703(a) of OPA 1990. OPA 1990 keeps the first two defences listed under CWA – act of god and act of war. However, OPA 1990 deletes the defence of negligence on the part of the US Government and addresses that an act or omission of a third party shall not be an act or omission of an employee or agent of the responsible party or a third party who has any contractual relationship with the responsible party.[27] Also, OPA 1990 limits the right of the responsible party to invoke the previously mentioned defences if it fails or refuses to report the incident as required by law or to provide all reasonable cooperation and assistance requested by a responsible official in removal activities or to comply with an order issued under Intervention

24. R. Force & J. Force, *supra* n. 4, 10.
25. See subs. (i) of CWA.
26. See § 9607(a) of CERCLA.
27. Contractual relationship includes land contract, deeds, easements, leases or other instruments transferring title or possession. See § 2703(d) (1) and § 2704(c) (1) (2) of OPA.

on the High Seas Act. Even more if the incident was proximately caused by the violation of an applicable federal safety or operating regulation by the responsible party, an agent or employee, the responsible party loses the right to limit its liability.[28] This provision is stricter than that of the CWA.

CERCLA[29] has almost the same provision on the defences for liability as OPA 1990.

3.5 Limitation of Liability for Marine Pollution

In its pre-OPA 1990 deliberations, Congress was emphatic and unequivocal about its burning desire to eviscerate the Limitation of Liability Act of 1851 for oil pollution purposes. Some commentators had pointed out long before OPA 1990 was passed, for example, that the Limitation of Liability Act could be used to shield shipowners from damages arising from 'pollution of navigable waters by oil and other harmful substances'.

Under subsection (f) of CWA, the liability is limited to: USD 125 per gross tonnes or USD 125,000, whichever is greater in the case of an inland barge, USD 150 per gross tonnes of any other vessels or for a vessel carrying oil or hazardous substances as cargo USD 250,000, whichever is greater. The limitation of liability for onshore and offshore facilities is USD 50,000,000.

According to section 2704(a) (1–4) and (b) of the OPA 1990, the limitation amounts are different for other tank vessels, other non-tank vessels, offshore facility, onshore facility and a deepwater port, as shown in Table 17-1.

It is not quite clear whether there is a limitation for liability provided under CERCLA. However, it does provide the limitation of financial responsibility for the owner or operator of each vessel over 300 gross tonnes as USD 300 per gross tonnes or USD 5,000,000, whichever is greater.[30] Also it mentions that the level of financial responsibility may be adjusted to protect against the risk under the discretion of the President after the promulgation of this law[31] which shall not preempt any state from imposing any additional liability with respect to the release of hazardous substances within such state.[32] It can be concluded that there is no limitation of liability for hazardous substances under CERCLA, with one exception – the guarantor may limit his liability within the limitation amount of the financial responsibility mentioned earlier.[33]

OPA 1990 contains liability limits that are significantly higher than the limits contained in the CWA or FWPCA and other previous legislation, and its limit for vessels is a 3,900% increase on the old liability limits of the FWPCA.[34] This

28. See 33 USC § 1471.
29. See § 9607 (b) of CERCLA.
30. See § 9608(a)(1) of CERCLA.
31. See § 9608(a)(4) of CERCLA.
32. See § 9614(a) of CERCLA.
33. See § 9608(d)(1) of CERCLA.
34. Awoyemi, *supra* n. 7, 341.

Table 17.1 OPA 1990 Limitation Amounts

Type of vessels or facilities	*Limitation amount*	*Limitation amount*	*Remarks*
Single hull tank vessel (including one single hull vessel fitted with double sides only or a double bottom only)	USD 3,000 per gross tonnes	USD 22,000,000, if the vessel is greater than 3,000 gross tonnes	Whichever is greater
		USD 6,000,000, if the vessel is 3,000 gross tonnes or less	Whichever is greater
Other tank vessels	USD 1,900 per gross tonnes	USD 16,000,000, if the vessel is greater than 3,000 gross tonnes	Whichever is greater
		USD 4,000,000, if the vessel is 3,000 gross tonnes or less	Whichever is greater
Other non-tank vessels	USD 950 per gross tonnes	USD 800,000	Whichever is greater
Offshore facility		All removal costs plus USD 75,000,000	
Onshore facility and a deepwater port		USD 350,000,000	
Mobil offshore drilling units	Respect to discharge, or the substantial treat of discharge of oil	Same as tank vessel	
	Respect to removal costs and damages from any incident	Same as offshore facility	

monumental increase in the limits on liability appears to be too drastic, considering the fact that Congress could have introduced gradual increases in the liability limits, which could be subject to upward or downward review depending on the prevailing circumstances at the time. Congress decided to do the opposite by allowing only upward adjustment of the liability limits to reflect the consumer price index. Section 2704(d)(4) provides that the President shall, by regulations issued not less often than every three years, adjust the limits of liability specified in subsection (a) to reflect a significant increase in the consumer price index. In 1991, the President by an executive order delegated this duty to the Secretary of Transportation.[35]

4 THE LEGISLATION OF CHINA ON MARINE POLLUTION AND DAMAGE

From the late 1970s, China has enacted a series of basic laws and regulations relating to marine environmental protection and marine activities.[36] They[37] are the Marine Environmental Protection Law (MEPL),[38] Regulations Concerning Prevention of Pollution of Sea Areas by Vessels (Vessel Pollution Prevention Regulation),[39] Regulation Concerning Prevention of Environmental Pollution by Ship-Breaking (Ship-Breaking Pollution Prevention Regulation),[40] Regulation Concerning the Environmental Protection in Offshore Oil Exploration and Exploitation (Offshore Oil Exploration and Exploitation Regulation),[41] Regulation Concerning the Dumping of Wastes at Sea (Dumping Regulation),[42] Regulation Concerning Prevention of Pollution Damage to the Marine Environment by Land-based Pollutants (Land-based Pollution Regulation),[43] Regulation of Preventing Pollution of the Changjiang River Water Areas from the Garbage of Ships and Solid Wastes along the River Bank,[44] Regulation Concerning Prevention of

35. Ex. Ord. No. 12777, 18 Oct. 1991, 56 F.R. 54757.
36. H. Shan, *Comparative Study of Int'l American and China OPLRs – To Establish China Comprehensive OPLR*, PhD dissertation (Tulane University School of Law, 2007), 20.
37. P. Guo & L.J. Zhao, 'The Legal Protection of the Marine Environment in China: Current Situation and Challenges', in *Prevention and Compensation of Marine Pollution Damage: Recent Developments in Europe, China and the US*, ed. M.G. Faure & J. Hu (Alphen aan de Rijn: Kluwer Law International, 2006), 288–292.
38. MEPL was originally enacted on 23 Aug. 1982, which is superseded by 1999 amendment.
39. It was enacted on 29 Dec. 1983.
40. It was promulgated by State Council on 18 May 1988.
41. It came into force from the same date on which the rule was enacted. The detailed rules had been promulgated on 20 Sep. 1990 by State Oceanic Administration with the title of 'Implementing Rules of the Administrative Regulation on the Environmental Protection on Ocean Oil Exploration and Exploitation'.
42. It was issued by State Council on 6 Mar. 1985 through Ord. No. 34 and been enforced from 1 Apr. 1985.
43. It was issued by State Council on 22 Jun. 1990 through Ord. No. 61 and been enforced from 1 Aug. 1990.
44. It came into force from 1 Mar. 1998. The title of the law is translated by the author from the original Chinese version of the law.

Pollution Damage to the Marine Environment by Coastal Construction Projections[45] (Coastal Construction Pollution Regulation), Regulation on Preventing and Curing Ocean Pollution from Ocean Construction Projects[46] (Ocean Construction Pollution Regulation) and so on.

Obviously, almost all of the laws and regulations mentioned, which are seldom concerned with civil liabilities of wrong-doers, focus on the administration of marine pollution protection mostly, and it is also clear that the laws and regulations were made according to various sources of pollution, which resulted in the different competent administrative authorities.

According to Article 5 of MEPL, the Ministry of Environmental Protection (MEP)[47] is in charge of the harmonized administration of national environmental protection and the environmental protection for preventing and curing maritime pollution damages from land sources and costal construction projects. The State Oceanic Administration (SOA) is in charge of administration of the ocean environment, arranging the investigation, monitoring, watch keeping, evaluation and scientific research of maritime environment and preventing and curing marine pollution damage from ocean construction projects and dumping wastes at sea. The Maritime Safety Administration (MSA) is liable for the administration of marine environmental pollution caused by non-military ships in the port of jurisdiction and ships other than fishery ships, military ships outside the port of jurisdiction, as well as for the investigation of pollution accidents. The State Fishery Administration is in charge of the administration of marine environmental pollution caused by non-military ships in the fishery port of jurisdiction and the fishery ships outside the fishery port of jurisdiction, as well as the investigation of pollution accidents. The administration and management of marine environmental pollution concerned with military ships, as well as the investigation of pollution accidents, is under control of the Department of Military Environmental Protection.

Obviously, there are different competent authorities involved under the various Regulations and Rules mentioned previously. Furthermore, even under the same applicable Regulations, there are various competent authorities to manage a case involving marine pollution damage. For example, the provision of Article 3 of Ship-Breaking Pollution Prevention Regulation indicates that all of the authorities under the power of MEP, MSA, State Fishery Administration and the Department of Military Environmental Protection may have the right to manage and control environmental protection against pollution damages from the dismantling of ships according to the division of the work.

45. It came into force 1 Aug. 1990, which had been amended on 25 Sep. 2007 by State Council through Ord. No. 507. The amendment came into force 1 Aug. 2008.
46. Translated by the author from the original Chinese version. It was issued by State Council on 19 Sep. 2006 through Ord. No. 475 and been in force since 1 Nov. 2006.
47. The new name was used from March 2008 instead of State Bureau of Environmental Protection according to Reform Plan of Chinese government.

5 THE MAIN PROVISIONS OF CHINESE LEGISLATION ON LIABILITY FOR MARINE POLLUTION

5.1 CHINA MARITIME CODE

The China Maritime Code (CMC) is silent on the liability for marine pollution, except for one clause that is concerned with the limitation of liability in Chapter XI of CMC.[48] If a claim is within the scope of one international convention on the civil liability for oil pollution damage that China enters, the responsible party may not limit his liabilities according to Chapter XI of CMC, but within the amount of limitation under such convention.[49] If one maritime claim is outside the scope of the 1992 CLC Convention, such as pollution damage caused by bunker[50] or harmful and noxious substance carried by ships, the liability of such pollution damage may be limited under Chapter XI of CMC if it is caused by ships defined in CMC.

5.2 MEPL

MEPL is made for protecting and improving the marine environment, protecting marine resources, maintaining ecological balance, preventing and curing pollution damage, ensuring human health, promoting sustainable development of economy and society. Therefore, it aims at the prevention and cure of marine pollution damage from different sources, such as land originals, coastal construction projects, ocean construction project, dumpling wastage, shipping or relative activities. There is only one general provision in Article 66, which holds that the State shall establish and enforce a legal system of civil liability for oil pollution arising from the vessel, set up an oil pollution insurance system and a compensation fund mechanism, according to the principle that the tanker owner and the cargo owner shall bear the oil pollution risks in common. MEPL also holds that the infringer who damages the ocean environment shall eliminate the harm and be liable for the losses or damages thereof.[51] If it was proved that the pollution damages were

48. The title of Ch. XI of CMC is 'Limitation of Liability for Maritime Claims'. China is not a country member to the 1976 LLMC, but the provisions of Ch. XI of the CMC are quite similar to those of the 1976 LLMC. Art. 207 is concerned with those maritime claims that may limit under CMC; however, Art. 208 is concerned with unlimited claims under Chapter XI. Art. 208(2) provides that the claims for oil pollution damage under the International Convention on Civil Liability for Oil Pollution Damage to which the People's Republic of China is a party shall not be limited according to the provisions in Ch. XI.
49. The convention refers to the 1992 CLC, and China is a country member of the CLC, which has effects in China since 5 Jan. 2000.
50. Although China has approved Bunker Convention 2001, because the latter has no independent system for limitation of liability, the limitation of liability for bunker pollution damage shall be ascertained according to relative provisions under Ch. XI of CMC in China.
51. Article 90 of 1999 MEPL.

caused completely by the intention or negligence of a third party, then the third party shall be liable for compensation liability. According to Article 92, the responsible party may invoke the following to exempt liability if the marine pollution damages are inevitable after taking timely and reasonable measures: (a) war; (b) inevitable nature power; (c) a negligent act or any omission of the public authorities that are in charge of a lighthouse and any other navigable aids in performance of their duties. It is clear that the civil liability does not exclude the application of an administrative penalty or the criminal liability of the responsible party.[52]

5.3 Vessel Pollution Prevention Regulation

The Vessel Pollution Prevention Regulation (hereunder: Old Regulation) was firstly enacted in 1983 according to the 1982 MEPL, while the latter was amended in 1999 and some authorities are preparing to amend the Regulation in past years. The Regulation was finally amended in 2009 and came into force from 1 March 2010. Since the amended Regulation (hereunder: New Regulation) changed a lot not only on the words of title of the Regulation,[53] but also on the provisions, this paper analyses on the basis of provisions of both old version and new amended one.

Under Article 4 of the Old Regulation, no ship in Chinese sea areas or ports is allowed to discharge oil, oily mixture, wastage or any other harmful and noxious substance against the provisions of MEPL, as well as this regulation. There are also a few provisions concerned with the civil liability of the infringer. The provisions of Articles 12 and 13 of Old Regulation concern providing security or liability insurance for pollution damage. If a pollution accident occurs, the shipowner or infringer shall provide financial security for possible pollution liability before the beginning of the voyage. For a tanker that is more than 2000 tonnes and engages in international trade this Regulation shall apply, as well as the CLC Convention, which China approved.

Chapter X of the Old Regulation is specifically concerned with damage claims arising from ship pollution accidents. Any entity and individual who suffered marine pollution damage caused by ships may litigate in a court according to the rules of the MEPL. However, it provides that disputes concerning liability and the amount of damages may be conciliated by the competent authority. If the claimant does not accept the decision of the competent authority, he may take further action in court. However, the claimant may take action directly in court or in an arbitration committee if a foreign factor is involved. If the claimant applied to the competent authority to settle the case, he shall report the pollution damage in writing according to the requirements under Article 41 of Old Regulation. If the shipowner hopes to invoke defence for exempting the liability for pollution damage, a report shall be

52. See Arts 93 and 94 of MEPL.
53. The title of New Regulation is Regulations Concerning Prevention and Cure of Pollution of Ocean Environment by Vessels.

made to the competent authority in writing to certify that the pollution damage is within one of the exceptional conditions under MEPL.[54]

There is only one article dealing with the removal cost of pollutants. However, it only provides that any entity and individuals who took part in removing pollutants shall report in a timely manner to the competent authority in writing, according to the requirement of the Vessel Pollution Prevention Regulation once the removing activities are finished if they claim for any removing costs.[55] It implies that the competent authority and not the responsible party shall pay for removing costs first. It is also not clear whether the claimant may claim removal costs directly from the responsible party or the competent authority may claim removal costs from the responsible party on behalf of the claimant.

However, the Chapter VII of New Regulation concerns specifically with pollution damage in vessel accident, which is amended according to 1999 MEPL, particular Articles 90 and 92 of 1999 MEPL have been provided in New Regulation fully and completely. In addition, Article 52 of the New Regulation provides clearly that the limitation of liabilities of vessel pollution damage shall be ascertained according to CMC. If the pollution damage occurs within the oceanic areas under the jurisdiction of Chinese government which resulted from the bulk consistent oil substances on board of the vessel, then the limitation of liabilities for pollution damages shall be governed by some International Conventions approved by Chinese government. Until now, China is the membership of CLC Convention 1992 as well as Bunker Convention 2001.[56] All sea-going vessels navigated in the ocean under the jurisdiction of Chinese government shall insure the pollution liability or provide available financial security, the only one exception is those vessels carrying on non-oil substance which gross tonnage is less than 1,000.[57] Further, Article 54 of New Regulation provides the shipowner of all Chinese vessels shall apply for civil liability insurance certificate or other financial security certificate for oil pollution damage to the authority of Maritime Safety Administration. And it is the first detailed provisions on pollution liability insurance for the shipowner in China.

In addition, Article 55 of New Regulation provides that the necessary costs for removing pollutants and for emergency disposing occurred by any entities organized by the Nation, shall has priority to other costs or damages happened in oil pollution accident. It is considered to be helpful for the Nation to take quick and reasonable measures to protect ocean environment.

As for the solving method for the disputes of pollution damages, Article 57 of New Regulation provides that the claimant may choose to apply for the mediation by the authority of SMC, or trigger the arbitration process or court litigation process separately and individually.

---

54. See Arts 41 and 44 of Old Vessel Pollution Prevention Regulation.
55. See Art. 42 of Old Vessel Pollution Prevention Regulation.
56. It should be noted that China also approved Fund Convention 1992 only on the behalf of Hong Kong, therefore, Fund Convention 1992 does not bind the mainland of China now.
57. See Art. 53 of New Vessel Pollution Prevention Regulation.

5.4 Ship-Breaking Pollution Prevention Regulation

The Ship-Breaking Pollution Prevention Regulation applies to all dismantling activities on shore or on water in China. Every entity or individual who suffers directly pollution damage arising from dismantling is entitled to claim for damages against the responsible party. Like the provisions of the Vessel Pollution Prevention Regulation,[58] the competent authority may settle the case under the demand of the claimant on the disputes of liability as well as the amount of the damages. Also, the claimant may[59] choose to sue directly in a court. According to Article 25 of the Ship-Breaking Pollution Prevention Regulation, if the pollution damage is caused by an irresistible natural disaster and may not be avoidable after necessary measures had been taken in a timely manner to prevent or minimize the loss or damage, the responsible party may exempt its compensation liability. Obviously the exemption of the Ship-Breaking Pollution Prevention Regulation is narrower than that of MEPL. There is no provision under the Ship-Breaking Pollution Prevention Regulation related to the removal cost for pollution damage.

5.5 Offshore Oil Exploration and Exploitation Regulation

If the pollution damages occurred by activities of oil exploration and exploitation at sea, the claimant shall demand the damage claim to the competent authority under Article 32 of Law on Environmental Protection (LEP) and Article 42 of MEPL. Furthermore, the claimant shall report to the competent authority in writing according to the requirements of this Regulation.[60] There is a provision on removal costs in this Regulation similar to that of 1983 Vessel Pollution Prevention Regulation.[61] On points of dispute settlement and exemption pollution liability, this regulation has almost the same provision as of 1983 Vessel Pollution Prevention Regulation.

On 20 September 1990, the Implementation Rules of Regulation concerning the Environmental Protection in Offshore Oil Exploration and Exploitation were promulgated. The provision of Article 28 of the Implementation Rules indicates that compensation liabilities include: (a) the damages of seawater quality and biological resources caused by pollution damage of marine environment by the action of operators, and the costs taken by the innocent party for removing and curing pollution; (b) loss of revenue of the innocent party caused by pollution damage of marine environment by the action of operators, the repairing costs for damaged equipment or instruments and the costs arising from necessary preventing measures taken by the innocent party; (c) the investigation costs for settling the disputes of pollution damages caused by ocean oil exploration and exploitation. It

58. It means 1983 Vessel Pollution Prevention Regulation.
59. Emphasis by the author.
60. See Art. 22 of Offshore Oil Exploration and Exploitation Regulation.
61. See Art. 23 of Offshore Oil Exploration and Exploitation Regulation.

is the first to clarify the scope of compensation liability for marine pollution damages in related Regulations and Rules. Further, it is the first to provide that the time-bar is three years, which shall begin on the date the claimant knows or might know the pollution damage in domestic laws. Except for the innocent party, each entity and individual involved in removing pollution operation may deliver the report to a competent authority. As for the settlement of disputes for marine pollution, the Implementation Rules of Offshore Oil Exploration and Exploitation Regulation have almost the same provision as other Regulations mentioned earlier.

The exemption provisions under the Implementation Rules of Offshore Oil Exploration and Exploitation Regulation are slightly different from others mentioned previously. The responsible party may invoke one of the following to exempt its liability if the pollution damage is caused by (a) war, (b) irresistible nature power or (c) the intention or negligence of third party completely. Also, he shall demonstrate that the marine pollution damage caused is unavoidable, although he had taken necessary measures.[62] If it is proved the pollution damage is caused completely by a third party, the third party shall be liable for compensation. The competent authority may decide whether the responsible party is exempted from liability on the basis of an investigation after the responsible party hands out the report to the authority for exemption.

5.6 Dumping Regulation

In order to control the dumping of wastes at sea and prevent marine environment pollution damage and maintain an ecological balance, the Regulation concerning the Dumping of Wastes at Sea was drafted in 1985. The Dumping Regulation excludes the ordinary operational discharge wastes from ships, planes or any other transportation instruments and equipments. The ships, planes, platforms and any other transportation instruments shall be exempted from compensation liability, if the action of dumping wastes at sea is within the scope of the exemption of MEPL. All dumping of wastes at sea shall be permitted by competent authority according to this Regulation. If illegal dumping of wastes at sea was made for saving life at sea or meeting the emergency circumstances, the person who actually dumped the wastes and the beneficiary of life-saving and emergency shall pay for pollution damage caused by the dumping. If it is proved that the pollution damage is caused by the negligence of a third party, the competent authority may command the third party to pay for pollution damages after the causation was confirmed.[63]

The competent authority may command the responsible party to pay the costs for removing pollutants and the losses or damages suffered by the innocent party. However, the competent authority may impose a warning or charge a fine of no more than Chinese Yuan (CNY) 100,000 to the infringer whose illegal dumping

62. See Art. 31 of Implementation Rules of Offshore Oil Exploration and Exploitation Regulation.
63. See Art. 15 of the Dumping Regulation.

wastes at sea caused marine pollution damage according to the situation and the degree of pollution damage.

Like the Implementation Rules of Offshore Oil Exploration and Exploitation Regulation, the Implementation Rules of Regulation concerning the Dumping of Wastes at Sea were promulgated in 25 September 1990. Almost all of the provisions under these Rules related to the time-bar, exemption of pollution liability, method to settle disputes thereof, third party liability and so forth are very similar to those of the Implementation Rules of the Offshore Oil Exploration and Exploitation Regulation,[64] except for the provision of compensation liability. Under the Implementation Rules of the Dumping Regulation, it is held that the compensation liability includes (a) the costs taken by an innocent party for removing and curing pollution and the costs arising from necessary preventing measures taken by an innocent party; (b) the economic losses of property caused by pollution and the damages of seawater quality and biological resources; (c) the investigation costs for settlement of the disputes of pollution damages caused by dumping wastes at sea. Comparing the Implementation Rules of Offshore Oil Exploration and Exploitation Regulation with the Implementation Rules of the Dumping Regulation, there are two points of difference. The latter provides that the economic losses of property, the former provide the losses of revenue of an innocent party; the former confirms that the repairing costs for damaged equipments or instruments caused by marine pollution, whereas the latter is silent on this point.[65]

5.7 Land-Based Pollution Regulation

The Land-Based Pollution Regulation regulates marine pollution damage caused by discharging land-source pollutants. There are few provisions concerned with civil liability for discharging land-source pollutants. According to Article 30, if the marine pollution damage from the discharging of land-source pollutants caused the serious economic losses, the competent authority may charge a fine calculated according to 30% of the direct losses of, which shall not be more than CNY 200,000. The penalty of the fine shall not exclude the responsible party to minimize the pollution and compensate for pollution damages.

5.8 Coastal Construction Pollution Regulation and Ocean Construction Pollution Regulation

The Ocean Construction Pollution Regulation is concerned with preventing and curing marine pollution from ocean construction projects. Article 27 provides that the entity for exploration and exploitation of oil, gas and mineral resources at sea shall insure civil liability for pollution damage. The responsible party shall

64. See Arts 38 and 39 of the Implementation Rules of the Dumping Regulation.
65. See Art. 37 of the Implementation Rules of the Dumping Regulation.

minimize the threat of pollution and pay for pollution damages caused by disobeying these Rules in ocean construction projects. If it is proved that the damage is caused by the intention or negligence of a third party, the third party shall be liable for compensation.[66]

On the other hand, the Coastal Construction Pollution Regulation is concerned only with preventing and curing ocean pollution from coastal construction projects. It hence does not deal with civil liability for pollution damage.

6 COMMENTS ON THE CHARACTERISTICS OF AMERICAN LAWS AND CHINESE LAWS

Comparison and analysis of US law and Chinese law on the field of marine pollution damage may reveal many differences between them.

6.1 CHARACTERISTICS OF US LAW

On the basis of studying and analyzing the US Federal Statutes on maritime pollution damage, we may come to the following conclusions:

(1) In the United States, there are special laws that are divided by different pollutants on marine pollution damage – oil and harmful and noxious substances.

(2) All of the US laws provide strict liabilities for marine pollution damage borne by the responsible party with 3 or 4 defences that are slightly different in CWA, OPA 1990 and CERCLA.

(3) In addition to the shipowner or operator of the ships or any other facilities, anyone who is involved in marine pollution damage shall be liable for compensation.

(4) CWA, OPA 1990 and CERCLA each provide that the responsible party may limit the liabilities for maritime pollution damages, although the amount of limitation under each statute is different. And also the federal statutes do not affect the right of each state to provide even higher amounts of limitation.

(5) The claims listed in each federal statute are clear and detailed. Although they are different in federal statutes, the claims include at least removal costs, the loss of or damage to natural resources, the damages suffered by the claimant, the expenses to respond according to emergency plan and accessing costs and so forth.

(6) The US Federal Statutes on marine pollution damage do not exclude each State promulgating State statutes and regulations to provide wider kinds of claims and higher amounts of compensation for oceanic environmental protection.

66. See Art. 56 of the Ocean Construction Pollution Regulation.

It is believed that the US legislation regime is not the best. Some deficiencies exist in US legislation, for example, possible conflicts among CWA, OPA 1990 and CERCLA; and it is unforeseeable for the responsible party to bear the pollution liabilities if each state law applies in addition to federal statutes. However, it is not a bad example to establish domestic legislation on marine pollution damage.

6.2 Characteristics of Chinese Laws, Regulations and Rules

There is no special law or regulation on the civil liability for environmental pollution damage, let alone on marine pollution damage. On the basis of studying and analyzing applicable Chinese laws, regulations and rules on maritime environmental protection, we may come to the following conclusions.

(1) Almost all Chinese laws, regulations and rules were promulgated for the purpose of the administration of oceanic environmental protection, on the basis of different sources, which caused marine pollution damage not on the basis of pollutants. Not only the civil liabilities are provided under various Chinese laws, regulations and rules on marine pollution damage but also the criminal liabilities and administrative liabilities are provided, despite that only a few provisions are concerned with civil liabilities.

(2) Except the litigation, it is the competent authority that may settle disputes arising from pollution damage and can ascertain the liability under certain regulations and rules, particularly those made before 1990. Nothing is mentioned on this point in the regulations recently amended or drafted.

(3) Under various regulations and rules, there are different competent authorities that do not facilitate marine environmental administration enforcement and may lead to some problems for a complete protection of the marine environment.[67]

(4) The kinds and the scopes of claims under different laws, regulations and rules are various. The removal costs and the damages suffered directly by the innocent party may be compensated without any doubts. As for the costs to prevent pollution, accessing costs in investigation and expenses in restoring and recovering of natural resources and so forth, the situation is not quite clear.

(5) All Chinese laws, regulations and rules impose a strict liability on the responsible party as well, like the US Federal Statutes. However, the defences that the responsible party may invoke differ between the MEPL and other regulations or rules.

(6) Except the provisions on the limitation of liabilities of shipowners, bareboat charterers and operators under CMC, or the limitation of liabilities of shipowner under CLC, there are no any other provisions on limitation of

67. Guo & Zhao, *supra* n. 37, 293–297.

liabilities in other Chinese laws, regulations and rules. It shows that the responsible party of pollution damage may only benefit the limitation for those marine pollution damages within the scope of CMC or the scope of CLC. If other Chinese laws and regulations applied, the liability of responsible party for pollution damages may be unlimited.

(7) As for the responsible party, it is only the shipowner who is liable for oil pollution damage under CLC, not others. However, if it is proved that the pollution damage is caused completely by the act or omission of a third party, then the third party is responsible party under CLC. From the view of limitation of liabilities of responsible party, if there are some conflicts between CLC and CMC, the provisions under CLC shall have priority to those of CMC only if a foreign factor is involved.[68] However, under other Chinese laws, regulations and rules, any entity or individual may be the responsible party if only their activities cause marine pollution damage.

7 CONCLUSION

On the basis of a comparison of legislation on marine pollution damage between US Federal Statutes and Chinese laws, regulations and rules, China has no special law on this point. Although it may be reasonable to have different regulations and rules on maritime environment protection in accordance with different pollution sources from the aspect of administration, it is not available for the claiming and compensation under different laws and regulations. Furthermore, the provisions under these various laws, regulations and rules are different, as mentioned earlier. Except that there are different competent authorities, there is some confusion as to application of different domestic laws and regulations.[69]

From the view of the international conventions on marine pollution damage, there are four international conventions. They are 1992 CLC, 1992 Fund Convention, 2001 Bunker Convention and 1996 HNS Convention. The 1992 CLC, the 1992 Fund Convention and the 2001 Bunker Convention have come into force now and China has approved the 1992 CLC, the 1992 Fund Convention[70] and the 2001 Bunker Convention. Therefore, China has to face not only the problems of having no unification for legislation on marine pollution damage liabilities, but also the possible conflict between the international conventions China has approved and Chinese domestic laws, regulations and rules on marine pollution damage.

It is obvious that the international regime on marine pollution damage is established on the basis of the pollutants, rather than on the basis of pollution

68. See Art. 268 of CMC.
69. If the pollution damage involved foreigner factors caused by a tanker under the scope of CLC, CLC may apply, whereas it shall be applied under different laws and regulations according to various pollution sources. Another factor that cannot be neglected is that China approved 2001 Bunker Convention in November 2008, and it came into force in March 2009.
70. The 1992 Fund Convention applies only to Hong Kong Special Administrative Region, not to mainland China.

sources. It is the same for the US Federal Statutes. In order to protect the maritime environment perfectly and completely, it is clever and reasonable to solve the disputes between civil liabilities of pollution damages on the basis of pollutants, because of the varied threats and dangers of pollution damage to the ocean environment from different pollutants. It is suggested that one special law should be made in China in the future on marine pollution damage to solve definitely and reasonably the problems discussed in this chapter. This future legislation should be made on the basis of all international conventions in this area as well as take into account the experience and the advantages of legislation of other countries, such as the US Federal Statutes.

References

Awoyemi, O.O. *The Anatomy of United States' Oil Pollution Act of 1990.* SJD dissertation. Tulane University School of Law, 2002.

Ferrey, S. *Environmental Law: Examples and Explanations.* 2nd edn. New York: Aspen Publishers, 2004.

Force, R. & J. Force. *Marine Pollution.* Vol. 1. Materials used for students, Fall 2007, Tulane University School of Law.

Grad, F.A. 'A Legislative History of the Comprehensive Environmental Response, Compensation and Liability (Superfund) Act of 1980'. *Columbia Journal of Environmental Law* 8 (1982): 1.

Guo, P. & L.J. Zhao. 'The Legal Protection of the Marine Environment in China: Current Situation and Challenges'. In *Prevention and Compensation of Marine Pollution Damage: Recent Developments in Europe, China and the US*, edited by M.G. Faure & J. Hu. Alphen aan de Rijn: Kluwer Law International, 2006.

Peterson, P. & D. Kopec. 'Crude Legislation: Liability and Compensation under the Oil Pollution Act of 1990'. *Rutgers Law Journal* 23 (1992): 599.

Schoenbaum, T.J. *Admiralty and Maritime Law* (Hornbook Series). 3rd edn. Egan, MN: West Group Publishing, 2001.

Shan, H. *Comparative Study of Int'l American and China OPLRs – To Establish China Comprehensive OPLR.* PhD dissertation. Tulane University School of Law, 2007.

Part V
China

Chapter 18

China and International Protection of Marine Environment

*Song Ying**

1 INTRODUCTION

The importance of the oceans and seas has long been recognized in their communication, defence and exploitation of the marine living and non-living resources. Therefore, competition and conflicting sovereign claims by states over the oceans and seas are as old as states. The legal rules of the law of the sea are as old as international law. Hugo Grotius's Mare Liberum (1609)[1] was considered an important early source of both the law of the sea particularly and the international law generally. Freedom of the sea was the main theme of customary law of his time.

China is a coastal state with a long tradition of communication and exploitation of marine resources. Because of its large population and economic growth, whatever China[2] does with regard to the seas will have an important impact on the world. As far as legal rules are concerned, China, externally, has participated in

* Peking University, China.

1. See B.E. Carter, P.R. Trimble & C.A. Bradley, *International Law*, 4th edn (New York: Aspen Publishers, 2003), 836; M.N. Shaw, *International Law*, 5th edn (Cambridge: Cambridge University Press, 2003), 490.
2. In 2008, the total population of China reached 1,325 billion, gross domestic product (GDP) to 7,903,235 million international dollars, see World Bank, World Development Indicators Database, 15 Sep. 2009, <http://worldbank.org> (visited 20 Sep. 2009).

Michael G. Faure, Han Lixin & Shan Hongjun, *Maritime Pollution Liability and Policy: China, Europe and the US*, 323–345.

most of the international treaties on the law of the sea.[3] Internally, China has already established the domestic legal framework for implementing those international norms. This chapter provides a brief and preliminary overview of the international and domestic legal framework for the protection of the marine environment.

2 OVERVIEW OF THE INTERNATIONAL LAW FOR THE PROTECTION OF THE MARINE ENVIRONMENT

2.1 HISTORICAL EVOLUTION

The legal rules of the law of the sea, including those for the protection of the marine environment, are generally considered as more advanced, more comprehensive and better developed than those of other branches of international law. The international legal framework for the protection of the marine environment is based on several groups of international treaties[4] on different subjects at both regional and global levels. Therefore, treaty law constitutes a major part of the legal rules in this area. As early as the mid-nineteenth century, bilateral treaties for the protection of marine living resources were concluded in Europe.[5] Until the mid-twentieth century, most attention in this area was paid to conservation and management of fishery resources. With the economic, scientific and technological development in the twentieth century, marine transportation and other exploitation activities (e.g., seabed activities) became not only possible but also more intense than before. Marine pollution, especially vessel-based pollution, triggered and stimulated further international law-making. In 1954, the International Convention for the Prevention of Pollution of the Sea by Oil was concluded. This was the first treaty to address oil pollution of the sea. In 1972, a treaty on ocean dumping was adopted to regulate dumping activities, the 1972 London Dumping Convention.[6] Because of the limited scope of the 1954 treaty, the 1973 International Convention for the Protection of Pollution from Ships and the 1978 Protocol were concluded to regulate vessel-based pollution, the International Convention for the Prevention of Pollution from Ships (MARPOL) 73/78.[7] From then on, the development of marine

3. A 'Selected List of China's Participation to Marine Environmental Treaties' is attached at the end of this article.
4. The term 'treaty' is used as a generic term embracing all kinds of international agreements in written form.
5. For example, the 1867 Convention between France and Great Britain relative to Fisheries (Paris), see P. Sands, *Principles of International Environmental Law*, 2nd edn (Cambridge: Cambridge University Press, 2003), xxxv.
6. Convention on the Prevention of Marine Pollution by Dumping of Wastes and other Matters, signed on 29 Dec. 1972 in London, entered into force on 30 Aug. 1975, and entered into force for China on 14 Nov. 1985.
7. MARPOL 73/78 was signed on 17 Feb. 1978; and entered into force (also for China) on 2 Oct. 1983.

protection treaties moved on a fast track. A series of treaties on pollution prevention and control were made, targeting various sources of pollution. Finally, after ten years of negotiations, the 1982 United Nations Convention on the Law of the Sea (UNCLOS) was concluded and entered into force in 1994.[8] Codifying most of the customary rules in this area, it was considered the 'constitution for ocean governance'[9] because of its general nature and wide scope of application. The legal framework based on UNCLOS was further developed and improved with several new treaties or the amendments to the existing ones.

2.2 Treaty Regime[10]

Marine environment treaties may be divided into categories according to different standards. From the aspect of the contracting parties, the treaties can be bilateral, regional or global. From the aspect of subject matter regulated, the treaties can be either specific or general. In combination of these two standards, I have divided the relevant treaties of this area into the following three categories – general framework treaty, regional arrangements, and treaties on specific subjects.

The first is the general framework treaty, such as the 1982 UNCLOS, which established the general legal framework for almost all aspects of the law of the sea. The main achievement of the treaty for the environment was the establishment of general obligations for states to protect and preserve the marine environment by addressing most of the environmental threats posed by marine pollution and over-fishing.[11]

The second category is regional arrangements, which address either general or more specific subjects. The best example is perhaps the United Nations Environment Programme (UNEP) Regional Seas Programme (RSP). The programme was first launched in 1974 and is one of UNEP's most significant achievements in the past thirty years.[12] RSP aims to address the accelerating degradation of the world's oceans and coastal areas through regional arrangements of sustainable management and use of the marine and coastal environment by engaging neighbouring countries in comprehensive and specific actions to protect their shared marine environment. It has accomplished this by stimulating the creation of RSP prescriptions for sound environmental management to be coordinated and implemented by countries sharing a common body of water. Today, more than 140 countries participate in 13 RSPs established under the auspices of UNEP. Six of these programmes are directly administered by UNEP. Nine regions have reached regional

8. UNCLOS entered into force for China on 7 Jun. 1996.
9. D. Hunter, J. Salzman & D. Zaelke (eds), *International Environmental Law & Policy*, 2nd edn (New York: Foundation Press, 2002), 659.
10. See the annex at the end of this chapter.
11. For example, the 1995 UN Fish Stock Agreement.
12. <www.unep.org/regionalseas/about/strategy/default.asp> (visited 3 Apr. 2009).

treaties.[13] A New Global Strategy was adopted in 2007, which incorporated an ecosystem management.[14] Another regional arrangement may be found with regard to the Antarctic, though the countries involved are from different continents. The Antarctic plays an important role in the global environment, especially with regard to climate change. The Antarctic Treaty was concluded in 1959.[15] Its main objective is to freeze national claims to sovereignty in the Antarctic continent. Under Articles I and II, the Antarctic is to be used for peaceful purposes only, including scientific investigation. Under this treaty framework, several more specific international agreements on the protection of the Antarctic environment were added, including the 1972 Antarctic Seal Convention, the 1980 Antarctic Marine Living Resources Conservation Convention, the 1988 Convention on the Regulation of Antarctic Mineral Resources Activities and the 1991 Antarctic Environmental Protection Protocol. Environmental objectives, at present, played an important part in the treaty regime of the Antarctic.

The third category are the treaties targeting more specific subjects of marine environmental protection, mainly under the auspices of International Maritime Organization (IMO). The subjects regulated may include those discussed in the following section.

2.2.1 Ocean Dumping

The major treaty on ocean dumping is the 1972 London Convention on the Prevention of Marine Pollution by Dumping of Wastes and other Matters (LC 1972).[16] The objectives of this treaty were to prevent the pollution of the sea by intentional dumping of waste and other matter liable to create hazards to human health, to harm living resources and marine life, to damage amenities or to interfere with other legitimate uses of the sea. The treaty prohibited dumping certain specific matter in the sea (e.g., radioactive waste) and established a permit obligation for dumping other matter.

2.2.2 Vessel-Based Pollution

MARPOL 73/78 established specific international regulations to implement the objectives of completely eliminating intentional pollution of the marine environment by oil and other harmful substances and minimizing accidental discharge.[17] The six annexes attached to the treaty effectively enlarged the scope of regulation.

13. No treaties were concluded for the two RSPs of East-Asian Seas and North-West Pacific to which China participated.
14. UNEP, *Global Strategic Directions for the Regional Seas Programme 2008–2012: Enhancing the Role of the Regional Seas Conventions and Action Plans*, UNEP (DEPI) RS.9/6, 31 Oct. 2007, <www.unep.org/regionalseas/> (visited 3 Apr. 2009).
15. The Antarctic Treaty was signed on 1 Dec. 1959 and entered into force on 23 Jun. 1961 and entered into force for China on 8 Jun. 1983.
16. The LC 1972 was signed on 29 Dec. 1972 and entered into force on 30 Aug. 1975; and entered into force for China on 14 Nov. 1985.
17. Sands, *supra* n. 5, 440.

2.2.3 Land-Based Pollution

Being the major source of pollution of the seas, land-based pollution to the marine environment mainly comes from the run-off from land, rivers, pipelines and other outfall structures and from or through the atmosphere. At present, apart from certain general principles in UNCLOS and other treaties and several 'soft law' documents, no general treaty has been concluded on this matter. In 1995, a Global Programme of Action for the Protection of the Marine Environment from Land-Based Activities (GPA) was adopted at a conference in Washington, D.C. Although this document was not legally binding, it provided recommendations to states, regional and international organizations on identification and assessment of problems, establishing priorities for actions; setting management of objectives for priority problems; identifying, evaluating and selecting strategies and measures to achieve objectives; and developing criteria to assess the effectiveness of the strategies and measures. The latest development was the 2006 Beijing Declaration, calling states to apply ecosystem approaches and to take action at the national, regional and international levels.[18]

2.2.4 Intervention in Accidents or Emergency of Oil Pollution

No matter how well-equipped vessels may be and how strict the standards for vessels may be, accidents of oil spill happen from time to time. After the 1967 *Torrey Canyon*, the 1978 *Amoco Cadiz* and the 1989 *Exxon Valdez* oil pills, over fourteen international treaties were concluded to provide the legal framework for international cooperation to combat emergency situations of oil spills. The principal treaties include the 1969 Intervention Convention,[19] the 1973 Intervention Protocol,[20] the 1989 Salvage Convention,[21] the 1990 OPRC Convention[22] and the 2000 HNS Protocol.

2.2.5 Liability and Compensation for Pollution Damages

In order to establish an incentive to prevent and reduce harm to the marine environment and also to provide timely, adequate and effective compensation to the

18. UNEP, 'Beijing Declaration on furthering the implementation of the Global Programme of Action for the Protection of the Marine Environment from Land-based Activities', 16–20 Oct. 2006.
19. The International Convention Relating to Intervention on the High Seas by Oil, signed on 29 Nov. 1969, entered into force on 6 May 1975 and entered into force for China on 24 May 1990.
20. The Protocol was signed on 2 Nov. 1973 and entered into force on 30 Mar. 1983; entered into force for China on 24 May 1990.
21. The International Convention on Salvage was signed on 28 Apr. 1989 and entered into force on 14 Jul. 1996; entered into force for China on 14 Jul. 1996.
22. The International Convention on Oil Pollution Preparedness, Response and Cooperation was signed on 30 Nov. 1990 and entered into force on 13 May 1995; entered into force for China on 30 Jun. 1998.

victims of pollution accidents, several treaties were concluded to establish rules of liability and compensation in relation to pollution and damages to the marine environment. At the global level, the principal treaty was the 1992 International Convention on Civil Liability for Oil Pollution Damage (1992 CLC),[23] the 1992 Oil Pollution Fund Convention, the 1996 Hazardous and Noxious Substances (HNS) Convention and 2001 Bunker Oil Convention.[24]

2.2.6 Conservation of Marine Living Resources

The international effort for conservation of marine living resources has a long history. Many of these treaties regulated the conservation of specific species. For example, the 1946 International Whaling Convention,[25] the 1993 Convention for the Conservation of Southern Bluefin Tuna and the 1995 UNCLOS Agreement of Straddling Stock.[26] The objectives of these treaties include protecting particular species or habitats of fisheries by limiting the total catch and restricting or prohibiting certain fishing methods (i.e., driftnet fishing) through the international cooperation.

2.3 Competent Institutions

The protection of the marine environment involves several international organizations. However, two of them played the most important role in this area. The first is the IMO, which was founded on 6 March 1948 in Geneva. Its objective mainly was to prevent marine pollution by ships. Many of the treaties were concluded and implemented under the auspices of the IMO, including the MARPOL73/78. The second organization is UNEP, which was established on 15 December 1972. Its objective mainly was to assess and solve major existing or emerging environment problems, promote international cooperation on environmental issues, promote international environmental law-making, encourage regional environment cooperation and so forth. RSPs, as mentioned earlier, were formulated and managed by the UNEP.

The following features might be identified from the overview of the international legal framework for the protection of the marine environment. First,

---

23. The Protocol of 1992 to amend the International Convention on Civil Liability for Oil Pollution Damage was signed on 27 Nov. 1992 and entered into force on 30 May 1996; entered into for China on 5 Jan. 2000.
24. The International Convention on Civil Liability for Bunker Oil Pollution Damage was signed on 23 Mar. 2001 and entered into force on 21 Nov. 2008; entered into force for China on 9 Mar. 2009.
25. The International Convention for the Regulation of Whaling was signed on 2 Dec. 1946 and entered into force on 10 Nov. 1948; entered into force for China on 24 Sep. 1980.
26. The UNCLOS Agreement on the Conservation and Management of Straddling Fish Stocks and Highly Migratory Fish Stocks was signed on 4 Aug. 1995 and entered into force on 11 Dec. 2001. China signed it on 11 Dec. 1996.

the framework is based on treaties. The treaty regime is both regional and global and specific and general. Second, the legal principles, tools and approaches reflect the evolution of international environmental law. For example, a fragmented approach targeting a particular problem has gradually given place to the ecosystem approach. Third, apart from the treaty regime, the role of the more flexible 'soft law' measures increased, such as GPA, with regard to the land-based pollution.

3 CHINA: CHALLENGES AND LEGAL RESPONSES

3.1 CHALLENGES OF THE COSTAL AND MARINE ENVIRONMENT

China is a coastal state, with an 18,000-kilometer coastline, 6,500 islands and nearly 3 million km^2 of marine area.[27] There are four costal seas, from the north to the south, respectively, the Bohai Sea, the Yellow Sea, the East China Sea and the South China Sea.[28] The eleven coastal provinces are the most populated and highly productive region – approximately 40% of the total population (550 million), 50% of the big cities and 60% of the national gross domestic product (GDP) (CNY 30,000 per person). In 2008, the total value of marine production was CNY 296.62 billion. This figure is expected to reach CNY 300 billion in 2010.[29] However, as a result of the high concentration of population and rapid economic growth in this region, the coastal and marine environment on the whole has quickly deteriorated since the 1980s. For example, in the Bohai Sea, the single inland sea of China, approximately 31% of the area is polluted.[30] The most challenging examples are discussed in the following section.

3.1.1 Polluted Marine Areas

The quality of the marine water of river mouths, bays and marine areas near big cities, where the water quality is worse than Grade I,[31] was 200,000 km^2 in 1998

27. State Oceanic Administration (SOA), *China Marine Environmental Quality Reports* (*zhongguo haiyang huanjing zhiliang gongbao*), 2000–2009, <www.soa.gov.cn/hygb/hyhjzlgb/> (visited during May 2009–July 2010).
28. The total marine area of the four costal seas is nearly 4.73 million km^2, where China has disputes on jurisdiction with the neighbouring states, including the South Korea, Japan, Viet Nam, the Philippines, Malaysia, Indonesia and Brunei. See J.S. Yang, 'Overview of Important Marine Political, Economic and Environmental Issues' (*Zhongda haiyang zhengzhi jingji he huanjing wenti zongshu*), in Z.G. Gao & H.W. Zhang (eds), *Collection of Papers on State Marine Policy* (*Haiyang guoce yanjiu wenji*) (Beijing: China Ocean Press, 2007), 187–188.
29. Internet News, 'The SOA Publish *the China Marine Development Report 2009*, <www.jmnews.com.cn> (visited 19 May 2009).
30. SOA (2007), *supra* n. 27.
31. The quality of the marine water is classified into four grades in accordance with 'Marine Water Quality Standard' (GB3097 – 1997) enacted jointly by the State Environmental Protection Administration and the State Oceanic Administration. Grade I, the best quality, is suitable for marine fishing and protected zones.

compared with 100,000 km^2 in 1992.[32] The polluted marine area doubled in only six years. With the consistent efforts over the last ten years, this figure has dropped to 145,000 km^2 and 137,000 km^2 in 2007 and 2008, respectively.[33] But the figure raised to 146,980 km^2 again in 2009.[34] The major cause of such significant polluted marine area was the rapid increase of discharge of pollutants (both industrial and municipal) into the marine environment from land-based sources. Up to 88.4% of the coastal outfall structures discharged pollutants directly into the seas in violation of the discharging standards in 2008.[35]

3.1.2 Eutrophication

During decade of 1990–2000, more than 200 cases of 'red tide' was reported. In the three-year period of 1997–1999 alone, there were forty-five cases; the total loss was approximately CNY 2 billion (United States Dollar (USD) 242 million).[36] A project on the marine environment of the Hangzhou Bay of China, sponsored by the World Bank, indicated that about 88% of nitrates and 94% of phosphates input into the coastal waters came from the Yangtze River, which carried agricultural fertilizers into the coastal waters as a result of water and soil erosion and flood.[37] The number of cases and the affected marine area increased to eighty-two cases and 11,610 km^2 in 2007 and sixty-eight cases and 13,700 km^2 in 2008.[38] Many of the incidents took place in the East China Sea.

3.1.3 Oil Spills

In 1998, more than twenty large-scale oil spills were reported in the coastal waters of China.[39] Nearly 30% of the marine area of the Bohai Sea exceeds the oil pollution standard permitted in marine water.[40] But in 2006, the total number of oil spills reached 124, with more than 1,216 tonnes of oil spilled in the coastal areas.[41] In 2008, the number of cased increased to 136, but the volume reduced to 155 tonnes as a result of stricter enforcement of the law.[42]

---

32. SOA (2000), *supra* n. 27.
33. SOA (2008, 2009), *supra* n. 27.
34. SOA (2010), *supra* n. 27.
35. SOA (2009), *supra* n. 27.
36. SOA (2000, 2009), *supra* n. 27.
37. Z. Xiao, 'Examining the Sick Ocean' (*bamai bing haiyang*), *China Environmental Newspaper* (*zhongguo huanjing bao*), 19 Jun. 1999, 3.
38. SOA (2008, 2009), *supra* n. 27.
39. State Environmental Protection Administration (SEPA), '1998 Report on the State of Environment', *1999 China Environmental Yearbook*, China Environmental Yearbook Publisher, 1999, 119.
40. SOA (2000), *supra* n. 27.
41. Ministry of Environmental Protection (MEP), '2006 Report on the State of Environment', *2007 China Environmental Yearbook*, China Environmental Yearbook Publisher, 2007, 208.
42. Ministry of Environmental Protection (MEP), '2008 Report on the State of Environment', <www.zhb.gov.cn/zkgb/2008zkgb/200906/t20090609_152536.htm> (visited 14 Sep. 2009).

3.1.4 Habitats Destruction

In the past twenty years, the total land area reclaimed from the sea and development from the tidal-flat area is more than 700,000 hectares. Approximately 57% of coastal wetlands and 73% of mangroves have been lost. Nearly 80% of coastal coral reefs have been destroyed to varying degrees.[43] The State Oceanic Administration (SOA) announced that the ecosystems of the coastal marine areas continued to deteriorate in from 2007 to 2009.[44]

3.1.5 Overfishing

Marine living resources are economically and socially important for China. In 1999, the total annual fisheries production of China was 40 million tonnes, among which 17 million tonnes was by capture and 22 million tonnes by aquaculture.[45] However, this figure increased to 92.1 million tonnes in 2008 for the annual total fisheries production, among which 74.9 million tonnes was for capture (inland and marine) and 17.2 million tonnes for aquaculture, according to the Food and Agriculture Organization (FAO).[46] It was reported that although the Northwest Pacific remained by far the most productive fishing area,[47] most stocks throughout the whole Asian and Pacific region were currently being fully harvested and exploited at unsustainable levels.[48] The major causes of concern are overfishing and the use of destructive fishing techniques in this region. Other causes of the decrease of marine fish resources include coastal erosion resulting from increased land subsidence from ground water extraction, sediment starvation as a consequence of inland dam and irrigation barrage construction and offshore mining of sand and marine traffic.[49]

3.2 Legal Responses: External

As indicated earlier, international legal protection of the marine environment is based on treaties. These treaties were basically the result of the competing and compromising interests of the members of the international community. The author conducted a survey on the relevant treaties to which China has participated (Table 18.1).[50]

---

43. SOA (2000), *supra* n. 27.
44. SOA (2008, 2009 and 2010), *supra* n. 27.
45. FAO, 'World Fisheries Production, by Capture and Aquaculture and by Country (1999)', <www.fao.org/fi/statist/summtab/default.asp>, 22 Jul. 2001.
46. FAO, 'The State of World Fisheries and Aquaculture' (2008), Rome, 2009, 4.
47. R. Grainger, 'Recent Trends in Global Fishery Production', <www.fao.org/fi/trends/catch/catch.asp>,22 Jul. 2001.
48. UNEP, *Global Environmental Outlook*, 'Chapter 2: Regional Perspectives' (Oxford University Press, 1997), 51.
49. *Ibid.*
50. See the annex for the details of these treaties.

Table 18.1 China's Participation in Selected Marine Environmental Treaties

	Before 1970	*1970s*	*1980s*	*1990s*	*After 2000*	*Total*
Marine Treaties	3	4	12	7	8	34
China	0	1	11	12	10	34

There are thirty-four relevant treaties[51] as indicated in Table 18.1. On the aspect of marine treaties, this Table 18.1 shows clearly that majority of the treaties (thirty-one treaties, 91% of the total thirty-four treaties) concerning marine environment mentioned earlier were concluded after 1970. This may be partly explained by the increased international environmental concern since the 1972 Stockholm Conference and its follow-up activities, though many treaties of law of the sea existed before then. In 1980s, the number of treaties reached a peak, and they started to decline in 1990s because many of the crucial issues were tackled during the 1970s–1980s. The international legal framework was fairly well established. After the 1990s, the number declined compared with that of 1980s. In addition, China also concluded more than a dozen of bilateral agreements on marine scientific cooperation.[52]

However, on the aspect of China's participation in these environmental treaties, China joined only one treaty before 1980s, which is the 1966 Convention on Load Line.[53] In fact, most of these treaties (twenty-three treaties, about 68% of the total thirty-four treaties) entered into force for China between the 1980s and 1990s. This implies that it takes time (sometimes a decade or even two decades) for China to decide to get involved or to join international efforts. Participation by China remained fairly stable between the 1980s and 1990s. After 2000, China continued to participate actively in the international efforts in this field.

Apart from the treaties, the most important effort to develop guiding principles or action plans for the protection of marine environment is Chapter 17 of Agenda 21, 'Protection of the Oceans, all kinds of seas, including enclosed and semi-enclosed seas, and coastal seas and the protection, rational use and development of their living resources', which was adopted at the UN Conference on Environment and

51. All the protocols in the list are considered as a separate treaty. The year indicated in the table is the year of entry into force for both the respective treaty and China. Because China only signed the 1995 Straddling Fish Stock Agreement on 11 Dec. 1996, it is not included in the thirty-three treaties.
52. There are fifteen bilateral agreements in the list provided by the SOA, <www.soa.gov.cn/hyjww/zwgknew/xxml/webinfo/2008/04/1208840263050619.htm> (visited 16 May 2009).
53. The Convention of the Load Line (LL 1966) was signed on 5 Apr. 1966 and entered into force on 21 Jul. 1968; enter into force for China on 5 Jan. 1971.

Development in 1992 in Brazil.[54] Although the Agenda 21 is not a legally binding document, it is the international community's action plan for the earth. Chapter 17 outlines proposals to guide subsequent international law-making, including the establishment of national coordinating mechanisms to develop land and water use and existing policies; integrated coastal and marine management; preparation of coastal profiles; environmental impact assessment and systematic observation; disaster contingency plans; improvement of coastal human settlement (especially treatment and disposal of sewage, solid wastes and industrial effluents); and conservation and restoration of altered critical habitats. Accordingly, China published the China Ocean Agenda 21 in 1996. It sets forth the basic strategies, strategic objectives and principal measures for achieving sustainable development of the oceans and the seas.[55]

3.3 Legal Responses: Internal

China's participation in the previously mentioned treaties triggered changes domestically. In accordance with the provisions of the relevant treaties, China promulgated laws, administrative regulations or measures and issued national policies, strategies or action plans for the purpose of marine and coastal environmental protection, management and development. Institutional arrangements have also been made or adjusted for this purpose with the establishment and development of the domestic legal framework.

For example, on 15 May 1996, the Standing Committee of the National People's Congress of China issued a declaration upon ratification of the UNCLOS.[56] The Declaration announced that, first, the legal regime concerning the territorial sea established in the 1992 Law on the Territorial Sea and Contiguous Zone should be maintained. Prior permission or notice is required for the passage of foreign warships in the territorial sea. Second, China claimed sovereign rights and jurisdiction over an exclusive economic zone (EEZ) of 200 nautical miles and the continental shelf. With any state whose coast is either opposite or adjacent to the Chinese coast, the issue delimitation shall be negotiated and settled in accordance with the 'equitable principles' of international law. On 16 July 1998, the Standing Committee of the National People's Congress promulgated the Law on the EEZ and Continental Shelf of the People's Republic of China. After that, several laws were implemented, including the 2001 Administrative Law of the Utilization of Marine Areas, the 2006 Administrative Regulation of Prevention and Treatment of

54. N.A. Robinson (ed.), *Agenda 21: Earth's Action Plan* (New York: Ocean Publication, Inc., 1993).
55. SOA, *China Ocean Agenda 21* (Beijing: China Ocean Press, 1996).
56. United Nations, *Report of the Secretary-General on the Law of the Sea*, (A/51/645), 1996, para. 34; *Report of the Secretary-General on the Law of the Sea: Law of the Sea*, (A/52/487), 1997, paras 64–67, <www.un.org/Depts/los/> (visited in April 2009).

Marine Pollution by Marine Construction Projects and the 2007 Administrative Measures of the Marine Function Zones.

3.4 Select Issues in the Legal Responses

With regard to the domestic legal responses to the challenges in marine protection, the following issues require further attention.

3.4.1 Land-Based Sources of Pollution

Land-based sources of pollution are the major contributor to marine pollution, so they are an important issue in China's environmental laws and regulations. In China, it is dealt with in several pieces of legislation because of the complexity of such a source of pollution. Then the issue is how to coordinate all of these relevant laws and institutions. In China, the relevant laws include the Environmental Protection Law (1989), the revised Marine Environmental Protection Law (1999 revised), the Regulation Concerning Prevention of Pollution Damage to the Marine Environment by Land-based Pollutants (1990), the Administrative Regulation of Prevention and Treatment of Marine Pollution by Marine Construction Projects (2006),[57] the Water Pollution Prevention and Control Law (2008 revised), the Water and Soil Conservation Law (1991), the Water Law (1988), the Agricultural Law (1993) and so on. The following points are noteworthy.

3.4.1.1 Environmental Impact Assessment

Environmental impact assessment (EIA), an important tool for environmental protection and management in international treaties, is also employed by China. In China, the Environmental Protection Law (1989) provided the requirement of an EIA report. In 1986, the Management Measures on the Environmental Protection for Construction Projects was jointly issued by the State Council, State Planning Committee and State Economic Committee. In 1998, these Management Measures were revised and incorporated into a more formal Management Regulation on the Environmental Protection for Construction Projects. The EIA Law was passed in 2002. In 2009, the State Council Regulation on the EIA of Plans was adopted and entered into force on 1 October 2009. The scope of plans subject to EIA requirement covers both general plans (e.g., regional plans and marine area development plans) and specific plans (e.g., industry, agriculture and nature development).

EIA is a useful tool for environmental protection in both the international treaties and domestic laws of many countries. In the past ten years, more than

57. The practice of Chinese environmental legislation is that the most important policy issues are dealt with in a framework law, such as the Marine Environmental Protection Law (1999 revised). In order to implement the framework law, a number of administrative regulations are made later.

90% of construction projects in China underwent EIA.[58] But the experience with EIA in China is mixed. First, some projects that have passed EIA still caused serious pollution, which called the effectiveness of the tool into question. Second, public participation was not widely and fully guaranteed, though there were some isolated cases of good public participation in the EIA in some big cities and in some highly controversial projects. Third, development plans are recently subject to EIA, which shows progress, but the question remains whether this kind of EIA can be effective because such plans, especially development plans, are usually administered by local governments, which are, at the same time, under tremendous pressure in regard to development, the local GDP target and personal promotion of the local officials. In many of these cases, environmental need is naturally and logically put in second place if the existing administrative evaluation system is not to be changed. The Campaign for Scientific Development, which started in 2007, hopefully will gradually change the vision on development for both the central and local governments.

3.4.1.2 Land-Use Planning

Land-use planning (or zoning) is to make an integrated and overall land use plan for various activities and facilities, that is, industry, agriculture, transportation, and so on, in order to achieve the objectives of balanced development, taking into account the local natural, social and economic conditions of a given place. It is an effective method to minimize the conflict of multiple land-uses and to prevent environmental pollution via rational geographical allocation. Due to the complexity of land-based pollution to the marine environment, the requirement of rational and effective land-use planning, especially in the coastal area, is important to conduct careful land-use planning. To a certain extent, the current heavy pressure of pollution on the coastal and marine environment is due to inappropriate land-use planning of the coastal provinces in the past.

Article 23 of the 2004 revised Land Administration Law of China provides that plans for all-round harnessing of rivers and lakes and for their development and utilization shall be developed with the overall plan for land utilization. Article 24 also provides that the government at all levels shall exercise close supervision over the plans for land utilization and keep control over the total area of land to be used for construction. Moreover, the revised Marine Environmental Protection Law (1999) was intended to provide more restrictions in this aspect. First, China implemented the system of 'function zones' (exploitation zones, treatment and protection zones, nature reserve zones and so forth) to conduct rational use of marine resources.[59] Second, the geographical allocation of discharging

58. W. Tao, 'Explanation on the Draft EIA Law', 12 Dec. 2000, *Collection of China Environmental Law and Regulations (2002–2003)* (Beijing: Chemical Industrial Publication, 2003), 10.
59. Article 6 of the revised Marine Environmental Protection Law (1999). See also G.Q. Li (ed). *Studies on China's Comprehensive Marine Administration* (Beijing: China Ocean Press, 1998), 275–281.

outlets along the coast should be subject to more stringent control (Article 30). Third, it is prohibited to build any new highly polluting plants (e.g., paper mill, chemical plant, electroplating plant) that are not equipped with sufficient pollution control and treatment facilities in the land near the coast (Article 45). But the fact that more than 88% of the discharging outlets were in violation of the law is a big problem and called the effectiveness of the law into question.

3.4.2 Water Resources and Water Pollution

In China, the regulations and rules concerning the protection and rational use of water resources are found mainly in the 1988 Water Law and in the 1991 Water and Soil Conservation Law. The relevant rules concerning water pollution are found mainly in the 2008 Water Pollution Prevention and Control Law (revised). Marine water pollution is regulated by another law. Therefore, water is regulated at least by four separate laws.

In accordance with the laws concerning water pollution in China, focus is on the control and prevention of waste effluents. The basic approaches toward water pollution include direct control and economic incentives. The principal measures include the following: (1) water quality standards and effluent control standards (at both national and local level), (2) permits, (3) discharge fees or dues and (4) penalties (civil and criminal). Chinese domestic laws require that waste effluents must be treated before being discharged into river, lakes, seas or other public waters. Apart from discharge standards, Article 18 of the 2008 revised Water Pollution Prevention and Control Law imposed a total quantity control on select pollutants discharged. The total quantity control is implemented on the provincial, municipal and county levels. This total emission and effluent quantity control approach represents a change and a new tendency toward pollution prevention and control in Chinese environmental law.

River basins are an important issue in China, so the 2008 Water Pollution Prevention Law provided a mechanism for solving problems in this aspect (Article 15). It improved the situation of transboundary pollution. But with regard to the marine environment, the revised law only states that it is a matter of the Marine Environmental Protection Law. However, no express link is made between the protection of fresh water and marine water. Under such circumstances, the ecosystem approach advocated and incorporated in the international treaties is not reflected in the domestic law. The 2006 Beijing Declaration and the China National Action Plan hopefully will bring changes to such administrative fragmentation.

Another challenge is how to improve the effectiveness of the relevant laws with regard to township and village enterprises (TVEs). For example, even though the previous concentration-based discharge standards applied in China are complemented by the total quantity control in the 2008 revision, Chinese TVEs that contributed to economic growth, in fact, caused serious pollution problems as a result of the insufficient technology and funds for pollution control and treatment. As a last resort, the central government shut down 999 small plants along the Huaihe River that could not possibly have met the water discharge standards by the given period of time of 1998.[60] Needless to say, the cost of these shutdowns was very high. The question remains as to how these small plants could possibly have been established and operated in the first place if the requirements of EIA and the Three Synchronisation Policy[61] were effectively carried out. The current financial crises hopefully will help change the development model for the TVEs.

3.4.2.1 *Other Sources of Pollution*

Other sources of pollution, for example, coastal development projects and air pollution, also contribute to marine pollution. These sources of pollution have been regulated by other laws in China. In the past years, China made considerable efforts to improve energy efficiency and reduce emission of greenhouse gases.

3.4.3 Vessel-Based Pollution

China is a party to the UNCLOS, MARPOL73/78 and other relevant treaties, as indicated in the annex. In implementing these treaties on vessel-based pollution, China promulgated laws and administrative regulations. The following issues may require further elaboration.

3.4.3.1 *Regulatory Authority*

In China's marine administration, the competent agencies include the SOA, MEP (formerly State Environmental Protection Administration (SEPA)), the Ministry of Transport (Maritime Bureau), the Ministry of Agriculture (MOA, for fishery) and the eleven coastal provincial authorities. Each of these authorities is responsible for the matter designated by law. The 1999 revised Marine Environmental Protection Law, though allocating more competence to the SOA, still followed the traditional decentralized and fragmented administrative structure. The coordination among the different administrative authorities was and perhaps is still a difficult issue.

3.4.3.2 *Implementation of MARPOL 73/78*

As a party to MARPOL 73/78, China has taken measures to regulate vessel-based sources of pollution. In 1983, the Regulation of the People's Republic of China Concerning the Prevention of Pollution of Marine Areas by Vessels was promulgated by the State Council. The protection measures include the requirement of installation of pollution control equipment in ships; the discharge permit; water quality standards; restrictions on certain activities, such as incineration at sea; emergency measures, and so forth. This regulation was later replaced by the

60. UNEP, UNEP (WATER) NOWPAP 1 G, 3/7, 1998, at 7.
61. It requires that the facilities for pollution control and prevention should be designed, constructed and operated along with the main development projects involving new construction, renovation, or reconstruction. See UNEP, *Global Environment Outlook*, *supra* n. 48, at 158.

2009 State Council Regulation on the Administration of Prevention of Pollution of Marine Environment by Vessels which entered into force on 10 March 2010. The new regulations, while maintaining the existing administrative framework, substantially made the legal standards more stringent.

3.4.4 Dumping of Waste and Other Matters

China participated in the 1972 London Convention. Ocean dumping of wastes and other matters is regulated in China. Dumping is permitted but only if it is in accordance with relevant laws and administrative regulations and rules.

In China, a special Administrative Regulation was issued by the State Council in 1985, the Regulation on the Management of Ocean Dumping. In 2007, the Interim Administrative Measures on the Dumping Zone was made. In accordance with the regulations, dumping is only permitted with a dumping permit. Wastes are divided into three categories, prohibited substances, substances with special permit and other wastes, following the provisions of the 1972 London Convention. The permit will specify the quantity and category of the wastes to be dumped and also the time limit and the method of dumping. Dumping should be done in the dumping sites designated by the competent authority. Foreign wastes are prohibited from dumping within the jurisdiction of China.

3.4.5 Exploitation of Marine Resources

Marine resources, living and non-living, gradually gained more attention as a result of the decrease of available land resources. The legal framework with regard to the exploration and exploitation of these resources is divided into two parts, as discussed in the following section.

3.4.5.1 Living Resources

Living resources have received increasing attention in China over the past few years because more elaborate and detailed rules have been made. In 2004, the Fishery Law was revised again. This law applies to both inland water and marine water of the EEZ under the jurisdiction of China. The basic approach of these regulatory rules is to set limits on the quantity of catch because overfishing is becoming increasingly worse in this region. A catch permit is now employed. In China, a period (normally two months) of prohibited fishing in all the adjacent seas is imposed. Certain fishing methods are also prohibited.

3.4.5.2 Non-Living Resources at the Sea-Bed

China is granted the status of 'pioneer investor' under UNCLOS. Relevant laws and regulations were passed in this regard. The regulatory measures include EIA requirement, installation of pollution control equipment, license, and so forth.

3.4.6 Protection of Critical Areas

China has designated many marine areas as 'National Park' or 'Protected (Conservation) Area' or other similar areas with special protection. Such designation may be made by the central or local government according to the significance of such areas. Within such areas, certain activities, for example, construction and mining, are restricted or prohibited. In 2006, twenty-nine sites were designated at the national level and 149 protected sites were also designated at the provincial or municipal level. The total protected area has reached 37,584 km^2.[62]

3.4.7 National Marine Environmental Policies and Strategies

National policies and strategies, though not binding legal norms, play an important role in environmental protection and management. Apart from the environmental laws and regulations, China issued various national policies and strategies relevant to marine environmental protection and coastal development. In the evolution of the environmental norms, some of them originally came from more soft documents, for example, declarations, policies and strategies. Therefore, it is important that attention should be given to these policy and administrative documents. Another reason for review of such documents is that these documents, especially the more comprehensive economic, social and environmental policies, produce a direct impact on the marine environment and coastal management and development.

3.4.7.1 General Economic, Social and Environmental Policies

In 1996, China published the Ninth Five-Year Plan for National Economic and Social Development and the Outline of the Long Term Target for the Year 2010 and the Ninth Five-Year Plan for State Environmental Protection and the Outline of the Long Term Target for the Year 2010. Both documents put environmental protection high on the agenda. In 1992, China set out China Agenda 21 – White Paper of Population, Environment and Development of China in the Twenty-First Century. In 2006, the State Council published the report on the China Environmental Protection (1996–2005), which gave a systematic review of the environmental protection in China.

3.4.7.2 Policies and Strategies for Marine Environment and Coastal Development and Management

The most comprehensive and basic policy document in China is the China Ocean Agenda 21 issued by the SOA in 1996.[63] The overall objective of this document is to construct a well-cycling marine ecosystem in order to formulate a scientific and reasonable ocean development system for the promotion of the sustainable

62. MEP, (2008), at 208.
63. SOA, *China Ocean Agenda 21* (Beijing: China Ocean Press, 1996).

development of the marine economy. The strategic principles include: (1) to take the development of the ocean economy as the core task, (2) to develop at a moderately fast speed, (3) to pursue an integrated development on both land and sea, (4) to develop the ocean by relying on science and education, and (5) to coordinate various developments.[64]

4 CONCLUSION

This overview demonstrates that the marine environmental challenges in China are ever more alarming. Eutrophication, oil spills, habitat destruction and overfishing are only a few examples of the deterioration of the marine environment. In order to tackle these challenges, both legal and policy measures have been taken. To date, China has already established the basic legal framework for the development and management of the coastal and marine environment. To a certain degree, the environmental norms in this area are more developed than those in other environmental areas. In this process, China has participated in most of the principal environmental treaties. These treaties produced positive input and were the driving force for China to establish and develop its domestic legal framework in this area. Most of the crucial issues have been covered to varying degrees. No obvious gap was left according to the information available at this stage. The characteristic of this domestic legal framework is a top-down approach, in which governmental authorities at all levels played a key role in ruling-making, implementation and enforcement. In fact, according to the 2008 State Council's White Paper on China's Efforts and Achievements in Promoting the Rule of Law, environmental laws, including the 1999 revised Marine Environmental Protection Law, are classified as the administrative law. This is also the reason why public participation in the whole process of law-making, implantation and enforcement is not well established and facilitated. There is an obvious need to improve the public awareness and public participation in this field.

The most important and urgent issue at present is how to improve the effectiveness of the law, especially with regard to land-based pollution. The fragmentation of the administrative authorities is a major obstacle. For a long time, the coordination of administrative authorities in this field has not been satisfactory. The central government has decided to improve the functioning of the government by merging relevant agencies and institutions, which are to be allocated more broad jurisdiction. This is certainly a positive move to improve the effectiveness, but more defined and clear procedural rules incorporated in the existing and new law are also important because regardless of the size of the agency, horizontal cooperation and coordination is still necessary.

Sustainable or scientific development for the marine environment is not a mere political statement. In order to achieve sustainable or scientific development,

64. *Ibid.*, 7–10.

environmental considerations should be integrated into the decision-making of agriculture, industry, trade, investment, research and development, infrastructure, finance and management and economy generally.[65] The root causes of environmental problems (exploitation of resources, energy generation, clean technologies and so forth) should be tackled. Environmental policies should move away from strictly sector issues to encompass broad social considerations in order to achieve lasting effects. China is still a developing country; therefore economic growth remains high on the agenda. Without economic support, environmental protection cannot be sustainable. Without environmental protection, economic growth and social development cannot be sustainable either. The best approach to improve the marine environmental status is to integrate the environmental objectives into the decision-making process of all economic, political and social issues. The development should be truly sustainable or scientific.

65. UNEP, *GEO-2000*, 1999; UNEP, GEO 4 – Environment for Development, 2007, <www.unep.org>.

Annex

Selected List of China's Participation to Marine Environmental Treaties

Treaties: General Legal Framework			
Title	Signature	Entry into force	Entry into force for China
UN Convention on the Law of the Sea (UNCLOS)	1982-12-10	1994-11-16	1996-06-07
- Agreement on Part XI	1994-07-18	1996-07-28	1996-06-07
- Agreement on the Conservation and Management of Straddling Fish Stocks and Highly Migratory Fish Stocks	1995-08-04	2001-12-11	1996-12-11 (signed)
Treaties: Regional Arrangements			
Title	Signature	Entry into force	Entry into force for China
Antarctic Treaty	1959-12-01	1961-06-23	1983-06-08
- Antarctic Environmental Protection Protocol	1991-10-04	1998-06-14	1998-01-14
- Antarctic Marine Living Resource Conservation Convention	1980-05-20	1982-04-07	2006-10-19
Convention on the Conservation and Management of Highly Migratory Fish Stocks in the Western and Central Pacific Ocean	2000-09-05	2004-06-19	2004-12-02
Treaties: Specific Subject-Matters (IMO)			
Title	Signature	Entry into force	Entry into force for China
Convention for the Safety of Life at Sea 1974 (SOLAS)	1974-11-01	1980-05-25	1980-05-25
- Protocol of 1978	1978-02-17	1981-05-01	1983-03-17
- Protocol of 1988	1988-11-11	2000-02-03	2000-02-03

Convention on Load Line (LL 1966)	1966-04-05	1968-07-21	1971-01-05
- LL Protocol 1988	1988-11-11	2000-02-03	2000-02-03
Convention on Tonnage Measurements of Ships 1969 (TONNAGE)	1969-06-23	1982-07-18	1982-07-18
Convention on the International Regulation for Prevention Collisions at Sea (COLREG 1972)	1972-10-20	1977-07-15	1980-01-07
International Convention for Safe Containers (CSC 1972)	1972-12-02	1977-09-06	1981-09-23
International Convention on Standards of Training, Certification and Watchkeeping for Seafarers (STCW 1978)	1978-07-07	1984-04-28	1984-04-28
MARPOL 73/78 (Annex I)	1978-02-17	1983-10-02	1983-10-02
MARPOL 73/78 (Annex II)	1978-02-17	1987-04-06	1983-10-02
MARPOL 73/78 (Annex III)	1978-02-17	1992-07-01	1994-12-13
MARPOL 73/78 (Annex IV)	1978-02-17	2003-09-27	2007-02-02
MARPOL 73/78 (Annex V)	1978-02-17	1988-12-31	1989-02-21
MARPOL Protocol 1997 (Annex VI)	1997-09-26	2005-05-19	2006-08-23
Convention on the Prevention of Marine Pollution by Dumping of Wastes and other Matters (LC 1972)	1972-12-29	1975-08-30	1985-11-14
- Protocol of 1996 (LC PROT 1996)	1996-11-07	2006-03-24	2006-10-29
International Convention Relating to Intervention on the High Seas by Oil	1969-11-29	1975-05-06	1990-05-24
- Protocol of 1973	1973-11-02	1983-03-30	1990-05-24

Protocol of 1992 to Amend the International Convention on Civil Liability for Oil Pollution Damage (CLC 1969)	1992-11-27	1996-05-30	2000-01-05
Convention on Limitation of Liability for Maritime Claims	1976-11-19	1986-12-01	1997-07-01
Athens Convention relating to the Carriage of Passengers and their Luggage by Sea (PAL 1974)	1974-12-13	1987-04-28	1994-08-30
- Protocol of 1976 (PAL PROT 1976)	1976-11-19	1989-04-30	1994-08-30
International Convention on Salvage (SALVAGE 1989)	1989-04-28	1996-07-14	1996-07-14
Convention on Facilitation of International Maritime Traffic (Facilitation 1965)	1965-04-09	1967-03-05	1995-03-17
International Convention on Oil Pollution Preparedness, Response and Cooperation (OPRC 1990)	1990-11-30	1995-05-13	1998-06-30
OPRC/HNS 2000	2000-03-15	2007-06-14	2010-02-19
International Convention on Civil Liability for Bunker Oil Pollution Damage (BUNKERS 2001)	2001-03-23	2008-11-21	2009-03-09

Sources: IMO: Status of Multilateral Conventions and Instruments in Respect of which the International Maritime Organization or its Secretary-General Performs Depositary or other Functions as at 30 April 2010, <www.imo.org>; United Nations Treaty Collection: <www.treaties.un.org>, April 2009–November 2009).

References

Carter, B.E., P.R. Trimble & C.A. Bradley. *International Law*. 4th edn (New York: Aspen Publishers, 2003).

Gao, Z.G. & H.W. Zhang, (eds). *Collection of Papers on State Marine Policy* (*Haiyang guoce yanjiu wenji*). Beijing: China Ocean Press, 2007.

R. Grainger, 'Recent Trends in Global Fishery Production'. <www.fao.org/fi/trends/catch/catch.asp>, 22 July 2001.

Hunter, D., J. Salzman & D. Zaelke (eds). *International Environmental Law & Policy*, 2nd edn. New York: Foundation Press, 2002.

Li, G.Q. (ed). *Studies on China's Comprehensive Marine Administration.* Beijing: China Ocean Press, 1998.

Sands, P. *Principles of International Environmental Law.* 2nd edn. Cambridge: Cambridge University Press, 2003.

Shaw, M.N. *International Law.* 5th edn. Cambridge: Cambridge University Press, 2003.

Tao, W. 'Explanation on the Draft EIA Law', 12 December 2000, *Collection of China Environmental Law and Regulations (2002–2003).* Chemical Industrial Publication, 2003.

Xiao, Z. 'Examining the Sick Ocean' (*bamai bing haiyang*). *China Environmental Newspaper (zhongguo huanjing bao)*, 19 June 1999.

Chapter 19

On the Application of CLC 92* in China

*Chen Qi***

1 INTRODUCTION

The convention on Civil Liability for Oil Pollution Damage (CLC) CLC 69 and CLC 92 are the major international conventions stipulating civil liability for oil pollution damage caused by ships. China acceded to CLC 69 on 30 January 1980, which became effective on 29 April 1982; and acceded to CLC 92 on 5 January 1999, which became effective on 5 January 2000, at which time CLC 69 became void.

Since CLC 69 and CLC 92 took effect in China, many disputes arose on the issue of whether these two conventions should be used in purely domestic[1] oil

* CLC 92 is the abbreviation of 'the International Convention on Civil Liability for Oil Pollution Damage of 1969'. CLC 92 is a Protocol to CLC 69. Subsequently, CLC 92 was amended in 2000; however, compared with CLC 92, the Amendment 2000 only increases the limitation of liability for claims. Although CLC 69 has been amended several times, the core content (strict liability) has never changed. Therefore, this chapter will discuss the topic mainly on the basis of CLC 92, unless it is necessary to mention other CLC versions.

** Dalian Maritime University, China.

1. According to Art. 2 of the CLC 92, the CLC 92 is only applied to oil pollution damages caused by any sea-going vessel and seaborne craft. Therefore, the 'purely domestic cases' in this chapter refer only to the cases involving oil pollution caused by sea-going vessels and seaborne crafts in Chinese territorial waters.

Michael G. Faure, Han Lixin & Shan Hongjun, *Maritime Pollution Liability and Policy: China, Europe and the US*, 347–357.

pollution cases in China? The Chinese academic circle has many arguments on this point.[2] Moreover, there are also completely conflicting decisions in similar cases in the Chinese judicial practices. In the oil pollution case of 'Min ran gong Number 2',[3] both the Guangzhou Maritime Court and Higher People's Court of Guangdong Province ruled that CLC 69 should be applied to purely domestic oil pollution cases, whereas in the oil pollution case of 'Yan Jiu You Number 2',[4] Qingdao Maritime Court and Higher People's Court of Shandong Province held that purely domestic oil pollution cases should be regulated by Chinese domestic laws, rather than CLC 69.

Those approving application of CLC to purely domestic oil pollution cases[5] argue that: (1) CLC 69 and CLC 92 explicitly state that they:

> 'Apply exclusively to pollution damage occurred within the territory including the territorial sea of a contracting state'

(The CLC 92 also includes the exclusive economic zone established by a contracting state or an area extending no more than 200 nautical miles from the baselines from which the breadth of a contracting state's territorial sea is measured.) So, when oil pollution damage is caused within the territory of a contracting state, CLC 69 or CLC 92 will apply, regardless of whether the oil pollution accident has foreign-related factors. (2) The Chinese government did not make any reservation to refuse the application of these two conventions to purely domestic oil pollution cases occurring in China. (3) According to the principle of 'treaty must be preserved', this convention should be applied to all kinds of oil pollution cases, including purely domestic ones.

---

2. There are many related papers. Such as, S. Yuzhuo, '*The Application of Laws on Compensation for Oil Pollution Damage Caused by Costal Transport Ships*' (original text in Chinese: Yan Hai Yun Shu Chuan Bo You Wu Sun Hai Pei Chang Fa Lv Shi Yong Wen Ti Yan Jiu) (June 2002), *Journal of Dalian Maritime University* (*Social Science Edition*), Nos 1–2; M. Faure and W. Hui, 'Financial Caps for Oil Pollution Damage: China and the International Conventions', in *Prevention and Compensation of Marine Pollution Damage: Recent Developments In Europe, China and the US*, ed. M.G. Faure & J. Hu (Alphen aan de Rijn: Kluwer Law International, 2006). G. Zhengyi and L. Pengjiu, 'On Several Legal Issues of Application of CLC' (original text in Chinese: CLC You Wu Min Shi Ze Ren Gong Yue Shi Yong De Ruo Gan Fa Lu Wen Ti), *Journal of Shenyang Normal University Science Edition* 31 (2007): 6; H. Lixin, *Study on Legal System of Compensation for pollution damages from Ships* (original text in Chinese: Chuan Bo Wu Ran Sun Hai Pei Chang Fa Lu Zhi Du Yan Jiu) (Beijing: Legal Press, 2007); Y. Xiaohan, *Case on Limitation of Liability for Oil Pollution Damage Compensation* (original text in Chinese), Guangzhou Maritime Court Cases, 24 Mar. 2003.
3. Case of the Fujian branch of China ship fuel supply Co. applying for limitation of liability, Guangzhou Maritime Court, 17 Feb. 2000. See <www.ccmt.org.cn&hs&news&show.php?cId=3185>.
4. Case of the Yantai Rescue & Salvage Bureau applying for limitation of liability, Qingdao Maritime Court, See H Lixin, *supra* n. 3, 3.
5. W. Zili, 'The Application of Laws on Establishment of Fund for Liability Limitation for Oil Pollution Cases', Conference Proceedings of the Eleventh Forum on Maritime Trials, September 2002, 249; Y. Xiaohan, 'Special Issues on Establishing Fund on Liability Limitation for Oil Pollution Damages', *Maritime Law Review* 1 (20001): 34–38.

However, the majority of academics[6] suggest that purely domestic cases should not be regulated by CLC 92[7] for the four following reasons:

(1) Although many Chinese laws and regulations implement that international treaties concluded or acceded to by China will take preference over domestic laws in case of conflict, most of these provisions are set down in the foreign-related part of relative laws.[8] Therefore, it can be deduced that purely domestic civil and commercial legal relations or civil procedures should be regulated by domestic laws rather than by international conventions.
(2) The purpose of establishing an international civil and commercial convention generally is to eliminate conflict of laws among different countries. Therefore, these conventions should be applied only to international or foreign-related civil and commercial legal relations.
(3) Most shipping enterprises in China are still not sufficiently developed, especially in the aspect of the conditions of tankers – many of them are relatively old and small. These shipping enterprises are apparently not ready for the high limitation of liability established by the international conventions. Therefore they will be overburdened if CLC is applied to purely domestic oil pollution cases.
(4) Applying CLC to purely domestic oil pollution cases will cause the Court to be confronted with many dissonances between Chinese law and CLC.

Article 141 of Minutes of the Second National Conference on Judicial Work of Trial of Foreign-Related Commercial and Marine Cases,[9] which was promulgated by the Supreme Court of the People's Republic of China (PRC) (Minutes) on 26 December 2005 provides:

> CLC 92 that China has acceded to shall be applied to claims for compensation of oil pollution damages caused by ships of contracting states with foreign-related factors, including claims for compensation of oil pollution damages caused by Chinese ships engaged in international shipping lines in Chinese territorial seas. Claims for compensation of oil pollution damages caused by Chinese ships which are not engaged in international shipping lines in Chinese territorial seas shall not be regulated by this convention.

This provision seems to have ended all controversies on this issue; however, considering this issue from the aspects of the basic principle of law, governmental policy and judicial practice, it should be concluded that we should apply CLC to oil

6. Such as, S. Yuzhuo, *Maritime Law Monograph* (original text in Chinese: Hai Shang Fa Zhuan Lun) (Beijing: Legal Press, 2007); Faure & Hui, *supra* n. 3; Lixin, *supra* n. 3.
7. Since the CLC 69 was mostly replaced by the CLC 92, unless otherwise mentioned, the following text of this paper will not discuss the CLC 69 any more.
8. Such as, Ch. 8 of General Principles of Civil Law, Part 4 of Civil Procedure Law.
9. Translated by the author from the original Chinese version.

pollution cases occurring in Chinese territorial seas regardless of whether the cases involve foreign-related factors.[10]

2 REASONS TO APPLY CLC TO CHINESE PURELY DOMESTIC OIL POLLUTION CASES

China should apply CLC to purely domestic cases for several reasons, as discussed in the following section.

2.1 The View That 'International Conventions Only Regulate Legal Relationships with Foreign-Related Factors' Is Inaccurate

One of the main reasons why scholars oppose the application of CLC to purely domestic oil pollution lies in the belief of the aim of international civil and commercial conventions is to clear up conflict of laws among different countries. As a result, these conventions should regulate only international or foreign-related legal relationships but not purely domestic ones.[11]

It is true that one of the aims of international conventions is to eliminate conflict of laws among countries, by way of enacting international uniform substantive law to unify civil and commercial laws, so as to resolve civil and commercial disputes fairly and reasonably. However, to view the conflict elimination as the only aim of international conventions is very lopsided.

Take the well-known Hague Rules,[12] for example; the original intent of these was clearly not to eliminate conflict of laws among countries. In fact, before the Hague Rules were enacted, almost all countries followed the principle of contractual freedom with regard to the contract of carriage of goods by sea covered or proved by a bill of lading. That is to say, conflict of laws was not a problem at that time. The problem was, instead, widespread abuse of the carriers' strong bargaining power – adding numerous unreasonable exemption clauses into bills of lading.

This kind of unbalanced situation existed not only in international transportation, but also in domestic transportation, that is to say, there was no conflict of laws among those countries with regard to bills of lading. The aim of the Hague Rules was actually to correct the imbalance of power between carriers and shippers, and to establish international uniform substantive law by which to set up the

10. See *supra* n. 2.
11. Lixin, *supra* n. 3, 64; Z. Lujun, 'Several Legal Issues on the Disposal of Marine Oil Pollution' (original text in Chinese: Hai Shang You Wu Chu Li Zhong De Ruo Gan Fa Lu Wen Ti), International Forum on Compensation Regime for Oil Pollution Damages in Shanghai, June 2001; S. Weili, 'On the application of International Conventions', *Annual of China Maritime Law* (2003): 195.
12. The Hague Rules is the short name of International Convention for the Unification of Certain Rules of Law Relating to Bills of Lading, 1924.

minimum statutory obligations and maximum rights for ocean carriers. Probably just because of this reason, the Hague Rules stipulates that 'the provision of this convention shall apply to all bills of lading issued in any of the contracting states' and did not limit the carriage of goods within the scope of 'international carriage'. Namely, when a bill of lading or other document of title was issued under a purely domestic carriage between two or more ports of a contracting state, it was still regulated by the Hague Rules.[13]

It is obvious from the example of the Hague Rules that preventing conflicts of law is not the only aim of international civil and commercial conventions, and some conventions are constituted to remedy unreasonable legal discrimination in a certain area or to fill gaps in laws. These international conventions can be applied not only to international civil and commercial legal relationships, but also to purely domestic relationships of contracting states if the constitutors of these conventions intend to do so.

If we review the historical background of CLC 69, we will find that this convention was established shortly after the serious marine oil pollution event involving the *Torrey Canyon*, in which the damaged parties received almost nothing for compensation. Due to the *Torrey Canyon* accident, the international community realized that the damaged parties might not be reasonably protected if shipowners were entitled to limit liability according to the then-valid convention – the International Convention relating to the Limitation of the Liability of Owners of Sea-Going Ships (1957). The existing law at that time could unreasonably prejudice the damaged party in an oil pollution incident and not sufficiently discourage harm to the ocean environment. As a result, it was necessary to establish an independent system of limitation of liability of oil pollution at that time. In fact, CLC 69 also stated its purpose in the beginning:

> conscious of the dangers of pollution posed by the worldwide maritime carriage of oil in bulk, convinced of the need to ensure that adequate compensation is available to persons who suffer damage caused by pollution resulting from the escape or discharge of oil from ships, desiring to adopt uniform international rules and procedures for determining questions of liability and providing adequate compensation in such cases.

The scope provision of CLC 69[14] indicates no limitation to foreign-related issues. On the contrary, the drafters intended that CLC 69 protect all contracting states from oil pollution – related damage, and the origin and ownership of the ship itself was irrelevant. Therefore, if China applies CLC to only foreign-related oil pollution cases, China will fail to meet the purpose of CLC.

---

13. Article 10 of the Hague Rules explicitly stipulated: 'the provisions of this convention shall apply to all bills of lading issued in any of the contracting states'.
14. See Art. 2 of the CLC 69: 'This Convention shall apply exclusively to pollution damage caused on the territory including the territorial sea of a contracting state and to preventive measures taken to prevent or minimize such damage.'

Moreover, many international civil and commercial conventions have clearly stated that they regulate international or foreign-related issues only at the very beginning of their establishment. For example, the United Nations Convention on Contracts for the International Sale of Goods, 1980 (CISG) provides:

> This Convention applies to contracts of sale of goods between parties whose places of business are in different States.

The Hamburg Rules[15] stipulate that

> This convention is applicable to all contracts of carriage by sea between two different States when these states satisfy certain conditions.[16]

Because these international conventions have explicitly restricted their scope to only foreign-related civil relationships, we certainly cannot apply them to purely domestic issues for the same reason. In contrast to the CISG and the Hamburg Rules, CLC has no such foreign-related limitation in the scope of application.

2.2 The So-Called 'Principle of Inside-Outside Distinction' Should Not Be the Basic Principle for Applying International Conventions in China

Some scholars allege that the 'principle of inside-outside distinction' should be followed when applying an international civil and commercial convention in China.[17] According to this principle, the international conventions concluded or acceded to by China should only regulate foreign-related legal relationships, and should never apply to any domestic relationships. Moreover, almost all application provisions of treaties are in the foreign-related chapter or part of Chinese civil laws,[18] although the Chinese Constitution does not provide explicit guidance on it. Based on the above reasons, these scholars point out that 'principle of inside-outside distinction' is the basic principle of Chinese civil laws.

However, the so-called 'principle of inside-outside distinction' might disobey the basic principle of 'the treaty must be preserved' which is generally accepted by the international community. According to the principle of 'the treaty must be preserved', every contracting state should perform all obligations of treaties

15. See Art. 1 of United Nations Convention on the Carriage of Goods by Sea, 1978.
16. See Art. 2 of United Nations Convention on the Carriage of Goods by Sea, 1978.
17. Many Chinese laws or regulations contain a chapter or part named 'Application of Law in Relation to Foreign-related Matters', for example, Ch. 8 of the Principle of Civil Law. In these chapters or parts, there are always some provisions about application of international conventions to foreign-related legal relationships. Therefore, due to the common use of such provisions in China, many scholars hold that all international conventions conclude or acceded to by China should only apply to foreign-related issues rather than purely domestic ones, based on this 'principle of inside-outside distinction'.
18. Such as Ch. 8 of General Principles of the Civil Law; Ch. 14 of Maritime Code; Part 4 of Civil Procedure Law.

concluded or acceded to in good faith unless there are contrary reservations. Therefore, in the absence of a relevant reservation, any action taken by the contracting state to exclude treaties concluded or acceded to from being applied to purely domestic legal relationships will break the principle of 'the treaty must be preserved' whether in legislation or judicial practice.

Although the provisions of the application of international treaties are in the chapter or part named 'Application of Law in Relation to Foreign-related Matters' or the similar in all relevant Chinese laws and regulations, this should not become the evidence of the establishment of the 'principle of inside-outside distinction' in China for two reasons: (1) The provisions stipulate that conventions should be applied to foreign-related legal relationships when there is no contrary reservation in Chinese laws does not imply that the treaties concluded or acceded to by China do not also regulate purely domestic relationships; (2) The Chinese Constitution and other Chinese laws do not clearly establish 'the principle of inside-outside distinction', that is, explicitly exclude international conventions from applying to any purely domestic civil and commercial legal relationships.

2.3 Applying CLC to Purely Domestic Cases Could Accelerate the Development of Chinese Shipping Industry

A point of view opposing application of CLC to purely domestic cases is that the high-level limitation of liability of CLC will heavily burden the Chinese oil shipping enterprises engaged in purely domestic shipping lines and consequently damage the Chinese shipping industry.[19]

Although, strictly speaking, whether applying CLC to purely domestic oil pollution cases will damage or benefit the Chinese shipping industry is a policy issue but not a problem of law, the above view still needs to be argued because holding a purely domestic carrier to a lower safety standard than foreign-related carrier has no theoretical or practical justification. In the reverse, applying CLC to purely domestic oil pollution cases will actually benefit the whole oil shipping industry of China in the long run.

From the opponents' point of view, in the current Chinese oil shipping market, most oil shipping enterprises are small companies and most of their oil tankers are relatively small and old, thus facing a higher accident and pollution rate. As a direct result, these small companies cannot bear much risk of accident. Applying CLC to purely domestic oil pollution cases require these small companies to spend more money on insurance fees, possibly driving them out of business and possibly leading to a shortage of shipping capacity, consequently damaging the whole Chinese oil shipping industry. However, the experience of the reform and opening-up practice of China might persuasively refute these arguments. China's thirty-one years of experience with the reform and opening-up practice have proved one truth: protecting the laggard productivity inhibits the development

19. Faure & Hui, *supra* n. 3; Lixin, *supra* n. 3, 64.

of advanced productivity. Just because there are many small, poorly managed companies and small or old tankers in the oil shipping market, oil pollution accidents have occurred frequently in recent years in Chinese territorial seas and qualified shipping enterprises and advanced oil tankers struggle in a crowded market with low shipping prices made possible by unsafe shipping. Therefore, applying CLC to purely domestic oil pollution cases will promote the optimum competition in Chinese shipping industry, especially the oil shipping industry. Another benefit is to speed up the elimination of old ships and the purchase of new oil tankers and thereby to generally speed up the development of the Chinese ship-building industry.

The present situation protects shipping interests at the expense of the victims and the marine environment. According to Article 142 of the Minutes, the liability of liable persons in a purely domestic oil pollution caused by ships should be determined in terms of China Maritime Code (CMC), Chinese Marine Environment Protection Law and other regulations, and liable person in oil pollution cases may be entitled to limitation of liability pursuant to Chapter 11 of the CMC.

This provision of the Minutes apparently might deviate from universally accepted principles of compensation for oil pollution damages – the strict liability principle and higher limitation of liability. It will not only cause insufficient compensation of domestic damaged parties but also will do harm to Chinese marine environment protection. In the long run, national interests will be harmed if we do nothing about it.

2.4 Applying CLC to Purely Domestic Oil Pollution Cases Could Resolve Some Practical Issues in Chinese Judicial Practice

Those opponents also argue that applying CLC to purely domestic oil pollution cases will cause Chinese courts to be confronted with many dissonances between Chinese law and CLC;[20] however, they fail to specify the dissonances they fear.

Applying CLC to purely domestic oil pollution cases will actually remove some practical issues in Chinese judicial practice. For example, applying CLC to all oil pollution cases[21] will eliminate all discrepancies of the understanding of 'foreign-related factors'.

According to Article 141 and Article 142 of the Minutes, Chinese Courts should apply CLC only to foreign-related oil pollution cases. As for purely domestic cases, Chinese domestic law, such as the CMC and Marine Environment Protection Law, should be applied. So, when a lawsuit demanding compensation for oil pollution damage goes before a Chinese Court, the Court must first determine whether this case contains foreign-related factors. It is generally accepted

20. Zhengyi & Pengjiu, *supra* n. 3, 102.
21. See *supra* n. 2, all oil pollution cases here refers to all cases involving with sea-going vessel and seaborne craft in Chinese territorial waters.

under Chinese laws that a civil relationship will be a foreign-related civil relationship as long as any element of it contains foreign-related factors – subject, object or factum juridicum – thus any case involving any foreign-related civil relationship is a foreign-related case. However, some scholars think that if both the injured party and the owner of the oil-spilling ship come from China, the case should be deemed as a purely domestic case – even if there are foreign-related factors present.[22] In other words, it is suggested that if the legal subjects of an oil pollution relationship contain no foreign-related factors, the CLC should not be applied to such an oil pollution case. Based upon this viewpoint, these scholars disagree with Article 141 of the Minutes, which prescribes that the CLC should be applied to oil pollution damage occurring in Chinese territorial seas caused by a Chinese ship engaged in international shipping lines.

Furthermore, some 'Chinese' geographical areas are not clearly defined as a foreign area or Mainland Chinese area in current Chinese judicial practice, such as Hong Kong and Macau. Consider ships transporting oil between ports of mainland China and Hong Kong, Macau and Taiwan. If such ship spills oil within Mainland China's territory, should the case be deemed foreign-related? At present, no conclusion has been drawn about this issue in both theory and judicial practice in China. However, if China uniformly applies the CLC to all cases arising from oil pollution damage, the aforesaid difficulty will be removed and the Chinese courts can entirely avoid the question of whether a case is foreign-related.

3 CONCLUSION

In short, CLC should also apply to purely domestic oil pollution cases occurring in Chinese territorial seas. Therefore, it is suggested that Article 141 of the Minutes should be modified accordingly. At an appropriate time in the future, the basic principle of the application of international civil and commercial conventions should be explicitly stipulated in high-status laws of China, such as the future Chinese Civil Code or even the Chinese Constitution.

It is worth emphasizing that the Regulations of the PRC on the Prevention of Vessel-Induced Marine Environmental Pollution[23] was just promulgated by the State Council of China on 2 September 2008. Article 52 of the said Regulation provides:

> The limitation of liability for claims for ship-resource pollution damages shall be determined in accordance with the provisions of the Maritime Code of the People's Republic of China relating to the limitation of liability for maritime claims. But when pollution damage is caused by persistent oil in bulk carried on ships in the territorial sea of the People's Republic of China, the limitation of liability for ship owners shall be determined in accordance with the related

22. See *supra* n. 20.
23. Author's translation from the original Chinese version.

> international convention concluded or acceded to by China. The persistent oil in the foregoing paragraph means any persistent hydrocarbon mineral oil.

According to the provision, as long as a pollution accident in Chinese territorial seas involved with persistent oil in bulk carried on ships, the limitation of liability for the claim shall be decided in accordance with the CLC, regardless of whether there is any foreign-related factor. This provision is distinctly different from Article 142 of the Minutes and the former is more consistent with the meaning of the CLC.

In addition, there is another inaccuracy existing in Article 141 of the Minutes, which reads as follows:

> The CLC 92 that China has acceded to shall be applied to claims for compensation of oil pollution damages caused by ships of contracting states with foreign-related factors, including claims for compensation of oil pollution damages caused by Chinese ships engaged in international shipping lines in Chinese territorial seas.

According to this provision, CLC will regulate oil pollution cases caused only by vessels of contracting states with foreign-related factors. This prescription also seems to conflict with the original aim of CLC. According to CLC paragraph 1 of Article 2[24] and Article 3,[25] CLC does not require the liable vessel to be owned by a contracting country. As long as the oil pollution occurred in the territorial seas of a contracting state, even if the liable vessel belonged to a non-contracting country, the claim for the oil pollution damages should also be regulated by CLC. Although according to the provisions concerning compulsory insurance in CLC, only vessels of contracting states will be considered qualified vessels, whereas in the aspect of compensation liability, there is no similar limit. All of these regulations are in line with the basic principle of international private law; where a tort or offence has been committed the law of the place of tort should be applied. For these reasons, Article 141 of the Minutes, which limits the vessel in the scope of contracting countries' vessels is obviously inappropriate.

---

24. See para. 1 of Art. 2 of the CLC 92: '1. Paragraph 1 is replaced by the following text: 1. 'Ship' means any sea-going vessel and seaborne craft of any type whatsoever constructed or adapted for the carriage of oil in bulk as cargo, provided that a ship capable of carrying oil and other cargoes shall be regarded as a ship only when it is actually carrying oil in bulk as cargo and during any voyage following such carriage unless it is proved that it has no residues of such carriage of oil in bulk aboard.'
25. See para. 1 of Art. 3 of the CLC 92: 'This Convention shall apply exclusively: (a) to pollution damage caused: (i) in the territory, including the territorial sea, of a Contracting State, and (ii) in the exclusive economic zone of a Contracting State, established in accordance with international law, or, if a Contracting State has not established such a zone, in an area beyond and adjacent to the territorial sea of that State determined by that State in accordance with international law and extending not more than 200 nautical miles from the baselines from which the breadth of its territorial sea is measured.'

References

Faure, M. & W. Hui. 'Financial Caps for Oil Pollution Damage: China and the International Conventions'. In *Prevention and Compensation of Marine Pollution Damage: Recent Developments in Europe, China and the US*, edited by M.G. Faure & J. Hu. Alphen aan de Rijn: Kluwer Law International, 2006.

Lixin, H. *Study on Legal System of Compensation for pollution damages from Ships* (original text in Chinese: Chuan Bo Wu Ran Sun Hai Pei Chang Fa Lu Zhi Du Yan Jiu). Beijing: Legal Press, 2007.

Lujun, Z. 'Several Legal Issues on the Disposal of Marine Oil Pollution' (original text in Chinese: Hai Shang You Wu Chu Li Zhong De Ruo Gan Fa Lu Wen Ti). Papers Collection of Shanghai International Forum on Compensation Regime for Oil Pollution Damages, June 2001.

Weili, S. 'On the Application of International Conventions'. *Annual of China Maritime Law* 14 (2003) 195.

Xiaohan, Y. *Case on Limitation of Liability for Oil Pollution Damage Compensation* (original text in Chinese), <www.maritimelawyer.cn/?p=520>.

Xiaohan, Y. 'Special Issues on Establishing Fund on Liability Limitation for Oil Pollution Damages'. *Maritime Law Review* 1 (2001): 34–38.

Yuzhuo, S. *Maritime Law Monograph* (original text in Chinese: Hai Shang Fa Zhuan Lun). Beijing: Legal Press, 2007.

Yuzhuo, S. *The Application of Laws on Compensation for Oil Pollution Damage Caused by Costal Transport Ships* (original text in Chinese: Yan Hai Yun Shu Chuan Bo You Wu Sun Hai Pei Chang Fa Lv Shi Yong Wen Ti Yan Jiu). *Journal of Dalian Maritime University* (*Social Science Edition*), Nos 1–2 (June 2002).

Zhengyi G. & L. Pengjiu. *On Several Legal Issues of Application of CLC* (original text in Chinese: CLC You Wu Min Shi Ze Ren Gong Yue Shi Yong De Ruo Gan Fa Lu Wen Ti). *Journal of Shenyang Normal University Science Edition* 31 (2007): 6.

Zili, W. 'The Application of Laws on Establishment of Fund for Liability Limitation for Oil Pollution Cases'. Conference Proceedings of the Eleventh Forum on Maritime Trials, September 2002.

Chapter 20

Compensation for the Domestic Oil Pollution in China's Coast: Which Law Shall Apply?

*Zhang Liying**

1 INTRODUCTION

For oil spill compensation, although China is a state party of the International Convention on Civil Liability for Oil Pollution Damage (CLC),[1] the scope of its application has always been controversial. It is of no doubt that CLC should be applied to foreign-related oil spills; the controversial issue is whether it should also be applied to domestic oil spills. In domestic oil spills, the polluter mainly flies a Chinese flag and the spill itself takes place in Chinese territorial sea. After a careful review of the domestic oil spill compensation practices by the various Chinese maritime courts, the author noted that the applicable laws are mainly the following: (1) the domestic law, *inter alia*, the Marine Environment Protection Law;[2] the

* Professor, *Scientiae Juridicae Doctor*, and Tutor for the doctoral students in China University of Political Science and Law.

1. China acceded to the International Convention on Civil Liability for Oil Pollution Damage, 1969 (hereafter 1969 CLC) on 30 Jan. 1980 and its 1992 Protocol on 5 Jan. 1999. These two conventions entered into force in China on 29 Apr. 1982 and 5 Jan. 2000 respectively. China acceded to the 2000 Amendments afterwards.
2. Y. Si, 'Research on Applicable Law in the Compensation of Oil Pollution Damages from the Coast Ship', *Journal of Dalian Maritime University (Social Science Edition)* 6 (2002): 1.

Michael G. Faure, Han Lixin & Shan Hongjun, *Maritime Pollution Liability and Policy: China, Europe and the US*, 359–369.

Maritime Code of the People's Republic of China,[3] and the General Principles of the Civil Law;[4] (2) the international conventions;[5] and (3) the international practice.[6] This diverse situation in the application of laws poses a negative effect on the stable protection of the interests of the two parties involved in the oil spill compensation. In order to ameliorate the governance of compensation in domestic oil spills, the clarification of the application of law is very important.

2 THE APPLICATION OF CLC IN THE DOMESTIC OIL SPILLS

Some scholars are of the opinion that domestic oil spills should apply CLC, according to which (specifically CLC1992) the application could cover all spills of the persistent oils (regardless of the tonnage) by all types of vessels (regardless of the flag state) in the territory (including the territorial sea) of a state party. The oil spill by a vessel in her flag state's territory is not an exception in the application of CLC. Consequently, considering that China is indeed a state party to the CLC and no reservation was announced at accession, the CLC shall be applied to all marine oil spills, foreign-related or domestic, before China can enact a special law dealing with this issue.[7] Nonetheless, in practice, many doubts have been raised concerning the idea of applying CLC to coastal oil spills. In the *MINRANGONG 1* case,[8] one of the opinions expressed by the court is Article 13 of the Regulations Concerning the Prevention of Pollution of Sea Areas by Vessels, which excludes application of CLC to the vessels with a gross tonnage not exceeding 2000 tonnes. According to Article 13:

> Vessels engaged in international trade with a bulk oil carrying capacity of 2,000 tons or more shall, besides observing these Regulations, be bound by the provisions of the International Convention on Civil Liability for Oil Pollution Damage, 1969.

Other scholars hold that the international conventions should be applied to foreign-related cases, but not to coastal domestic cases; those of the opposite opinion maintain that according to the Environmental Protection Law, international conventions should take precedence over domestic law in environmental protection

3. The Oriental Ambassador Case, please see Z. Zhang & Y. Yin, 'Law Is the Key for the Marine Environment Protection: Scholars Meets in Tianjin', *Tianjin Daily* (2 Oct. 2002).
4. For the Athens Horizon case, see *supra* n. 2.
5. For more information about the *MINRANGONG 2* case, please see G. Xu & Z. Li, 'Study on Several Legal Issues in the Maritime Cases', <www.ccmt.org.cn/hs/explore/exploreDetial.php?sId=2353>.
6. The *South Ocean* case, see *supra* n. 3.
7. *The Limits of Liability and the MINRANGONG 2 case*, the internet version is available at <http://sina.chinalawedu.com>.
8. L. He & H. Wu (eds), *The Analysis of the Selective Maritime Cases* (Xiamen: Xiamen University Publishing, 2004), 247.

cases. CLC is a convention related to environment protection, and China did not announce any reservations during ratification, hence the CLC should be applied in China.[9]

The author is of the opinion that the precedence and the direct application of the international treaties mainly applies to foreign-related civil relations. As for its application in domestic relations, there is no clear and unified provision in our constitution. Judging from several domestic laws relating to the international treaties, the direct application, the parallel application and the domestic incorporation of the international treaties exist simultaneously. In the meantime, there are also several laws without provisions in relation to the international treaties. In civil matters, the international treaties that China has ratified or acceded to may take the precedence directly, whereas in the other relations, the manner of the application of the international treaties shall be decided on a case-by-case basis. Article 142 (2) of the General Principles of the Civil Law states that:

> If any international treaty concluded or acceded to by the People's Republic of China contains provisions differing from those in the civil laws of the People's Republic of China, the provisions of the international treaty shall apply, unless the provisions are ones on which the People's Republic of China has announced reservations.

Article 238 of the Civil Procedural Law also has a similar provision. These are the legal basis for the application of international treaties in civil cases in which their provisions are different from the domestic law of China. Outside the civil relations, as a result of the inconsistent practice and the various concerns, especially the lack of a constitutional basis, it cannot be generally concluded that the direct application or precedence of the international treaties has already or will become the only way for their application in China. Consequently, the application of non-civil international treaties (including human rights treaties) in China can be decided only after careful research into the relevant domestic law and the provisions of the treaties. Article 142 of the General Principles of the Civil Law is placed under Chapter VIII Application of the Law in Civil Relations Involving Foreign Parties; the similar clause in the Maritime Code has also been placed under the chapter of the foreign related matters. This means that the precedence of the treaties here is specifically targeting the 'Civil Relations with Foreigners'. According to Article 178 of the Opinions of the Supreme People's Court on Several Issues Concerning the Implementation of the General Principles of the Civil Law,

> Civil relations with foreign elements means one or two parties of the civil relations are foreigner, stateless persons or foreign legal persons; or the object of the civil relations is outside of the People's Republic of China; or the legal facts which establish, change or terminate the civil relations are occurred outside of China.

---

9. Z. Wen & H. Yu, 'The Application of the Liability Limits of 1969 CLC on the Domestic Coastal Tankers', in *Annual of China Maritime Trial*, ed. Z. Jin (Beijing: Court Press, 2000), 515–516.

Consequently, for civil matters without foreign elements, there should not be an issue of the direct application of the treaties. Based on the literal interpretation of the legislation, the author holds that for cases in which the oil spill occurred in China's territorial water and both of the parties are under the Chinese flag, there is no foreign element involved; hence it would not involve the issue of direct application of CLC.

In addition, considering the fact that unlike most of the Western developed countries, Chinese laws separate foreign-related civil relations from domestic civil relations; that is, the applicable law for the foreign related relations and non-foreign-related relations is different. The limits of liability for passenger claims is a perfect illustration of this. China acceded to the Athens Convention Relating to the Carriage of Passengers and Their Luggage by Sea, 1974 and has incorporated its limits of liability into the China Maritime Code (CMC); however, the limits of liability do not apply in coastal passenger carriage. The liability for the international passenger carrier is Special Drawing Right (SDR) 46,666 per passenger;[10] whereas the coastal passenger carrier has a much lower limit of only China New Yuan (CNY) 40,000 per passenger.[11] As for the liability for marine claims, there is also a huge difference between the international limit and the coastal limit. The international limit is the same with the Convention on Limitation of Liability for Maritime Claims (LLMC) 1976;[12] whereas the coastal limit shall comply with the regulations of the Department of Transportation.[13] This rule has significantly reduced compensation for coastal claims. Its rationale is connected to the special national conditions of China – an under-developed economy and a lower level of life quality. Although China has joined several international conventions, in the foreseeable future, the application of the international standards in the domestic cases would still be impossible. Chinese coastal vessels, especially those tankers with smaller tonnages, cannot afford the liability limits in the convention, even if the 1992 Protocol to CLC has already entered into force for China. During the drafting process of the international conventions, this situation had already been taken into consideration. The convention clearly states the application does not apply to domestic relations,[14] allows the states to make reservations on this issue[15] or provides for

10. See The Maritime Code of the People's Republic of China, Art. 117.
11. See Provisions of the Limitation of Liability With Respect to the Carriage of Passengers by Sea Between the Ports of the People's Republic of China (Promulgated by Decree No. 6 of the Ministry of Communications), Art. 3.
12. *Supra* n. 10, Ch. XI.
13. See Provisions Concerning the Limitation of Liability for Maritime Claims for Ships With a Gross Tonnage Not Exceeding 300 Tonnes and Those Engaging in Coastal Transport Services As Well As Those for Other Coastal Operations (Promulgated by Decree No. 5 of Ministry of Communications on 15 Nov. 1993), Art. 4.
14. See Athens Convention Relating to the Carriage of Passengers and Their Luggage by Sea, 1974, Art. 2 (1) and United Nations Convention on the Carriage of Goods by Sea (The *Hamburg Rules*) *Hamburg*, 1978, Art. 2 (1).
15. International Convention on Salvage, 1989.

direct application of domestic law,[16] so as to leave room for those monist[17] countries. The author noted that the major features of the coastal carriers are small-tonnage, old age, high incident rates and inability to compensate. Comparing with the international carriage, the technical condition and seafarer's professional skills of the coastal carriage are significantly lagging behind. Under this circumstance, it would be unfeasible to apply the CLC standard regardless of the actual conditions of the domestic marine carriage industry.

3 THE REASONS FOR THE APPLICATION OF THE DOMESTIC LAW

After careful review of all of the cases after 2001 on the official website of China Foreign-Related Commercial and Maritime Trial, the author noted that most of the cases applied Chinese domestic law, with only one case – the *China Marine Bunker (Petro China) Co., Ltd* case (also known as the *MINRANGONG 2* case) applied the liability limit of CLC 1969. However, this judgment was later overruled by the court of the second instance on the grounds of the Maritime Code. The MINRANGONG 2 was a bunker liner owned by China Marine Bunker (Petro China) Co. She sank after the collision with the empty tanker *Donghai 209* near the port of Guangzhou. China Marine Bunker (Petro China) Co. applied for the liability limit prescribed in CLC 1969 and Protocol 1992 before the Guangzhou Maritime Court. Zhuhai Environmental Protection Bureau and the Marine & Aquatic Product Office of the Guangdong Province filed an objection to this application. The two objectors held that *MINRANGONG 2* was not a ship that fell under the jurisdiction of the CLC, and no foreign elements were involved in this case, hence CLC regulations should not be applied in this case. The Guangzhou Maritime Court concluded that according to the Environmental Protection Law, international conventions should take precedence over domestic law in environment protection cases. CLC 1969 is a convention related to environment protection, and China did not announce any reservations during ratification; consequently, CLC 1969 should be applied to the oil spill compensation of domestic carriers not exceeding 2,000 tonnes. The Guangzhou Maritime Court thus ruled in favour of the applicant.[18]

The prima facie issue of the application inconsistency in domestic oil spill cases is the application of law; however, the real issue at stake is the balance of interest between the victim and the polluter. In this regard, the liability limits of CLC 1969 and its Protocol are either too high or too low for China. This is one of the main reasons why the maritime courts choose to apply the domestic law. The liability limits in CLC 1969 are very low, with the highest approximately CNY 4,000,000. According to the Maritime Safety Administration, for the period from

16. See International Convention on Certain Rules Concerning Civil Jurisdiction in Matters of Collision, 1952, Art. 8.
17. H. Kelsen, *Principles* of *International Law* (Huaxia Publishing, Beijing, 1989), 132.
18. Z. Wen & H. Yu, *supra* n. 9.

1996 to 2001, the average damage of the domestic oil spill was CNY 5,050,000. The liability limits of CLC 1969 cannot even cover the average damage. Moreover, 80% of the tankers in operation are small vessels with a total tonnage not exceeding 1,000 tonnes.[19] In order to compete for oil carriage, domestic companies began to build or buy big tankers;[20] however, the larger the tanker is, the more oil it can carry. There is increased potential for incidents that will definitely incur huge amounts of compensation, and if CLC 1969 is applied, the liability limits will not be able to protect the interests of the victims. CLC 1992 significantly increased the liability limits, and the 2000 Amendments further lifted CLC 1992, by another 50.37%.[21] This trend reflects that the society is paying careful attention to the interest of the victims. However, in this case, even a vessel carrying only 200 tonnes could be liable for compensation higher than CNY 36,000,000 in one incident, if the 1992 CLC is applied, and even more if the 2000 Amendments are applied. This will probably pose a bankruptcy threat to coastal carriers and will definitely cripple the coastal industry in the long term. The liability limits of CLC are either too high or too low to suit the need of the domestic oil spills. This is the reason most Chinese courts choose to apply the domestic law. The domestic law, for example, the Maritime Code, although not entirely perfect, is at least less extreme than CLC. Its liability limits are somewhere in between the CLC 1969 and CLC 1992.

4 THE RATIONALE FOR THE APPLICATION OF DOMESTIC LAW

Although many scholars are in favour of the application of domestic laws in domestic oil spills, there is a slight difference as to which domestic law shall apply specifically. As mentioned earlier, the main suggestions are the Marine Environment Protection Law, the General Principles of the Civil Law and the Maritime Code.

4.1 The Limitations of the Marine Environment Protection Law

The Environment Protection Law (EPL) was enacted in 1989 and the Marine Environment Protection Law (MEPL) in 1982 and amended in 1990. Article 90 of the MEPL prescribes that:

> Those who cause pollution damage to the marine environment shall eliminate the damage and compensate the losses; in case of pollution damage to the

19. H. Liu, 'Speed up the establishment and implementation of China's Oil Spill Compensation Mechanism' (2003) *CMLA Newsletter* 1.
20. G. Xu, *Study on the Legal Regime for the Vessel-Induced Oil Spill Compensation* (Peking University Press, Beijing, 2006), 159.
21. Please see the sixteenth public notice of the Ministry of Transport re the 1992 CLC, the 2000 Amendments and the 1992 Fund Convention 1992, <www.moc.gov.cn/05zhengwu/haishijl/t20031203_24831.htm>.

> marine environment resulting entirely from the intentional act or fault of a third party, third party shall eliminate the damage and be liable for the compensation. If the State suffers heavy losses from the damages to marine ecosystems, marine aquatic resources and marine nature reserves, the departments invested by this law with the power of marine environment supervision and administration shall, on behalf of the State, put forward compensation demand to those who are responsible for the damages.

The cleanup cost is deemed a civil liability and is included in this chapter in the scope of oil spill compensation. Although both EPL and MEPL have general articles for oil pollution compensation, neither contains specific regulations concerning the scope of compensation, the system of compensation and the compensation limits. For example, there is no provision as to whether the (middle-) long-term damage caused by the oil spill should be included in the scope of compensation.

The generalized wording of the pertinent MEPL provisions rendered the MEPL almost inoperable. As a result of that, it is rarely used as the applicable law. Of all of the cases after 2001 on the China Foreign-related Commercial and Maritime Trial website, only one case used MEPL as the applicable law in deciding 'whether the recourse should be against the defendant' [translation].[22] This is vivid proof of the application limitations of the MEPL.

4.2 The Strict Liability and Compensatory Damage in the General Principles of the Civil Law Is Not Applicable to Domestic Oil Spill Disputes

Of all of the cases in which the General Principles of the Civil Law (Civil Law) is applied, the most quoted provisions are Articles 117(2),[23] (3)[24] and 124.[25] The *YANJIUYOU 2* case is one in which the Civil Law is applied. The incident occurred in August 1994; the competent court, Qingdao Maritime Court, found that the shipowner should not be entitled to the liability limits. The court applied the Civil Law and ordered a compensation of CNY 5,108,200.[26] Article 124 of the Civil Law applied strict liability to environmental pollution, according to which,

22. See *Hailong Co. Ltd v. Shilong Waterborne Transportation Co. Ltd*, Case No. 106 [2002] Guangzhou Maritime Court, <www.ccmt.org.cn/hs/writ/judgementDetial.php?sId=766>
23. See the General Principles of the Civil Law, Art. 117 (2): Anyone who damages the property of the state, a collective or another person shall restore the property to its original condition or reimburse its estimated price.
24. *Supra* n. 23, Art. 117(3): If the victim suffers other great losses therefrom, the infringer shall compensate for those losses as well.
25. *Supra* n. 23, Art. 124: Any person who pollutes the environment and causes damage to others in violation of state provisions for environmental protection and the prevention of pollution shall bear civil liability in accordance with the law.
26. For the *YANJIUYOU 2* Case, please see the Civil judgment by the Appeal Court of Shangdong Province Case No. 32 [1997].

any person who pollutes the environment and causes damage to others in violation of state provisions for environmental protection and the prevention of pollution shall bear civil liability in accordance with the law. The oil spill damage, which is a special tort, will have to use strict liability if this provision is applied. As for the scope of liability, Article 117 of the Civil Law sets out the principle of a compensatory damage. If all of the oil spill cases apply the principle of compensatory damage each time there is a major oil spill, a liner company will go bankrupt because no insurer will ever accept an unlimited liability for the oil spill insurance. This will significantly harm the development of the marine oil carriage and is completely against the international trend of shipowner's limited liability. In practice, due to the limited repayment ability of the shipowners, most of the domestic oil spill compensation orders have not been executed entirely. The previously mentioned *MINRANGONG 2* case is a typical example. The liable party in this case has the ability to execute only 5.6% of the total compensation ordered by the competent court.[27] Facing such a low rate, the international community and some countries established special funds as a complementary entity. The Fund contributes for the part that exceeds the shipowner's ability and the oil spill pollution without a liable party. This is why internationally the execution rate of the compensation order is relatively higher. Consequently, it will be inappropriate if one only uses the 'fairness' principle of the Civil Law when judging the oil spill disputes. The special features of the oil spill require a *lex specialis* for its regulation.

In addition, to apply the compensatory damage in domestic oil spills will cause imbalances within the different subdivisions of the maritime law. As mentioned earlier, the difference in liability limits between international and coastal passenger carriage are huge. This is the same situation with liability limits of maritime compensation. However, if the domestic oil spill is regulated by Civil Law, the oil spill compensation will not only be higher than the seagoing carriage, but also unlimited. Considering the unmatchable advantages of seagoing carriage – the technical conditions and professional abilities of the seafarers – the application of the Civil Law incurs manifest unfairness and creates incoherence with the entire limited liability system of maritime law.

4.3 The Application of the Maritime Code in the Coastal Oil Spills Compensation

Based on the this analysis, the author holds that the domestic oil spill should not involve the application of international conventions, yet as for the application of the domestic law, as discussed above, there are also some problems. Although the Maritime Code still contains certain imperfections, yet in the absence of a *lex specialis* for the compensation of the domestic oil spills, it remains a wiser choice.

27. Guangzhou Maritime Court ordered a total compensation of CNY 10,600,000, yet the shipowner only carried out CNY 600,000 and 20% of the cleanup costs.

Comparing to the Civil Law, the Maritime Code is a *lex specialis*, in view of the *lex specialis derogat priori generali* principle; it shall certainly take precedence over the Civil Law. Most cases in which the Maritime Code is applied are oil spills caused by vessel collision.[28] This indeed reflects the limitation of the Maritime Code in dealing with coastal oil spills. In the relevant cases, the most cited provision of the Maritime Code is Article 169(2), which states:

> The ships in fault shall be liable for the damage to the ship, the goods and other property on board pursuant to the proportions prescribed in the preceding paragraph. Where damage is caused to the property of a third party, the liability for compensation of any of the colliding ships shall not exceed the proportion it shall bear.

The court normally cites this provision as the legal basis for oil spill compensation and deems the oil spill to be the 'other property' caused by the collision. But the oil spill is a special maritime tort. It is different from the ordinary maritime tort in the sense that the damages it caused are more widespread, more resistant, more expensive and difficult to clean. It would not be reasonable if it is put together with the other ordinary torts. This caused limitations in the application of the Maritime Code in the oil spill disputes.

Apart from this, there are also certain problems in the limits of liability. Article 210 of the Maritime Code prescribes the liability limits for coastal carriage, according to which, the limits shall be in accordance with the Provisions Concerning the Limitation of Liability for Maritime Claims for Ships With a Gross Tonnage Not Exceeding 300 Tons and Those Engaging in Coastal Transport Services As Well As Those for Other Coastal Operations (Promulgated by Decree No.5 of Ministry of Communications on 15 November 1993). According to the data gathered by the Chinese classification association in 1996, 42% of Chinese tankers fall under the tonnage range of 2,000 to 10,000.[29] Within this range, the liability limit is SDR 5,100,000 under the 1992 Protocol; SDR 1,753,500 under the Maritime Code of China; SDR and 876,800 under the provisions of the Department of Transportation. The liability limits in the 1992 Protocols mainly target tankers in the international shipping line, which is obviously too high for the domestic tanker in coastal carriage. On the other hand, considering the immeasurable long-term damage to ocean and the victims, the liability limits in the provisions of the Department of Transportation are clearly too low. Its application will not increase the shipowners' attention in preventing the oil spill incidents, and in the end not will not be favourable for the protection of the marine environment. Consequently, the ideal solution

28. Please *see Guangdong Provincial Oceanic and Fishery Administration v. Yuyang Shipping Co. Ltd, Tianshen International Shipping Co. Ltd and China Ship Owners Mutual Assurance Association*, Case No. 89 [2001] Guangzhou Maritime Court, <www.ccmt.org.cn/hs/writ/judgementDetial.php?sId=1187>.
29. R. Zhou, 'The Application of Law in the Coastal Vessel-Induced Oil Spill Compensation', <www.ccmt.org.cn>.

will be the establishment of a *lex specialis* specifically governing the domestic oil spills.

5 THE SOLUTION: THE ESTABLISHMENT OF A *LEX SPECIALIS* AND A COMPENSATION FUND

As discussed above, if the limit of liability is too high, the shipowner will be exposed to the danger of bankruptcy, and if it is too low, protection of the victims will be put at risk. In order to guarantee compensation for the victims of the oil spill, China should establish a compensation fund. It is unrealistic to simply impose the huge compensation to the shipowners, the result of this practice can only reduce the vitality of the oil carriage industry. The essential part of the establishment of *lex specialis* is the introduction of the compensation fund. China is not a state party of the International Convention on the Establishment of an International Fund for Compensation for Oil Pollution Damage (Fund Convention) 1971 (hereafter 1971 FUND), and it is one of the countries without the mechanism of a compensation fund. This is why often the victims of an oil spill cannot obtain their compensation, while the shipowners are facing huge compensation. This situation is not good for the development of the marine oil carriage industry. The fact that the biggest beneficiaries of this industry are not sharing the damage does not help solve the problem.[30] If the mechanism of the compensation fund can be introduced rationally, it will, on the one hand, execute compensation for the victims, and, on the other, significantly reduce the burden of the shipowners. Even the oil industry will be a beneficiary of this mechanism.

6 CONCLUSION

The inconsistent application of law in domestic oil spill disputes will affect the protection of the rights of the two parties and the environment. It also weakens the consistency and authority of the maritime judgment. Hence, the clarification of applicable law in domestic oil spill disputes is of vital importance. The author is of the opinion that although China is a member of the CLC, and has stipulated both in the Civil Law and Maritime Code the precedence of the international treaties, these provisions all govern the civil relations with foreign elements. As for the compensation of the domestic oil spills, because there is no foreign element involved, CLC would not come into force. The domestic law should be applied. However, the provisions of the MEPL are too generalized to be operable; the strict liability and compensatory damage in Civil Law are not suitable for the marine oil spill; as the *lex specialis*, the Maritime Code is relatively more suitable for the domestic oil spill, yet there are certain limitations in the sense that the oil spill is different from

30. C. Yan, 'The Development of the Ideology and System of the Maritime Law in the Environmental Era', <www.law-walker.net/detail.asp?id=2715>.

the ordinary tort in maritime law. Consequently, to establish a *lex specialis* specifically governing the domestic oil spill would be the best choice. In order to essentially solve the compensation problem, the pertinent provisions of the *lex specialis* would need to include the introduction of a compensation fund.

References

He, L. & H. Wu (eds). *The Analysis of the Selective Maritime Cases.* Xiamen: Xiamen University Publishing, 2004.

Kelsen, H. *Principles* of *International Law.* Beijing: Huaxia Publishing, 1989.

Liu, H. 'Speed up the Establishment and Implementation of China's Oil Spill Compensation Mechanism'. *CMLA Newsletter* (2003): 1.

Si, Y. 'Research on Applicable Law in the Compensation of Oil Pollution Damages from the Coast Ship'. *Journal of Dalian Maritime University (Social Science Edition)* 6 (2002): 1.

Wen, Z. & H. Yu, 'The Application of the Liability Limits of 1969 CLC on the Domestic Coastal Tankers'. *Annual of China Maritime Trial*, edited by Z. Jin. Beijing: Court Press, 2000.

Xu, G. *Study on the Legal Regime for the Vessel-Induced Oil Spill Compensation.* Beijing: Peking University Press, 2006.

Yan, C. 'The Development of the Ideology and System of the Maritime Law in the Environmental Era', <www.law-walker.net/detail.asp?id=2715>.

Zhang, Z. & Y. Yin, 'Law Is the Key for the Marine Environment Protection: Scholars Meets in Tianjin'. *Tianjin Daily* (2 October 2002).

Zhou, R. 'The Application of Law in the Coastal Vessel-Induced Oil Spill Compensation'. <www.ccmt.org.cn>.

Chapter 21

Compulsory Liability Insurance on Carriage of Liquid and Dangerous Cargo by Coastal and Inland Waters in China

Chu Beiping & Zhang Jinlei***

1 INTRODUCTION

Early in 2004, the Ministry of Transport considered planning to apply a 'compulsory liability insurance system' to the liquid and dangerous cargo transportation by amending the Ordinance of Transportation in Coastal and Inland Waters.[1] The core of this system is that in order to run business, corporations shall apply for the License of Operation, and before they can apply for such license, corporations must obtain a certificate of liability insurance. However, there were different opinions regarding the necessity of such system. Some believed it is necessary to issue compulsory liability insurance on the liquid and dangerous cargo transportation in the Ordinance on Prevention of Marine Environmental Pollution by

* Chu Beiping, Professor of Dalian Maritime University, Ph.D., supporting member of LMAA, vice-director of Institute of Maritime Law of DMU. Zhang Jinlei, senior lecturer of Dalian Maritime University, Ph.D.

** Dalian Maritime University, China.

1. Author's translation from the original Chinese text.

Michael G. Faure, Han Lixin & Shan Hongjun, *Maritime Pollution Liability and Policy: China, Europe and the US*, 371–384.

Ships.[2] Presently, although the previously mentioned two Ordinances are under amendment, the regime of compulsory liability insurance on liquid and dangerous cargo transportation does not seem to be outlined clearly. This chapter will begin by explaining the importance of marine compulsory liability insurance in legislation to prevent pollution, then analyse the defects in the present legislation and necessity and obstacles of compulsory liability insurance on transport of liquid and dangerous cargo transportation and finally present suggestions for the amendment of the above ordinances.

2 THE FUNCTIONS OF THE COMPULSORY LIABILITY INSURANCE

Essentially speaking, insurance is a legal act by which the contracting parties will enter into an insurance contract voluntarily. Nevertheless, because the social division of labour and exchanges are strengthened by social and economic development, the foundation of the principle of liberty of contract has been shaken. In some circumstances, such freedom of contract is distributed and restricted by laws for the public interest. The public interest of the whole society has gradually become the standard to supervise and review the lawfulness of private acts and private interest. Today, people are increasingly involved in more highly dangerous actions, and the anxiety of the impossibility to anticipate future safety needs is intensified. Thus, compulsory insurance is being separated from traditional liability insurance, and evolves into a government-directing insurance that the subject with particular liability has to undertake.[3] Accordingly, the function of the compulsory insurance shall be constructed from many aspects, such as society, economy, law and so forth.

2.1 The Function of Strong Social Administration

The compulsory insurance regime requires the insured to purchase the liability insurance with particular liabilities. It is because the acts of different persons may subject other people to large risks. The subject of insurance under a compulsory insurance system is mainly the liability of the insured to indemnify the victims. This system aims to distribute and pass over the earlier mentioned liability of the insured to a third party. Compulsory insurance is for the interest of such third party and also for the interest of the whole public. It is a kind of insurance that has strong social administrative function, as a result of which the compulsory insurance plays an important role in solving the civil dispute, easing social conflicts and sustaining social stability.

2. Author's translation from the original Chinese text.
3. G. Feng & H. Xiaoke, 'The Study on the Compulsory Insurance', *China Commercial Law Annual* (2007): 522.

Compulsory insurance not only transfers the risks in business but also shares the burden of government. More important, it removes public anxiety about indemnification for loss caused by professional errors or public accidents. Accordingly, it increases the sense of safety and public confidence.

Meanwhile, compulsory insurance shares the workload of the government in dealing with accidents and also eases the financial burden. The government is no longer the party that always makes the payment at the end of the day. Now, the government is able to distribute and transfer the social administrative risk, which can increase the risk resistance capacity of the whole society and sustain social stability.[4]

2.2 The Function of Extensive Economical Indemnification

Compulsory insurance has a function of extensive economic indemnification. By purchasing compulsory insurance, corporations and the individuals are able to transfer various liability risks in their business. It can reconcile the conflict between the huge liability of indemnity and the small capacity of making the payment. Thus, the insolvency and interruption of production resulting from incapability of indemnifying can be avoided, which may sustain the stability of the production, operation of a certain corporate and the stability of the whole economy.

The occurrence of the dangerous incident relates directly to the following factors: the unregulated market, the potential faults in the operation of production and transportation corporations, the small size of the corporations and the lack of safety measures. Therefore, in order to prevent dangerous accidents and strengthen the capacity to indemnify such accidents, it is important to establish better administration on the qualification and improve the obedience of law and capability of operating the business in those corporations.

There is great complexity and flexibility in designing the particular types of compulsory liability insurance. However, the scope of compulsory liability insurance shall be linked to economic development. Whether one should purchase such insurance is determined by many factors, for example, the economy, legislation or culture of a country or region; the awareness of the incident; and the risk to the public. These factors should also be considered in determining the construction of the compulsory liability insurance system on all liquid and dangerous cargo transportation, especially under the current legal system in China.

---

4. Y. Chen & Z. Jinhai, 'The Trend of the Future Development of the Compulsory Insurance and Its Inspirations', *Financial Accounting* 1 (2009).

3 NECESSITY OF COMPULSORY LIABILITY INSURANCE ON CARRIAGE OF LIQUID AND DANGEROUS CARGO TRANSPORTATION

The risks insured under the compulsory liability insurance are mainly from those of the holder of the dangerous cargo and the operator involved in the dangerous act. All of those risks are related to the public interest and therefore should be distributed by the combined power of society through insurance. Hence, whether compulsory liability insurance shall be applied to liquid and dangerous cargo transportation should be determined by the factor of its influence on public interest.

Liquid and dangerous cargos are generally highly inflammable, explosive and noxious. The incidents caused by dangerous chemicals in waters can spread over a large range, and are of great difficulty for rescue and disposition. The leakage, diffusion, explosion or other incidents caused by the dangerous cargoes in the process of the vessel transportation and loading and discharging at the wharves will not only result in excessive casualties and property loss but will also cause environmental pollution and ecological disasters, which will affect significantly the safety of water used in the daily life of residents as well as in industry and agriculture. Moreover, those incidents can also induce blockage of the navigation route and obstruct water transportation.

After the occurrence of these incidents, lack of a mechanism to provide for pollution compensation makes indemnification of victims promptly quite difficult. The liability of the accused corporations to indemnify the victims and restore the environment is rarely realized. In the case of civil litigation or administrative mediation, victims not only need to undertake the high cost of legal actions but also to face the various difficulties when making the claims. In this circumstance, it is crucial to change the situation in which the corporations gain the profits through polluting the environment while the society pays for the damage. Now, is the right time to establish an effective system of compulsory environmental liability insurance. The functions of the social administration and the economic indemnification of the marine compulsory liability insurance on the liquid and dangerous cargo are to benefit the public.

In fact, in order to provide directions on the constitution of the environmental liability insurance system, the State Environmental Protection Administration of China and the China Insurance Regulatory Commission jointly declared the Opinions on Directing the Compulsory Insurance of Environmental Pollution on 4 December 2007, and formally confirmed the route chart to establish 'Green Insurance'[5] system. According to these directions, corporations that are storing

5. 'Green Insurance' is a figurative name for liability insurance for the environmental pollution. Broadly speaking, in such a system a corporate entity purchases insurance in an insurance company for the risk for environmental pollution incident that might happen, and the insurance company will indemnify the victims when the pollution incidents occur. In that way, the corporations insured can avoid the risk for huge amounts of indemnification, while the victims can be indemnified effectively and promptly.

and transporting dangerous cargo are categorized into three kinds of corporations experimenting the compulsory insurance.

Today, liability insurance for environmental pollution has already drawn great attention from the government of China. The study on the establishment of liability insurance for environmental pollution is put forward in many government documents, for example, Certain Guidance on Innovation and Development of Insurance Industry Issued by State Council of China on 15 June 2006.

Foreign experience with liability insurance for environmental pollution provide good reference. The regime of liability insurance for environmental pollution is established in many developed countries, for example, the United States, Germany, the United Kingdom, France, Sweden and so on and developing countries such as India. Germany and India have even enacted special laws in order to enforce a system mainly containing compulsory liability insurance.

4 GLANCE ON THE CURRENT LAWS ON THE PREVENTION OF POLLUTION TO THE OCEAN

4.1 International Conventions

The international conventions on preventing pollution of ocean mainly include the International Convention for the Prevention of Pollution from Ships 1973, as modified by the Protocol of 1978 relating thereto (MARPOL73/78); International Convention for Civil Liability for Oil Pollution Damage 1969, as modified by the protocol of 1992 (CLC 1992); International Convention on Liability and Compensation for Damage in connection with the Carriage of Hazardous and Noxious Substances by Sea, 1996 (HNS Convention 1996) and the International Convention on Civil Liability for Bunker Oil Pollution Damage, 2001 (Bunker Convention 2001).

4.1.1 MARPOL73/78

MARPOL73/78, which has been ratified by China, aims to achieve the complete elimination of intentional pollution of the marine environment by oil and other harmful substances and to minimize the accidental discharge of such substances. Though the harmful substances mentioned here fit a wide definition,[6] MARPOL73/78 is merely an instrument that provides for entire technical management without a compulsory liability insurance regime being established.

6. This substance means any substance if introduced into the sea, is liable to create hazards to human health, to harm living resources and marine life, to damage amenities or to interfere with other legitimate uses of the sea and includes any substance subject to control by the present Convention.

4.1.2 CLC 1992

CLC 1992 aims to adopt uniform international rules and procedures for determining questions of liability and to provide adequate compensation in such cases. In this convention a comprehensive compulsory liability insurance regime on persistent oil damage has been established. The oil mentioned in this convention means any persistent oil such as crude oil, fuel oil, heavy diesel oil, lubricating oil and whale oil, whether carried on board a ship as cargo or in the bunkers of such a ship. The Chinese government acceded to this convention in 2000, and it applies to seagoing vessels with gross tonnage of more than 2,000 tonnes. As for coastal vessels and other seagoing vessels that carrying persistent oil cargoes with tonnage less than 2,000 tonnes, other regulations will be issued by competent authorities in the future.[7]

4.1.3 HNS Convention 1996

The Hazardous and Noxious Substances (HNS) Convention 1996 provides a compensation and liability regime for incidents involving these substances. The HNS Convention 1996, when it comes into force, will ensure that ships carrying hazardous and noxious liquid substances are covered by liability insurance or will be covered by regimes similar to those already in existence for oil incidents. HNS mentioned in this convention means oils carried in bulk[8]; noxious liquid substances carried in bulk[9]; dangerous liquid substances carried in bulk[10]; dangerous, hazardous and harmful substances; materials and articles in packaged form[11]; liquefied gases[12]; liquid substances carried in bulk with a flashpoint not exceeding 60°C[13]; solid bulk materials possessing chemical hazards[14] and residues from the

7. In China, seagoing vessels and coastal or inland water vessels are usually treated in a different way in respect to application of laws. In the Notice of Performing the CLC1992 issued by MSA of China Transport and Communication Ministry in 2000 (No. [2000] 15). The MSA of China declared that CLC 1992 will not apply to the coastal vessels and inland water vessels registered in China.
8. As listed in appendix I of Annex I to the International Convention for the Prevention of Pollution from Ships, 1973, as modified by the Protocol of 1978 relating thereto, as amended.
9. As referred to in Appendix II of Annex II to the International Convention for the Prevention of Pollution from Ships, 1973, as modified by the Protocol of 1978 relating thereto, as amended, and those substances and mixtures provisionally categorized as falling in pollution category A, B, C or D in accordance with regulation 3(4) of the said Annex II.
10. As listed in the International Code for the Construction and Equipment of Ships Carrying Dangerous Chemicals in Bulk, 1983, as amended.
11. As covered by the IMDG Code, as amended.
12. As listed in Ch. 19 of the International Code for the Construction and Equipment of Ships carrying Liquefied Gases in Bulk, 1983, as amended and the products for which preliminary suitable conditions for the carriage have been prescribed by the Administration and port administrations involved in accordance with para. 1.1.6 of the Code.
13. The degree shall be measured by a closed cup test.
14. As covered by Appendix B of the Code of Safe Practice for Solid Bulk Cargoes, as amended, to the extent that these substances are also subject to the provisions of the International Maritime Dangerous Goods Code when carried in packaged form.

previous carriage in bulk of these substances. As can be seen from this definition, the HNS Convention has covered nearly all substances that might be dangerous to the environment.

4.1.4 The Bunker Convention 2001

The Bunker Convention 2001, a general regime and liability frame, is very similar to both the CLC 1992 and the HNS Convention 1996. It provides that the registered owner of a ship having a gross tonnage greater than 1,000 registered in a state party shall be required to maintain insurance or other financial security, such as the guarantee of a bank or similar financial institution, to cover the liability of the registered owner for pollution damage in an amount equal to the limits of liability under the applicable national or international limitation regime. However, in all cases, the insurance or financial security should not exceed an amount calculated in accordance with the Convention on Limitation of Liability for Maritime Claims, 1976, as amended. Bunker oil, as mentioned in this convention, means any hydrocarbon mineral oil, including lubricating oil, used or intended to be used for the operation or propulsion of the ship, and any residues of such oil. This convention came into force on China 9 March 2009. According to the Notice issued by MSA China, the Convention will also apply to coastal vessels from 1 July 2009. From this appointed time, the owners of such coastal vessels having gross tonnage greater than 1,000 tonnes is required to buy a compulsory liability insurance policy for bunker oil damage.

4.2 Domestic Laws

While the Ministry of Transport of China is drafting the Coastal and Inland Water Transportation Ordinance, a compulsory liability insurance regime for liability for carriage of liquid and dangerous cargo and passengers by coastal and inland waters set forth a requirement, according to which carriers are required to obtain a valid insurance certificate or financial security issued by first level insurance company while they apply before the competent authority for the Certificate of Operation in Coastal and Inland Waters. The carriers who failed to obtain liability insurance will be forbidden to carry liquid and dangerous cargoes in coastal and inland waters. Because the liquid and dangerous cargo as defined in the draft Ordinance merely includes the liquated gas, liquid chemicals and oils (inclusive of asphalt), there is a dispute regarding whether the Ordinance will also be applied to HNS other than dangerous cargo and oils. A prevailing point of view is that HNS as defined in the HNS Convention 1996 is wider in scope than that of the Ordinance. Therefore, it is suggested that definition of liquid and dangerous cargo shall be amended to a wider scope by reference to the HNS Convention 1996.

There is also a controversy as to which ordinance among Coastal and Inland Water Transportation Ordinance and Prevention of Marine Environmental Pollution by Ships Ordinance shall contain the compulsory liability insurance provision.

One argument is that the compulsory liability insurance will benefit the compensation for environmental damage, the compensation for the personal injuries or loss of life and compensation for the loss of property resulting from the blast, fire or other accidents due to the specification of liquid and dangerous cargo. This seems to justify the establishment of compulsory liability insurance in the Coastal and Inland Water Transportation Ordinance other than in Prevention of Marine Environmental Pollution by Ships Ordinance, because the latter one should mainly aim to ensure prevention of oceanic pollution. On the other hand, it is asserted that the Marine Environment Protection Law of China 1999, as the law of upper level of the Prevention of Marine Environmental Pollution by Ships Ordinance, has provided that

> The State Council shall enact detailed regulations with respect to the oil pollution liability insurance or oil pollution compensation fund.[15] Further, as a matter of fact, (1) most of the conventions relating to marine pollution have authorized the ship registration authority to perform the checking, expertise and manage the compulsory liability insurance; (2) It will be more convenient for MSA to issue the insurance confirmation certificate and supervise the validity thereof; (3) China have already accepted the CLC1992, the insurance certificate will be issued and supervised by MSA for and on behalf of the port state'[16]

Therefore, it is more reasonable to further set forth the compulsory liability insurance provisions in the Ordinance on Prevention of Marine Environmental Pollution by Ships.

On 5 October 2008, the Chinese government released the Ordinance on Prevention of Marine Environmental Pollution by Ships (reviewed draft) for collecting public comments. Article 60 of the draft ordinance provided that

> the ships that will sail in the waters in the jurisdiction of China shall obtained the liability insurance or financial security for civil liability for oil damage, but the ships not carrying oil substances and having tonnage less than 1,000 tons shall be excluded.

Pursuant to this provision, all vessels carrying oil cargo and all coastal vessels having gross tonnage greater than 1,000 tonnes that are navigating in coastal or inland waters are obliged to have liability insurance. Regarding the matter of how to ensure compensation for the victims suffering oil pollution damage from vessels carrying oil cargo will be dealt with through a regime similar to that in CLC 1992, although the Ordinance resolves only the problem of compulsory liability for tanker and bunker leaving other liquid and dangerous cargo damage liability unmentioned.

On 16 November 2006, the Ministry of Transport of China launched the Notice Regarding Enhance of Management of Entrance Approval of Carriage of

15. See Art. 60 of Oceanic Environment Protection Law of China 1999.
16. Authors' translation from the original Chinese text.

Passenger and Dangerous Cargo in Coastal and Inland Waters, in which no requisite requirement of an insurance certificate was set forth. Thereafter, the local government had consequently deleted the requirement of an insurance certificate in the approval procedure.[17] The Administrative Rules on Safety Supervision over the Carriage of Dangerous Cargo by Ships dated 1 January 2004 has provided in Article 20 that the ships carrying dangerous cargo shall be affected with relevant liability insurance in accordance with regulations on ship safety and prevention of pollution by ships and shall obtain the insurance certificate or financial security accordingly. However, should there be a lack of provisions regarding compulsory liability insurance either in the Ordinance on Transportation in Coastal or Inland Waters or Ordinance on Prevention of Marine Environmental Pollution by Ships, the compulsory liability insurance in the previously mentioned area shall not be performed under the Administrative Approval Act of China.[18]

This analysis illustrates that the laws in China relating to compulsory liability insurance do not cover every corner of the liability for damage caused by liquid and dangerous cargo carried by vessels. It is undoubted that prevention of oceanic pollution and water pollution shall play an important role in the battle against environment pollution, and the compulsory liability insurance shall be popularized and developed. However, it must be borne in mind that several obstacles in the way of popularization of liability insurance shall be removed.

5 OBSTACLES

5.1 The Scope of Liquid and Dangerous Cargo Is Difficult to Define

It can be seen from the suggested plan that liquid and dangerous cargo mainly means the liquid gas, liquid chemical, oil cargo (contain asphalt) and so forth. The reason to stipulate compulsory liability insurance is that liquid and dangerous cargo can be a significant threat to the public interest. However, there are similarities among the types of dangerous cargo,[19] dangerous goods, dangerous

17. It is noted from websites of several local governments that the insurance certificate shall be requested during the approval procedure for carriage of liquid and dangerous cargo by the individuals other than corporate. However, according to the Notice of Transportation and Communication Ministry dated 16 Nov. 2006, the vessels newly engaged in (inclusive of the converted vessel, purchased vessel and management-transferred vessel) as well as the vessel under an expired entrusted contract already engaged in carriage of passenger or liquid and dangerous cargo shall be forbidden to operate in the method of entrusted operation. Other vessels under a valid entrusted contract shall be ceased to operate under such way on or before 31 Dec. 2008, and shall be operated by newly registered corporate entity with qualification or by means of bareboat chartering.
18. Order of Transportation and Communication Ministry of P.R. China [2003] No. 10.
19. Including HNS or goods as mentioned in Chapter VII of International Convention for the Safety of Life at Sea, 1974, supplementary Arts I, II and III of MARPOL 73/78 and other stipulations on in conventions or regulations that China has joined.

chemicals[20] and dangerous substances.[21] Sometimes, it is difficult to distinguish those cargos for classification as liquid or dangerous cargo. Therefore, even if compulsory liability insurance was stipulated in the laws of China, there would still be an obstacle when deciding which cargo should be governed.

5.2 Companies' Lack of Strong Sense of Responsibility for Obtaining Public Liability Insurance

Many companies involved in the coastal and inland water carriage industry know little of the reason for and the effect of liability insurance and prefer to avoid effecting insurance because they suppose there is little chance of causing an accident. Some companies have no knowledge of the relevant laws or the concept of honesty. They are not willing to be covered by any liability insurance, and they will try to escape their responsibilities when accidents occur. Besides, the victims have no strong sense of self-protection and they might abandon their rights. This will leave the opportunity for companies to escape from their liability of compensation. Finally, the government does not use insurance as a method to deal with economical and social issues often enough, so the function of the market has not been fully developed.

5.3 Disadvantages of Compulsory Liability Insurance

Attention should be given to the fact that the compulsory liability insurance might increase the costs for companies that might have already been on the edge of deficit. If compulsory liability insurance were stipulated as a precondition of administrative approval, such as the regulation of 'certificate of operation of vessel' applied by the shipping company, it definitely will take a toll on company efficiency and may lead to insolvency issues in the current shipping market, although the power of the government to improve the insurance will be obviously strong. Accordingly, it is vital for the government to research this point.

It should be mentioned that China is still a developing country; the HNS convention has not come into force, and even some developed countries have not stipulated a compulsory liability insurance regime into the regulations of HNS or

20. Substances as stipulated in Art. 3 of Regulation of dangerous chemical registry management of China: explosive substance, compressed gas and liquid gas, inflammable liquid, inflammable solid, nature substance, goods inflammable when meeting humidity, oxidizer, organic peroxide, poison substance, corrosive substance, other simple substance, compound or mixture and other chemicals that are be proved to be dangerous by material as stipulated in the standard of division of Art. 3 of Regulation of safety management of dangerous chemical and (GB13690-1992) division and sign of common dangerous chemical.
21. See Distinguishing major dangerous source GB 18218-2000. Hazardous substance means a kinds or mixture of kinds of substance that has the danger of easily causing of fire, explosion or poison because of its characteristics in chemistry, physics or toxicity.

dangerous liquid cargo. Under these circumstances, it is still too early for China to implement strict compulsory liability insurance in transportation of all liquid and dangerous cargo through coastal and inland waters. The policy of China may be described as a combination of compulsory insurance and voluntary insurance.

5.4 Limitation of Insurance Product

Even if the law could establish a compulsory liability insurance regulation on liquid dangerous cargo, the implementation may not be easy, because the plan for compulsory insurance is still unclear. The implementation of compulsory liability insurance on road traffic took one year after the regulation was born, which is an example of the same kind.

One of the characters of liability insurance is that its liability and insured amount are less clear than in common property insurance. Liability insurance has a long 'tail', that is to say, there is a big time gap between effect of damage and the insured accident occurs. Therefore, the insured amount is difficult to ascertain. For example, consider a US case involving insurance compensation for asbestos shingles. Asbestos shingle was first used as construction material until it was found to be contaminated several decades later; as a consequence the injured workers began to claim. The insurance must cover several decades.[22] Therefore, liability insurance is of long term and the insured amount is unclear. It requires to a large extent that the insurance company be more competitive and product-creative, which is a difficult issue in the development of liability insurance. From the perspective of insurance companies in China, they are still in a preliminary level of development. They are still lacking management skills, technology, products development skill, intensive computation and data. However, the maturity of an insurance product is based on adequate collection of data. Today there are a very limited number of compulsory liability insurance systems in China, except for the compulsory liability insurance for road traffic. Therefore, the collection of data is not sufficient for a smooth development of the product.

5.5 Obstacles in Law

The legal environment in China with respect to compulsory liability insurance needs to be further improved. The compulsory liability insurance on liquid dangerous cargo cannot be isolated. It should be combined with relevant laws together to play its important role. Without a clear combination of laws, regulation of compulsory liability insurance cannot exert its power and may even be useless. Therefore, the combination of legal regulation can be considered from several aspects, as follows: (1) to establish direct action against underwriters; (2) to

22. G. Zuojian, 'Combine with Special and General Aspects, Improve Liability Insurance', 2 Aug. 2007, <http://finance.itus.cn/insurance/opinion/n75584.shtml>.

consider the limitation amount for liability carefully; (3) to make clear who shall be responsible for compensation liabilities of liquid and dangerous cargo, because it will directly affect who shall be qualified as the insured in the policy under compulsory liability insurance.

6 PROSPECT

6.1 The Joint Effort of the China Insurance Regulatory Commission and the Ministry of Transport of China

It has been suggested that the sense of insurance should be cultivated and the need for insurance should be guided. From this point, the current development of liability insurance needs the support of the government, which means that a good policy and the administrative competence of the department are very essential. It also depends on local governments' use of the idea of risk management to actively promote liability insurance. After the publication of Certain Guidance on Innovation and Development of Insurance Industry Issued by State Council of China on 15 June 2006, the result of this implementation is quite satisfying. Another important work is to follow the principle of 'government lead, policy guide, experiment well and finally expands' to stimulate the potential need of insurance by coordination of work among different departments of competent authorities. For example, the China Insurance Regulatory Commission and Ministry of Public Security of China can make policy together to promote the compulsory liability insurance for fire in public places; the China Insurance Regulatory Commission and Ministry of Construction of China can jointly promote compulsory liability insurance for construction quality; the China Insurance Regulatory Commission and Ministry of Security Supervise of China do joint research into compulsory liability insurance for employers in high-risk industries. Once it has been proved that compulsory liability insurance of liquid dangerous cargo is really needed, it will be implemented by the joint effort of the China Insurance Regulatory Commission and Ministry of Transport of China.

6.2 Experiment on Selected Industries

As discussed earlier, the compulsory liability insurance of liquid dangerous cargo should be put into the concept of 'environmental liability insurance'. However, regulation of 'environmental liability insurance' is a new theory in China. This policy of environment and economy directly relates to the direct or latent environmental risk. It should be examined by experimental application on the main industries that cause the most serious pollution in the past few years, including the companies involved with the production, management, storage, transportation and use of dangerous chemicals. The companies in petrochemical industry and disposal of cast dangerous chemical are the most dangerous to the environment. Therefore, the compulsory liability insurance of liquid dangerous cargo could be examined experimentally in these kinds of companies first.

During implementation, the principle of 'focus on the key point – first deal with the easy one, then the hard one' should be followed. Considering that China joined and implemented CLC 1992 and the Bunker Convention 2001, the concept of compulsory liability insurance of oil has been preliminarily formed. Shipping companies transporting liquid dangerous cargo in provinces and cities along the river and who have already experienced these kinds of accidents should have a good sense of insurance and a good standard of shipping operation. Therefore, these companies should be selected for experimenting with the voluntary liability insurance. Besides, jointly with the government, the China Insurance Regulatory Commission has promoted the compulsory liability insurance of companies in public places such as shopping malls, hotels and cinemas. With the expansion of voluntary and compulsory liability insurance, the rising needs of shipping companies and the experience of insurance companies, the compulsory liability insurance for environmental protection and public interest will satisfy the conditions of implementation. By that time the government can implement compulsory liability insurance of liquid dangerous cargo by administrative regulation throughout China.

6.3 Development of Insurance Product

Insurance companies insist on the rule of 'start from experiment to collect experience and then expand' in their business because they are the industries who manage risks. Only by first guarantying their own safety of operation they can successfully manage risks. Under the circumstances of a deficiency in experience and lack of data, attention should be drawn to both the effect and the speed by developing and testing methods in various fields to meet different needs.

The China Insurance Regulatory Commission has already attached great importance to the transformation and creation of insurance product. They try to promote matching supply and demand to a higher level and therefore encourage, guide and promote insurance companies to develop higher level products and explore new policies in respect to protection of intellectual property so as to protect the creative product.

7 CONCLUSION

The compulsory liability insurance of liquid dangerous cargo has obvious effects on social management and economic compensation, and the Chinese government should provide proper guidance and promote this kind of insurance. As to implementation, however, there are still obstacles at the current stage. First, the scope of liquid dangerous cargo is too wide to be defined exactly. After establishing regulation of the compulsory liability insurance of liquid dangerous cargo, which should be applicable to almost every kind of liquid or dangerous cargo, the competent authority should work on distinguishing or defining the ships transporting liquid dangerous cargo. This will have an enormous effect.

However, a completed and systematic legal system of compulsory liability insurance has not been formed yet in China. A few problems arise from various kinds of liquid dangerous cargo, such as direct action, limitation amount for liability and issues as to who shall be responsible for the liability. These difficulties cannot be solved by a general regulation of compulsory liability insurance in the Ordinance on Prevention of Marine Environmental Pollution by Ships. Even if both the law enacting department and competent authority make preparations to implement the compulsory liability insurance, there still will be obstacles linked to the issues of rate of premium and conditions and terms of policy. Therefore, the draft Ordinance on Prevention of Marine Environmental Pollution by Ships can first stipulate the compulsory liability insurance for the compensation of oil pollution as an experiment through which method some relevant experience can be accumulated and then expand the range of compulsory insurance to carriage of liquid dangerous cargo.[23]

References

Chen, Y. & Z. Jinhai, 'The Trend of the Future Development of the Compulsory Insurance and Its Inspirations'. *Financial Accounting* 1 (2009).

Feng, G. & H. Xiaoke, 'The Study on the Compulsory Insurance'. *China Commercial Law Annual* (2007): 522.

Zuojian, G. 'Combine with special and general aspects, improve liability insurance'. <http://finance.itus.cn/insurance/opinion/n75584.shtml>, 2 Aug. 2007.

23. Print Note: On 15 Sep. 2009, the China State Council launched the Ordinances on Prevention of Marine Environmental Pollution by Ships, which will come into force on 1 Mar. 2010.

Chapter 22

The Factors Affecting Judicial Authentication on Maritime Oil Pollution and Relevant Solving Measures

*Sun Guang**

1 INTRODUCTION

With the development of the maritime transportation industry, the world fleet continues to expand. Especially with respect to crude oil transportation, enlarging of the scale of international oil tankers has become a means to reduce the cost of crude oil transportation. Vessels may, during the voyage, run against rocks, collide, strand, sink and catch fire as well as experience other accidents that damage cargo oil tanks and fuel tanks, causing severe maritime pollution incidents. In addition, vessels in operation probably give rise to oil spill pollution by the discharge of sewage, such as bilge water from washing water and ballast water containing oil. Release of harmful substances from vessels is increasing daily, causing sustained and difficult-to-remove pollution, further damaging the maritime environment. Having a long coastline, China faces inevitable threat and damage from oil spill

* Dalian Maritime University International Law PhD Candidate; Judge of Dalian Maritime Court, Mail Address: No. 280, Luxun Road, Zhongshan District, Dalian City, Liaoning Province, China. Tel: 0411-82759808, Fax: 0411-82759816, E-mail: sunguang2028@163.com.

Michael G. Faure, Han Lixin & Shan Hongjun, *Maritime Pollution Liability and Policy: China, Europe and the US*, 385–395.

pollution by ships. 'As to first five years of twentieth century, oil-spill pollution events of ton-ships are more than eighty cases in Chinese waters.'[1] 'Oil pollution events from ships occurred about 124 cases around the national coast, with total oil spill volume of 1,216 tonnes in 2006.'[2] Preventing pollution from ships to protect Chinese maritime environment is not only a subject studied by shipping enterprises and administrative departments but is also a matter of focus to which the maritime justice pays attention. The number of civil cases on compensation for maritime oil pollution damage has been increasing. Oil spill pollution authentication is a necessary procedure to determine responsibility. Through cross-examination, authentication conclusion can become a convincing proof to protect legitimate rights and interests. However, the present situation of maritime oil-pollution authentication is not so systematic and functioning, which brings significant influence on the reliability of oil pollution authentication and leads to much confusion in the trial of civil cases of marine pollution.

2 BASIC LEGAL PROVISIONS OF JUDICIAL AUTHENTICATION

China has enacted and promulgated the Decision on the Administration of Judicial Authentication ('the Decision'), the Measures for the Administration of Registration of Judicial Authentication Institutions, the Measures for the Administration of Registration of Judicial Authenticators, and the General Rules on the Procedures for Judicial Authentication ('the General Rules') since 2005, to regulate activities of administration and proceedings of judicial authentication institutions and judicial authenticators. According to the previously mentioned regulations, judicial authentication refers to authenticating and judging activities regarding professional issues involved in the litigation as well as providing authentication reports, which are performed through scientific and technological means or professional knowledge by authenticators in the lawsuit. Judicial Authentication Licenses and Practice Licenses of Judicial Authenticators are necessary for authentication institutions and judicial authenticators. In the judicial authentication procedure, litigants are entitled to apply and commission to judicial authentication; courts enjoy the power to decide the authentication; judicial administration departments enjoy the power to supervise; and authentication institutions enjoy the right to implement. The purpose of this judicial authentication mechanism, which includes division of labour with individual responsibility, mutual coordination and restraint, is to guarantee the fairness of judicial authentication procedures in terms of subjects.

---

1. M. Cai-xia, 'The scope of maritime environment damage compensation in oil pollution by ships incidents' (translated by author), <www.riel.whu.edu.cn/show.asp?ID=4115>, 5 Jul. 2006.
2. C. Yan-hong, 'Chinese Marine Pollution Incident Doubling Every Year & Environmental Forensics unknown' in a legal system network (translated by author), <www.envir.com.cn/info/2008/5/529404.htm>, 29 May 2008.

Opinions of the authenticator who is an independent litigant participant have a direct effect on the judgments that the judge holds upon professional issues of specific cases, thereby affecting the confirmation of legal facts. However, authentication reports not only depend on the factors with comparative scientificity and objectivity, such as employing specialized knowledge and utilizing equipment and facilities, but also the knowledge and skill of individual specialists, which means that they may be influenced by individual characteristics and subjectivity to a certain extent. Therefore, both the Decision and the General Rules stipulate the authenticator accountability system and the basic principle upon authentication, which is legitimate, independent, objective and fair.

In order to guarantee the fairness of judicial authentication, generally speaking, there are four requirements: (1) Independence. Authenticators should employ science, technology or professional knowledge to authenticate and determine the special issues involved in the litigation independently. Judicial authentication institutions bear the responsibilities and liabilities independently while drawing the conclusion in their own name. (2) Neutrality. Judicial authentication institutions should maintain a neutral stance to all litigants and should not subject their judgments on the judicial authentication to the opinions of courts or any other organs, units and individuals. (3) Public welfare. Judicial authentication permitted by the state is a special activity related to public safety and interest. Accordingly, the fees they gain must be appropriate rather than profitable to cover working funds for equipment and material consumption. (4) Service. Judicial activities, belonging to the scope of legal services, are based on the need for litigation. Only by complying with the previously mentioned requirements, can judicial authentication institutions put themselves in a scientific and objective position to issue authentication statements on specialized technical problems involved in the legal proceeding.

Maritime oil pollution authentication should conform not only to the requirements but also to certain technical standards. According to Article 22 of the General Rules, while a judicial authenticator is authenticating, he should comply with the following sequences as well as employ the technical standards of the professional field: (1) national standards and technical specifications; (2) industry standards and technical specifications formulated by the competent authorities and industry organizations of authentication or other relevant industry competent authorities; and (3) technical standards and specifications accredited by the majority of experts in the professional field. If there are not any technical standards and specifications stipulated in the preceding paragraph, the judicial authenticators may use the relevant technical specifications formulated by their own judicial authentication institution. Oil pollution authentication belongs to the field of environmental forensics. In 1989, the *Exxon Valdez* tanker oil spill pollution that occurred in US waters led to the emergence of environmental forensics. Since then, environmental forensics has mostly focused on oil spill authentication and traceability research, finding great use in emergency treatment of pollution. With reference to the latest developments in international environmental forensics, the Specifications for Identification System of Spilled Oils on the Sea of China were implemented on 1 April 2008 as a national standard in China. The specifications stipulate methods of collecting, storing, preserving and authenticating oil spill samples, providing scientific approaches for authenticating the pollution source of oil spill and defining the troublemaker of an oil pollution event accurately.

3 FACTORS AFFECTING IMPARTIALITY OF JUDICIAL AUTHENTICATION

Although there have been detailed and unified authentication norms, in practice the same oil pollution events, authenticated by different institutions with the same norms, may be concluded differently. Apart from effects of a reasonable range of estimating, many other factors also lead to this result. There are objective factors, such as authentication institutions' outdated and backward instruments and equipment, authenticators' low professional and technology level as well as lack of legal knowledge. Subjective factors are mainly that authenticators do not have a strong sense of responsibility, making the authentication conclusions favour authentication principal, and that authentication conclusions are not certain, changing greatly sometimes. Moreover, the previously mentioned objective and subjective factors often interact as both cause and effect and link to each other inextricably. The fundamental determinants shall be self-financing operation of authentication institutions. Most authentication institutions are self-supporting legal entities or other organizations, whose primary purpose is to make profit, pursuing economic efficiency and ignoring social benefit. Now receiving the authentication fee, authenticators would inevitably incline to the principal subjectively, displaying more views in favour of rather than against the principal. In order to make profit, an authentication institution will also try to lower costs as much as, for example, to update equipment and instruments at a low rate, while the demands of oil-pollution authentication activities for equipment and technology are high. It is difficult for authentication institutions to achieve the status of public welfare and develop into an advanced authority technologically in the business for all of the factors discussed previously.

The profit of authentication institutions is mainly derived from authentication fees. Article 8 of the General Rules provides that judicial authentication institutions shall uniformly charge judicial authentication fees, and the other relevant provisions of the state shall apply to the charge items and standards. The reason the government regulates uniform specifications or standards on authentication fees are to avoid arbitrary charges. However, the Ministry of Justice has failed to issue those so far, perhaps because it is difficult to balance public welfare and advancement. In 2008, the charge specification on partial judicial authentication business was promulgated in Liaoning Province, China, which clarified that judicial authentication fees belonged to the service charges. According to that specification, the fees are generally low. However, Article 3 of that specification provides other charges besides the existing authentication charge standards, including: (1) translation fees and travel expenses related to authentication. (2) Costs incurred by additional inspection and testing beyond the business scope of equipment configured by authentication institutions according to rules. (3) Expenses arising from court testimony. (4) Other related costs incurred besides costs resulting from services of authentication activities. That specification provides that the charging amount and charging methods of these costs are to be negotiated by authentication institutions and their clients. This makes the amount of authentication fees flexible. Consider oil pollution authentication as an example (authentication institutions charge much higher than the amount in the specification, from thousand to tens of thousands SDR, so that it is beyond affordability by the parties.

In the *Tasman Sea* oil pollution case, the State Oceanic Administration expended about SDR 222,936 on survey, investigation, assessment and bioremediation research. The Bureau of Fishery Management Superintendence spent approximately SDR 43,636 on investigation and assessment.[3]

Whether these costs are reasonable is hard to determine. Inevitably, it is difficult to supervise authentication charging, and the specification of judicial authentication charging loses its meaning.

In order to solve the problems about the neutrality and independence of authentication institutions, some scholars suggest to follow the Hong Kong Government Laboratory and the American Public Laboratory. Both of them are directly invested by the state to establish independent and authoritative authentication institutions to ensure equipment and instruments are in good order, the standard of technology is sophisticated, the structure of staff is reasonable, the charges are appropriate and the legal action neutral and independent and to guarantee that they have a high level of credibility and authority. This approach seems to be good, yet after the earlier mentioned authentication institutions set up, the other kind of authentication institutions would shrink gradually because of lack of competitiveness. Then those set up by the state would also, in the absence of competition, cause corruptness and inefficiency. If the state establishes two competing institutions, this will bring about other problems such as waste of resources.

4 CONTROVERSY ON THE SCOPE OF MARITIME OIL POLLUTION AUTHENTICATION

Controversy on the scope of authentication, in addition to the previously mentioned factors, will affect maritime oil-pollution judicial authentication. According to Article One paragraph six of the International Convention on Civil Liability for Oil Pollution Damage (1992) ('the CLC'), to which China has acceded, pollution damage means loss or damage caused outside the ship by contamination resulting from the escape or discharge of oil from the ship, wherever it may occur. However, compensation for impairment of the environment (other than loss of profit from such impairment) shall be limited to the costs of reasonable measures

3. W. Xuan-mi, 'For the Azure Blue of Bohai Bay' (translated by author and the given DATA is converted from RMB to SDR by author), <www.cunews.edu.cn/html2006/hddq/hddqxw/092801431.htm>, 10 May 2006.

of reinstatement actually undertaken or to be undertaken and the costs of preventive measures and further loss or damage caused by preventive measures. The CLC regulates environmental damages in principle.

The United States promulgated the Oil Pollution Act in 1990 ('OPA 1990'), listing natural resources in the categories of environment, including land, fish, wildlife, biota, air, water, ground water, drinking water supplies and resources alike.[4]

OPA 1990 provides that persons liable can be claimed for compensation on losses to natural resources caused by oil pollution, which breaks the original compensation principle significantly, performing a comprehensive protection for victims. As prescribed in Article 95 paragraph one of the Marine Environmental Protection Law of the People's Republic of China 'the Marine Environmental Protection Law', pollution damage to the marine environment means any direct or indirect introduction of substances or energy into the marine environment that results in deleterious effects such as harm to marine living resources, hazards to human health, hindrance to fishing and other legitimate operations at sea, impairment of utilization, quality of sea water and degradation of environment quality.

The Marine Environment Protection Law does not limit the compensation of environmental damage to the costs of reasonable measures of reinstatement actually undertaken or to be undertaken. Therefore, when oil spill pollution occurs in sea areas of China, the State Oceanic Administration ('the SOA'), on behalf of the state, will claim for losses in the marine environment; the Bureau of Fishery Management Superintendence, on behalf of the state, will claim for medium-and-long term losses of natural fishery resources; and fishermen will bring lawsuit for losses of farming and fishing. The action is sometimes for as much as several tens of thousands SDR or even several tens of millions SDR. These cases invoke much dispute, not only because of the large sums involved but also because the parties hold different views on legal issues, such as whether the losses in the marine environment and medium-and-long term losses of natural fishery resources are both environmental damage or whether this applies only to the former. In addition, different parties have different views on authentication as follows, as discussed in the following section.

Authentication institutions, commissioned by the SOA and the Bureau of Fishery Management Superintendence separately, issue authentication reports on losses in the maritime environment and medium-and-long term losses of natural fishery resources, respectively, in accordance with their authority. The main method of authentication is to investigate sources of pollution, the fields and areas of pollution, sea conditions, water quality and plankton resources and so forth to refer to historical maritime data and to survey the affected seafood market. Subsequently, the losses to the maritime environment are estimated through spilled oil drift diffusion models and shadow engineering methods, and the value losses of maritime environmental capacity and the foundation losses of maritime ecological service are determined. As to medium-and-long term losses of natural fishery

4. H. Li-xin, *Study on the Law of Compensation for Pollution Damage Caused by Ships*, 1st edn (Beijing: Law Press China, 2007), 192.

resources, authenticators adopt trawls to measure fishery resources density on the spot and draw a conclusion on the direct fishery resources losses by comparing the density of fishery resources after the incident with average density of that over the same period of three years before the incident. Then, authenticators calculate the amount of medium-and-long term losses of natural fishery resources in the polluted areas in the light of three times of direct fishery resource losses in accordance with the Provisions on the Methods of Calculating Fishery Losses from Water Pollution Accidents of the Ministry of Agriculture, which prescribe that the calculation of economic losses of natural fishery resources should not be less than three times of the amount of loss of seafood from that of direct economic losses.

However, shipowners or their insurers, as defendants, do not agree greatly with the foresaid authentication conclusions, stating that these calculation methods are not scientific, because in many cases, fish do not die from oil pollution, but do transfer to other sea areas. Generally, the longest impact of oil pollution on the waters is not more than five months, and then marine life will migrate back after pollutants disappearance thus the plaintiffs' losses are less. The period for which authenticators monitor the fields and areas of pollution, water quality and losses rate of fishery resources should be three years at least. There are not huge losses in the maritime environment plaintiffs' claims, provided that tracing monitoring shows that the fishery resources have been restored to its original condition in the short term or the range of losses is not wide. Moreover, the medium-and-long term losses of natural fishery resources and to the marine environment are both environmental damage and should not be calculated repeatedly. Therefore, shipowners usually refer to other authenticators to issue authentication reports and apply for people with expertise (also known as 'expert assessors') to appear in court, raising doubts about the authentication reports commissioned by the SOA and the Bureau of Fishery Management Superintendence. Shipowners also consider that the environmental damage defined in the report is beyond the limitation of the Convention, and the SOA and the Bureau of Fishery Management Superintendence have no right to claim environmental losses, regardless of undertaking and planning to undertake any reasonable measures of reinstatement the Convention prescribed.

The main reason for the various disputes over oil pollution by ships is the potential ambiguity in the concept of environmental damage. In addition, environmental damage is not direct losses or easy to monetize, so that scientific rationality of authentication methods is challenged. Whether the medium-and-long term losses of natural fishery resources are classified as environmental damage is dependent on the meaning of environmental damage. In terms of Article 95 paragraph one of the Marine Environmental Protection Law, pollution damage to the marine environment is a generalized conception, including harm to marine living resources, hazards to human health, hindrance to fishing and other legitimate operations at sea, impairment of utilization quality of sea water and degradation of environment quality and so on. They not only have direct personal and property losses but also experience adverse effects as a result of the marine environment suffering damage. The medium-and-long term losses of natural fishery resources belong to pollution damage to the marine environment. Moreover, the environmental damage mentioned in the CLC differs from the above-mentioned pollution damage to the marine environment, excluding direct personal and property losses, which cannot be estimated in the form of actual loss, and the compensation is limited to the costs of reasonable measures of reinstatement actually undertaken or to be undertaken. As to the medium-and-long term losses of natural fishery resources, the authors think that these are not actual losses, but a sort of loss resulting from adverse effects on natural fishery resources caused by maritime environmental pollution in the future. Like the losses to the marine ecology, they are estimated losses and belong to the environmental damage stipulated in the CLC. Both losses to the marine environment and medium-and-long term losses of natural fishery resources should be claimed by a competent department on behalf of the state. This department is the SOA.

Someone holds the view that we may learn from OPA 1990 and the report concluded in the *Tasman Sea* case by the North China Sea Environment Monitoring Centre of the SOA to monetize environmental damage. That is to say, the scope of maritime ecological environment should include but not be limited to environmental damage of marine water quality and restoration costs, marine sediment damage and restoration costs, damage of a variety of biological resources and restoration costs, assessment expenses on determination of losses and monitoring of recovery and the costs spent on investigation and so on, without violating the international conventions and relevant laws.[5]

The scope of damage in this perspective actually exceeds that in the CLC, which leads to many disputes in regard to legal issues. Regulations should be drawn up in China to clarify the scope of compensation of environmental damage.

The ecological environment provides a kind of service, whose value is hard to quantify. Any rights in this regard will have to be monetized on certain standards eventually, when entering into areas of civil compensation, which will be difficult yet inevitable.[6]

Environmental damage cannot be proved by evidence directly, which needs to be determined by authenticators, through surveying on the spot, looking up information, and other reasonable means, in accordance with the Specifications for Identification System of Spilled Oils on the Sea of China and other national standards or internationally accepted methods. The results of authentications are not absolutely accurate, but are relatively reasonable.

5 DIFFICULTIES OF DETERMINING AUTHENTICATION REPORTS AND RELEVANT SOLVING MEASURES

Through cross-examination and certification, the court need decide whether to admit and accept oil pollution authentication reports as evidence that is used

5. M. Cai-xia, *supra* n. 1.
6. W. Nan-wei & Y. Xiao-han, 'On the Reality and Rational Consideration of Maritime Oil Pollution Compensation', <www.lwwzx.com/Freepaper/22386.htm>, 24 Aug. 2006.

as the basis for a decision in the case. However, there are many difficulties in doing that.

First, authentication reports are hard to determine because of loopholes. The contents of cross-examination on authentication reports involve authenticity and legitimacy. The former mainly includes whether samples are in good condition, whether the quantity of samples is adequate, whether the apparatus and equipment together with authentication methods are advanced and appropriate, whether authentication procedures and methods are in accordance with legal standards or other industry standards; whether analysis and demonstration of authentication conclusions are logical and scientific, and so forth. The latter mainly includes whether authentication institutions are established by law; whether authenticators have the legal qualifications or have interests in the case or have any relations with litigious parties; whether sources of the samples are legal; whether format and content of the authentication conclusions are in line with law; and whether authenticators appear before the court to cross-examine in accordance with the law.[7]

Loopholes always exist to varying degrees in authentication reports issued under the commission of the injured because of objective and subjective factors. That makes it difficult for courts to admit the legitimacy of the authentication reports and determine losses according to the reports.

Second, oil pollution authentication is so specialized that the judges can only review it in form or review it based on their own knowledge, yet it is hard for them to review the foundation of authentication, the usage of technical means and any other professional issues substantively. Facing the authentication reports provided by plaintiffs and defendants, as well as the tit-for-tat points supported by authenticators and expert assessors, who appear in court according to the application by plaintiffs and defendants, judges find it hard to determine which is more accurate or reasonable. In addition, there are no more authoritative authentication institutions to determine the effectiveness of those reports.

As to the two problems mentioned earlier, we may learn from the technical advisory systems used by other countries. There are two types of technical advisors: one is offering service for clients, the other is offering service for the court. In the Italian technical advisory system, each party can appoint his own technical advisor before or after authentication. However, technical advisors in the United Kingdom are giving service to the court only. China could consider establishing both of the technical advisors at the same time. That is to say, the parties shall be permitted to employ a technical advisor who will participate in and supervise the whole process on their own while one party or some parties apply for authentication and bring forward reasonable requests to avoid or reduce loopholes. Meanwhile, a technical jury system shall be set up so that the collegial panel can have

7. Y. Yu-juan, 'The Faith of Procedural Justice and the Reconfiguration of Judicial Authentication System', (translated by author), <www.cqvip.com/onlineread/onlineread.asp?ID=20307624&SUID=EGBNBFDHDNCBCOPIBOLGEMCBCDOIPNFP#>, accessed on 15 Jun. 2008.

technical experts to provide support for judging the points held by respective parties and determining whether authenticators are in fault in terms of procedure and substantiality from the prospective of technology. The party can prosecute and claim compensation in light of the degree of fault if case authenticators are in fault, to achieve authenticator accountability.

Third, it may lead to re-authentication because of the reasonable objection from defendants, but there are also plenty of problems in that area. For instance, the re-authentication may not proceed because of inadequate sampling or no sample preserved by original authenticators or the non-existence of an accident scene. That presents the court with a dilemma – the original authentication reports cannot be admitted while the re-authentication cannot be carried out. Although the plaintiffs do suffer from losses, the court finds there is no basis on which to make a determination. If no determination on the losses is made, it does not comply with the principles of fairness and justice to protect the legitimate rights and interests of injured parties. Therefore, judicial authenticators have the obligation to collect adequate samples when sampling at the scene and keep those until the end of final judgment. In addition, Article 29 paragraph two of the General Rules provides the qualifications of judicial authentication institutions that accept the re-authentication application must be higher than the original one. However, Article Eight paragraph one of the Decision provides that there is no hierarchical relationship between the authentication institutions. An authentication institution is not subject to any geographical restriction in accepting entrustment to engage in judicial authentication. The different regulations contradict each other, resulting in confusion upon choosing a judicial authentication institution when it comes to re-authentication. The competent administrative departments should clarify foresaid problems.

This chapter suggests that a competent administrative department draw up regulations on the upgrade of both judicial authentication institutions and authenticators. For example, those regulations prescribe that: (1) Competent administrative departments carry out an annual assessment on the previous year's authentication work of all judicial authentication institutions and authenticators. If the previous year's performance was outstanding and had a good social reputation, they would be evaluated for excellence and get their qualification upgraded. If the previous year's performance was bad and had a poor social reputation, they would be evaluated for failure and have their qualification downgraded. (2) In addition to accepting initial authentication applications, high-level authentication institutions and authenticators may undertake re-authentication in respect to authentication reports issued by low-level ones and deny effectiveness of low-level authentication reports. (3) Assessment methods may include voting by the judges, plaintiffs and defendants in the cases, in addition to examining the instruments and other hardware equipment, the number of authenticators and the qualification of authenticators. What is more, the author advises that proper standards and limitation of authentication costs shall be established and costs of re-authentication shall be higher than those of initial one. This will create a good mechanism for competition, which is favourable for the orderly development of judicial authentication and for the court to find the facts of cases to determine civil liability, protect legitimate rights and protect the interests of the parties.

In summation, with the implementation of relevant provisions of the judicial authentication, the procedures of judicial authentication standardize gradually. At the same time, the Specifications for Identification System of Spilled Oils on the Sea ensures the authentications of losses in maritime oil pollution have technical standards. All of these play an active role in protecting our maritime environment and those injured by oil pollution. However, some problems still exist in regard to oil-pollution authentication, with both procedures and substantiality, that greatly affect judicial guarantee of legitimate rights and interests. We should identify and solve them in a timely manner and properly to better protect our maritime environment.

References

Cai-xia, M. 'The Scope of Maritime Environment Damage Compensation in Oil Pollution by Ships Incidents' (translated by author). <www.riel.whu.edu.cn/show.asp?ID=4115>, 5 July 2006.

Li-xin, H. *Study on the law of Compensation for Pollution Damage Caused by Ships.* 1st edn. Beijing: Law Press China, Beijing, 2007.

Nan-wei, W. & Y. Xiao-han. 'On the Reality and Rational Consideration of Maritime Oil Pollution Compensation'. <www.lwwzx.com/Freepaper/22386.htm>, 24 August 2006.

Xuan-mi, W. 'For the Azure Blue of Bohai Bay (translated by author and the given data are converted from RMB to SDR by author). <www.cunews.edu.cn/html2006/hddq/hddqxw/092801431.htm>, 10 May 2006.

Yan-hong, C. 'Chinese Marine Pollution Incident Doubling Every Year and Environmental Forensics Unknown', in a legal system network (translated by author). <www.envir.com.cn/info/2008/5/529404.htm>, 29 May 2008.

Yu-juan, Y. 'The Faith of Procedural Justice and the Reconfiguration of Judicial Authentication System', (translated by author), <www.cqvip.com/onlineread/onlineread.asp?ID=20307624&SUID=EGBNBFDHDNCBCOPIBOLGEMCBCDOIPNFP#>, accessed on 15 June 2008.

Part VI

Case Studies

Chapter 23

Legal Issues from the *Hebei Spirit* Oil Spill Incident

*Hu James Zhengliang**

1 INTRODUCTION

The incident of oil spill from M/T *Hebei Spirit* of 7 December 2007 is the most serious incident of oil pollution from ships in the history of the Republic of Korea and was declared by the Korean government to be a national disaster. What attracts the attention of the international community is not only the significant damage caused by the incident, but also the myriad legal disciplines involving issues of criminal, civil and administrative law.

This chapter introduces the incident, including the pollution damage caused, the investigations into the incident, the legal issues arising therefrom and the assessment and settlement of claims for pollution damages. It also analyses what China, as one of the largest oil-importing countries, may learn from this incident and the handling thereof.

* Ph.D., Professor of Law, Director of Institute of Maritime Law, Shanghai Maritime University; Lawyer & Partner, Shanghai Wintell & Co Law Firm; E-mail: james.hu@wintell.cn.

Michael G. Faure, Han Lixin & Shan Hongjun, *Maritime Pollution Liability and Policy: China, Europe and the US*, 399–412.

2 THE INCIDENT[1]

On 7 December 2007, while at anchor about five miles off Taean on the west coast of the Republic of Korea (Korea) and waiting for a pilot from Daesan Port, the tanker *Hebei Spirit* of 146,848 gross tonnes, flying the flag of the Hong Kong Special Administrative Region, was struck by the ocean crane barge *Samsung No.1* with lifting capacity of 3,000 metric tonnes. The crane barge was being towed by two tugs, *Samsung T-5* and *Samho T-3*, before collision. The crane barge, the two tugs and an anchor boat, *Samsung A-1*, formed a marine spread. During bad weather conditions, the towing line of Samsung T-5 broke and the crane barge was blown by the strong wind into the tanker, puncturing three of her port side cargo tanks (Nos 1, 3 and 5).

Hebei Spirit was laden with about 209,000 MT of four different crude oils. Approximately 12,547 kilolitres or 10,500 tonnes of crude oil escaped into the sea. The oil polluted, in varying degrees of contamination, three of the four provinces along the west coast of Korea. The shoreline of some 375 kilometres composed of rocks, boulders and pebbles, as well as long, sandy amenity beaches and port installations in the Taean Peninsula and on the nearby islands, were polluted. Over a period of several weeks, mainland shorelines and islands further south also became contaminated by emulsified oil and tar balls. A considerable number of commercial vessels were also contaminated.

Large scale oil cleanup operations at sea and on shore coordinated by the Korean National Coast Guard Agency, a department of the Ministry of Maritime Affairs and Fisheries (MOMAF) were necessary.

At sea, more than 100 vessels of the Korean Coast Guard, the navy and the Korean Marine Pollution Response Corporation (KMPRC) and over 1,500 fishing vessels were deployed to carry out cleanup operations. The Coast Guard applied dispersants from vessels and later by helicopter over patches of floating oil. Tens of kilometres of oil booms were also deployed at sea and along coastal areas.

On shore, the Korean Coast Guard tasked twenty-one licensed cleanup contractors as supported by local authorities and fisheries cooperatives to undertake shoreline cleanup operations. Local villagers, army and navy cadets and volunteers from all over Korea also participated in the cleanup operations at numerous locations along the west coast. In excess of one million work-days were worked during the first two months.

Table 23.1 shows figures of emergency cleanup operations, which were finalized by 10 October 2008.

1. See IOPC Fund documents 92FUND/EXC.44/7, 92FUND/EXC.44/7/1, 92FUND/EXC.44/7/Add.1, <www.iopcfund-docs.org/cgi-bin/doc-server/iopcf-docs.cgi>.

Table 23.1

Total amount of resources used and waste oil recovered

	Human resources & equipment used							Waste oil recovered	
	Personnel (Volunteers)	Vessels (unit)	Helicopters (unit)	HCE* (unit)	Oil fence (km)	Absorbent (tonne)	Dispersant (kl)	Contaminated residue (tonne)	Waste oil (kl)
Total	2,132,322 (1,226,730)	19,864	346	28,973	46.77	493	298	32,074	4,175

* HCE: heavy construction equipment.

Besides the significant oil cleanup costs, the oil spill caused serious damage to fisheries and the aquaculture industry and tourism businesses in the affected areas. The total amount of pollution damages arising from the oil spill is currently estimated by experts at Korean Won (KRW) 566–601 billion or United States Dollar (USD) 411–437 million.

The Protection and Indemnity Club (P&I) Club of *Hebei Spirit* and the International Oil Pollution Compensation Fund 1992 (1992 IOPC Fund) appointed a team of Korean and international surveyors to monitor the cleanup operations and investigate the potential impact of the pollution on fisheries, aquaculture and tourism activities.

3 INVESTIGATION INTO THE INCIDENT AND LITIGATIONS

3.1 Investigation into the Incident

3.1.1 Investigations Conducted by KMST

The investigations into the causes of the incident were conducted by the Korean Maritime Safety Tribunal (KMST) according to the Korean Act on the Investigation and Inquiry into the Marine Incident and aimed at improving safety of life at sea and preventing the re-occurrence of similar incidents in the future. KMST has two tiers: the local maritime safety tribunal, which is the Incheon Maritime Safety Tribunal (IMST) in this matter and the Central Maritime Safety Tribunal (CMST).

On 4 September 2008, the IMST issued its investigation report in the form of a Decision, in which it found that both the marine spread and *Hebei Spirit* were to blame. In particular, it found that:

> (a) On the part of the marine spread, (i) it failed to take early measures against bad weather conditions; (ii) it drifted with wind and waves after losing its towing ability; (iii) it failed to transmit signals to nearby vessels and perform safety measures such as emergency anchoring while drifting; (iv) the towing line of 'Samsung T-5' broke while the crane barge was too close to 'Hebei Spirit' and eventually caused the collision.
> (b) On the part of *Hebei Spirit*, (i) she failed to discover the approaching marine spread earlier because of a lack of due care; (ii) she took passive measures to seek shelter, such as putting the engine to dead slow astern; and (iii) the fact that emergency countermeasures were not properly implemented as a result of lack of awareness of the seriousness of marine pollution all contributed to the pollution.

The Decision issued by IMST was appealed to CMST by the owners of *Hebei Spirit.* After a total of eleven trials during the period from 22 October to 4 December 2008, during which crew members of the marine spread and

Hebei Spirit, as well as witnesses, were interrogated,[2] on 4 December 2008, CMST issued its Decision, in which it affirmed the conclusions drawn by CMST that both the marine spread and *Hebei Spirit* were to blame. In particular, it found that:

> On the part of the marine spread, (i) it failed to take early actions in response to weather changes; (ii) it encountered bad weather and loosing towing ability, but continued navigation without taking any safety measures such as warning the other vessels nearby or performing emergency anchoring etc.
>
> On the party of 'Hebei Spirit', (i) she was negligent in performing her duty while at anchor and failed to take early and active measures; (ii) her main engine was not operable due to negligence in engine readiness.

CMST concluded, in particular, that the marine pollution was caused by oil spilled into the sea as a result of the puncturing of cargo tanks of *Hebei Spirit*, but the expansion of pollution was due to inappropriate emergency measures taken by *Hebei Spirit* after the collision, which increased the speed at which the oil spilled, and passive actions to prevent the oil spill.

The CMST Decision was appealed to the Korean Supreme Court by the owners of *Hebei Spirit* and others. The finding of negligence committed by the master and chief officer on *Hebei Spirit* in the CMST Decision was also criticized by some nautical experts and organizations such as International Association of Independent Tanker Owners (Intertanko) in the world shipping industry, and deemed 'technically flawed', 'biased and certainly lacking good faith' and susceptible contravention of the International Maritime Organization (IMO) guidelines that require cooperation and consultation with all interested parties before a final report is made.[3]

3.1.2 Investigation Conducted by MARDEP

The Marine Incident Investigation Section of the Marine Department of Hong Kong Special Administrative Region (Hong Kong MARDEP) as the flag authority of *Hebei Spirit* also conducted investigation into the causes of the incident according to the IMO Code for the Investigation of Marine Casualties and Incidents[4] for the purpose of improving the safety of life at sea and avoiding similar incidents in future.

The investigation revealed that: (i) the decisions for the marine spread to commence the towing voyage when adverse weather had been forecast is the main probable cause of incident; (ii) during the voyage the towing capacity of the marine spread could not overcome the weather conditions; (iii) the breaking of towing line caused the crane barge to drift uncontrollably and contact directly with *Hebei Spirit.* Other contributory factors are the towing voyage was not carried in accordance with towing survey certificate; towing line was not properly maintained; the

2. See 1992 IOPC Fund document 92FUND/EXC.44/7/1.
3. Fairplay International Shipping Weekly – Newswatch, 18 Dec. 2008.
4. Promulgated under IMO Assembly Resolution A.849(20).

tugs did not alert local VTS and nearby vessels when they lost control of the navigation. In other words, the marine spread is 100% to blame and *Hebei Spirit* is innocent.

3.2 Criminal Litigations in Korea

3.2.1 The First Trial

After the incident happened, the Daejeon District Prosecutor in Korea investigated the incident. On 21 January 2008, the Prosecutor submitted the indictment to Daejeon District Court, Seosan Branch against Samsung Heavy Industries (SHI), owners of *Hebei Spirit* and crew members of the marine spread and *Hebei Spirit.*

On 23 June 2008, judgment of first trial was issued, by which (i) the master of the tug *Samsung T-5* was sentenced to prison and imposed a fine, the master of the tug *Samho T-3* was sentenced to prison, and SHI was imposed a fine; (ii) the master and chief officer of *Hebei Spirit* and her owners were found innocent.

3.2.2 The Second Trial

The judgment of first trial was appealed by the Prosecutor, SHI, the master and other parties who were found guilty. On 10 December 2008, the 1st Criminal Division of Daejeon District Court issued the judgment of second trial, which reversed the original ruling of the court for first trial and by which (i) the master of the crane barge *Samsung No.1*, the master of the tug *Samsung T-5*, the master of the tug *Samho T-3* were sentenced to prison, the master of *Hebei Spirit* was sentenced to one year and six months of imprisonment and imposed a fine of KRW 20 million and the chief officer was sentenced to eight months of imprisonment, on the grounds of destruction of a vessel as a result of criminal negligence while on duty and violation of the Korean Marine Pollution Prevention Act and/or Seafarers Act; (ii) the owners of *Hebei Spirit* and SHI were imposed a fine.

3.2.3 The Third Trial

The judgment of second trial was appealed to the Korean Supreme Court by the Prosecutor and the parties who were found guilty. On 23 April 2009, the Korean Supreme Court issued its judgment, by which (i) the charges against the master of *Samsung No.1*, the charges against the master of the tug *Samsung T-5* related to destruction of a vessel as a result of criminal negligence while on duty and violation of the Korean Marine Pollution Prevention Act, the charges against the master and chief officer of *Hebei Spirit* related to destruction of a vessel as a result of criminal negligence while on duty were reversed, but the relative parts thereof were to be remanded to the court of the second trial; (ii) the remaining appeal of the master of the tug *Samsung T-5*, the master and chief officer of *Hebei Spirit*, SHI and the owners of *Hebei Spirit* were rejected.

In other words, the original ruling of the second trial on the destruction of vessel was overturned, but the other charges were upheld. As far as the master, chief officer and owners of *Hebei Spirit* are concerned, the sentence of imprisonment on the master and chief officer was overruled, but the fines imposed on all of them were maintained, and the master and chief officer would still stay out on bail in South Korea for a few more months.

3.2.4 Strong Reactions from the International Community

After the issuance of the judgment of the second trial by which the master and the chief officer of *Hebei Spirit* were sentenced to imprisonment on the grounds of destruction of a vessel as a result of criminal negligence while on duty and violation of the Korean Marine Pollution Prevention Act, the international shipping community reacted very strongly against the judgment. The owners and manager of *Hebei Spirit*, the Indian governmental institution and maritime groups, Hong Kong Shipowners' Association (HKSOA) strongly protested against and condemned the jailing of the two seafarers. There were a massive rally, gatherings and protest in India. The Indian maritime groups intended to call for a general boycott of Korean products and Korean ports by Indian seafarers. Almost all the non-governmental maritime organizations, including the International Group of P&I Clubs, Intertanko representing over 80% of the world's tanker owners, Baltic and International Maritime Council (BIMCO), International Transport workers Federation (ITF), International Ship Managers' Association (InterManagers), International Chamber of Shipping (ICS), International Shipping Federation (ISF), International Maritime Employers' Committee (IMEC) and Nautilus UK, severely criticized the criminal charges against the master and chief officer of *Hebei Spirit* and launched an international campaign to support the master and chief officer of *Hebei Spirit* for the purpose of their release.

The jailing of the two seafarers were described as 'very regrettable', 'inexplicable', 'wrongful', 'unfair', 'inhumane treatment', 'miscarriage of justice', 'justice denied', 'an unjust trial, a pure case of victimization', 'aggravated and aggrieved the entire shipping industry', 'grievous blow to seafarers around the world', 'part of efforts to assuage Korean public opinion and reduce Samsung's responsibility for the clean-up operation', 'to appease the public of Korea', and 'a classic case of legal face saving'.

The International Group of P&I Clubs said the judgment 'will send deeply disturbing signals to all those considering a career at sea and to their families'. ICS and ISF commented:

> 'The decision by the Korean Court to imprison the two officers . . . further adds to the concerns that the shipping industry has about the criminalization of seafarers in general.'

> 'This incident also highlights the fact that, in the course of providing this important service, seafarers are exposed to the vagaries of different national law, with some jurisdictions imposing criminal sanctions for oil pollution even when there is no intentional or wilful misconduct. There is a real need for uniformity based upon the internationally agreed MARPOL standards so that seafarers know where they stand. All nations should be deeply concerned about the impact of this decision on recruitment to the maritime profession at a time where there is a growing shortage of well-trained and qualified sea-going personnel.'

Nautilus UK pointed out:

> Whilst we sympathize with Korea for the environmental effects of the incident, we strongly believe that the two seafarers were unfairly caught up in efforts to address the crisis caused by the oil spill.[5]

3.3 Civil Litigations in Korea

3.3.1 Claims for Pollution Damages against SHI Etc. and SHI's Application for Limitation

A number of claimants filed lawsuits with Seoul Central District Court against SHI, Hyundai Oilbank as the receiver of cargo oil carried onboard *Hebei Spirit* and Korean government for oil pollution damages.

SHI filed an application with Korean court for limitation of liability in the sum of about USD 3.3 million under Korean law. After the application, thousands of Korean fishermen and other victims protested and urged Samsung Group to promise 'unlimited' accountability and compensation.[6] The claimants alleged that limitation of liability should not be applicable as the crane barge *Samsung No.1* is not a 'ship' or 'vessel'. The Korean court has not yet decided this issue.

3.3.2 Owners' Application for Limitation of Liability

Korea is a party to the 1992 Civil Liability Convention (CLC 1992) and the 1992 Fund convention. Therefore, the limitation amount of the owners of *Hebei Spirit* under CLC 1992 is Special Drawing Rights (SDR) 89.77 million. On 15 January 2008, the owners of *Hebei Spirit* filed an application with the 2nd Civil Division of Daejeon High Court for limitation of liability. Before the opening of the limitation proceedings, on 30 July 2008, the Court agreed to the establishment of a limitation fund through a security provided by the P&I Club of *Hebei Spirit* and deposited into the Court.

---

5. 'Nautilus protests over criminalization case', <www.nautilusuk.org/ngen_public/printable.asp?aid=3792>.
6. <www.straitstimes.com/Breaking%2BNews/Asia/Story/STIStory_320065.html>.

SHI and some claimants filed objections to the limitation of liability of the owners of *Hebei Spirit.*[7] However, such objections were dismissed by the relevant Korean court on 7 July 2009.

3.4 Recourse Litigations in China[8]

In January 2009, the owners of the *Hebei Spirit* and the 1992 IOPC Fund separately commenced recourse action against Samsung C&T Corp. as the owners of the crane barge *Samsung No.1* and the tug *Samsung T-5* and SHI as the operator/bareboat charterers thereof in the Ningbo Maritime Court of the People's Republic of China in an amount of USD 200 million respectively. The recourse litigations are ongoing.

4 ASSESSMENT AND SETTLEMENTS OF OIL POLLUTION CLAIMS

The total amount available for compensation to the victims of the oil spill under CLC 1992 and the 1992 Fund convention is SDR 203 million, including the amount under the CLC 1992. In anticipation of receiving a large number of claims, and after the First Cooperation Agreement concluded among the owners and the P&I Club of *Hebei Spirit*, KMPRC and MOMAF, the P&I Club of *Hebei Spirit* and the 1992 IOPC Fund opened a Claims Office (the Hebei Spirit Centre) in Seoul to assist claimants in the presentation of their claims for compensation.

As on 12 March 2009, 3,377 claims with a total amount of about USD 338 million were received by the Claims Office, among which 224 are claims for cleanup and other preventive measures, 16 are claims for damages to fishing boat hulls, nets and other structures, 413 are claims for damages to the fisheries and aquaculture, 2,723 are claims for damages to tourism and other economic sectors and one is a claim for damages to the environment. It is expected that a large number of significant claims will be further submitted by claimants in the near future. In the tourism sector, the experts appointed by the P&I Club and the 1992 IOPC Fund had been informed that several thousand individuals and organizations were preparing supporting documentation and were expected to submit claims in the near future for losses allegedly suffered as a consequence of the contamination.

As on 12 March 2009, 234 claims were assessed, of which 93 claims were paid, while 482 claims were rejected. The remaining claims were being assessed or additional information had been requested from the claimants.

In July 2008, a Second Cooperation Agreement was concluded, under which the P&I Club undertook to pay claimants 100% of the assessed amounts up to the

---

7. See IOPC Fund document 92FUND/EXC.44/7/1.
8. 'Hebei Spirit: Recourse action against Samsung Heavy Industries (SHI)', see <www.iopcfund.org/pr-pdf/Hebei%20Spirit%20recourse%20action_e.pdf>.

shipowner's limit of liability under CLC 1992, namely SDR 89.77 million. When this amount is paid out, the amount available from the 1992 IOPC Fund up to SDR 203 million, including the amount under the CLC 1992 will be used for compensation. However, in view of the uncertainty as to the total amount of the admissible claims, the level of payment was limited to 60% at the beginning and later was reduced to 35% of the amount of the damage actually suffered by each claimant, as assessed by the experts appointed by the P&I Club and the 1992 IOPC Fund.

5 MEASURES TAKEN BY KOREAN GOVERNMENT

5.1 Special Law for the Support of the Victims of the *Hebei Spirit* Oil Pollution Incident

In March 2008, the Korean National Assembly approved this Special Law, which entered into force as of 15 June 2008, to help the local residents in the affected areas and to restore the marine environment. Under the provisions of the Special Law, the Korean government was authorized to make payments to claimants, based on the assessments made by the P&I Club and the 1992 IOPC Fund, within fourteen days from the date they submitted proof of assessment to the government. If the P&I Club and the 1992 IOPC Fund paid claimants compensation on a pro-rata basis, the Korean government would pay the claimants the remaining percentage.[9] Therefore, claimants could receive compensation in full for the losses suffered as a result of the incident based on the assessments of claims by the experts appointed by the P&I Club and the 1992 IOPC Fund.

A new scheme in the application of the Special Law was later approved by the Korean government to enable the victims to receive loans in the amounts fixed in advance, if they did not receive an offer of compensation from the P&I Club and the 1992 IOPC Fund within six months from the date of submission of their claims.

5.2 Korean Government's Decision to 'Stand Last in the Queue'

To enable speedy compensation payments for the victims, the Korean government announced its willingness to 'stand last in the queue' in receiving compensation from the 1992 IOPC Fund for cleanup and recovery costs and other expenses incurred by the central and local governments.

5.3 Hardship Payments Made by the Korean Government

Also to help the local residents in the affected areas, the Korean central government and its local authorities paid moneys to them as hardship payments. It has been

9. See IOPC Fund document 92FUND/EXC.41/11.

reported in the press that in Taean County, which is one of the most affected areas, a total of 18,757 households received payments between KRW 746,862 (about USD 500) and KRW 2,916,600 (about USD 2,000). These payments were made as donations to the affected residents and therefore did not constitute payment of compensation for pollution damages and would not fall within the scope of compensation under of the 1992 Fund Convention.

6 WHAT MAY BE LEARNT IN LEGAL PERSPECTIVES BY CHINA?

6.1 Oil Water-Borne Transport and Legislations in China

Traditionally, China was an oil exporting country. Such a situation changed in the year of 1993, because from that year China became an oil-importing country and has now become one of the largest oil-importing countries like the United States and Japan. In recent years, the annual quantity of imported oil carried by sea amounted to more than 100 million tonnes and the quantity of oil or oil products in domestic water transport amounted to more than 200 million tonnes. Every day, more than 200 tankers navigate in Chinese waters.

According to statistics announced by the China Maritime Safety Administration (MSA), there were about seventy to eighty incidents of oil spills from ships in each year in recent years, among which several involved a considerable quantity of oil. For examples, the spill of light crude oil from the M/T *Tasman Sea* off Tianjin in 2002, the spill of bunker oil from the M/V *Chang Yang* in Huangpu River in Shanghai in 2003, the spill of bunker oil from the container vessel *MSC Ilona* at the mouth of the Pearl River in Guangzhou in 2004, all of which were the result of ship collisions. Another conspicuous example is the spill of crude oil from the M/T *Arteaga*, a Very-Large Crude Carrier (VLCC) flying the Portuguese flag and carrying 119,574 million tonnes of crude oil off Dalian in 2005 as a result of grounding, which involved more than 100 claimants with a total claim amount of nearly CNY 100 million, with legal proceedings for compensation lasting more than three years.

Fortunately, there has never been any disaster of oil spill in the waters within Chinese jurisdiction. Nevertheless, some Chinese nautical experts have predicted that, because of the very large volume of oil transport and increasingly congested traffic, occurrence of a very serious incident or even a disaster of oil spill from tanker in Chinese waters will be inevitable in the future, even if great efforts are continuously made by the Chinese government in avoiding such occurrence.

Considerable achievements have been made in formulating the Chinese legal system to cope with prevention of oil pollution from ships and compensation for oil pollution damages. Besides ratification and implementation of the international conventions or other instruments including, *inter alia*, MARPOL 73/78, International Convention for the Safety of Life at Sea (SOLAS) including ISM

Code, 1969 Intervention Convention, International Convention on Oil Pollution Preparedness, Response and Co-operation, 1990 (OPRC), CLC 1992, 1992 Fund Convention and 2001 Bunker Convention, China has enacted domestic laws and regulations, such as the Marine Environmental Protection Law of 1999, Regulations Governing Administration on Preventing Pollution to Marine Environment from Ships of 2009 and various technical requirements for the purpose of preventing oil pollution damages from ships. However, the application of 1992 Fund Convention is limited to Hong Kong Special Administrative Region. In addition, noticeably, CLC 1992 and 2001 Bunker Convention are not applicable to cases in which no foreign element is involved.[10]

The author is of the view that, from the perspective of legislation, China is not well prepared for coping with a very serious incident or even a disaster of oil spill from tanker such as the *Hebei Spirit* incident, once it occurs in the Chinese waters, mainly because of the lack of an oil pollution compensation fund.

6.2 The Necessity of Setting up a Domestic Oil Pollution Compensation Fund

As is commonly recognized in China, it will be economically inadvisable to apply the 1992 Fund Convention in Mainland China, because China has become one of the largest oil-importing countries and currently the living standards in the Mainland remains relatively low compared with the developed countries. It is also recognized that, in the case of a serious oil spill incident, let alone a very serious one or a disaster, the limit of liability available under CLC 1992 where foreign element is involved, or a lower limit under China Maritime Code (CMC) where foreign element is not involved, will be beyond or far beyond insufficiency of compensation for oil pollution damages. Therefore, a domestic oil pollution compensation fund similar to the Oil Spill Liability Trust Fund (OSLTF) in the United States[11] should be established as soon as possible.

After lengthy discussions, substantive steps were taken by the Chinese government in 2003. In January of that year, the former Ministry of Communications and the Ministry of Finance jointly submitted an application for setting up a compensation scheme for oil pollution damages from ships to the State Council. In August of the same year, the two Ministries formed draft Regulations Governing the Levy of Compensation Fund for Oil Pollution Damages from Ships and Its Use. The two Ministries then called for comments and suggestions from other ministries

10. International maritime treaties ratified and acceded to by China shall be applicable in China by virtue of para. 1 of Art. 268 of China Maritime Code (CMC) of 1992 which provides that such an international treaty shall prevail over the Code in case of repugnancy, except where reservation has been made. However, Art. 269 is contained in Chapter XIV 'Application of Law in Foreign-related Relations'.
11. See Subtitle I 'Trust Fund Code' of Title 26 'Internal Revenue Code' and Subchapter A 'Establishment of Trust Code' of Chapter 98 'Trust Fund Code', United State Code (USC).

and the oil industry. At the same time, the former Ministry of Communications appointed a research institute to draft documents similar to the *Claims Manual* of the 1992 IOPC Fund. However, the draft Regulations have not yet been approved by the State Council.

6.3 Improvement of Statutory Provisions Regarding Compensation for Oil Pollution Damages from Ships

The prevailing Maritime Code of 1992, Marine Environmental Protection Law of 1999 and Regulations Governing Administration on Preventing Pollution to Marine Environment from Ships of 2009 do not contain detailed provisions in regard to the party or parties liable for oil pollution damages from ships, the scope of compensation and so forth. There are no binding provisions with respect to assessment of damages, especially cleanup costs and damages to fishery resources and the marine environment. The need for improving the prevailing statutory provisions regarding compensation for oil pollution damages from ships, especially to help the victims in an effective and efficient way remains.

6.4 Improvement of Law Regarding Criminal Liability of Offending Seafarers

In recent years in China, increasingly more seafarers are found criminally liable, mainly in maritime incidents in which serious loss of life or damage to properties such as bridges is caused. It seems clear that the courts applied rules applicable to road incidents to maritime incidents in several cases, without regard to the fact that maritime activities are normally more risky.

As the *Hebei Spirit* incident proves, criminal liability of seafarers is not only a matter of criminal law but also may involve the issue of policy. The policy consideration may be of more importance in China, because China is a shipping power and availability of seafarers has become a problem. Such a situation requires improvement of Chinese statutory rules regarding criminal liability of seafarers as a result of maritime incidents including oil spill from ships.

6.5 Improvement of Emergency Response System

As a result of effective implementation of the related IMO standards in China, considerable achievements have been made with respect to establishment of national or local oil spill response plans. Nevertheless, it seems clear that further efforts are needed to improve the emergency response system, especially the availability of professional response resources, because the *Hebei Spirit* incident proves that a very serious oil spill incident may require large scale of oil cleanup operations.

7 CONCLUSION

Based upon this discussion of the *Hebei Spirit* oil spill incident and analysis, the following conclusions may be drawn:

The *Hebei Spirit* oil spill incident has not only caused very serious oil pollution damage in Korea, but also given rise to various legal proceedings in or outside Korea involving complicated issues of civil, criminal and administrative law;

A lot of effective and efficient measures have been taken by the government of the Republic of Korea to cope with the disaster, especially in helping the victims, with the assistance of the P&I Club and the 1992 IOPC Fund;

The KMST investigations into the incident and the Korean courts' finding of criminal liability of seafarers of *'Hebei Spirit'* has incurred strong criticisms from the international community, which implies the need for policy consideration in the issue of criminal liability of seafarers, especially where foreign seafarers are involved, and for international uniformity as much as possible in this regard;

In consideration of the prevailing conditions in China, the *Hebei Spirit* oil spill incident may help to prove the necessity of setting up domestic oil pollution compensation fund, improving the statutory provisions regarding compensation for oil pollution damages from ships and criminal liability of seafarers as well as improving the oil spill emergency response system in China.

Chapter 24

A Review and Critical Analysis on the *MSC Ilona* Oil Spill Incident at Pearl River Estuary

Li Zhonghua & Zhou Zhujun***

1 INCIDENT REVIEW

At 21:35, 7 December 2004, Panamanian container vessel *Hyundai Advance* collided with German container vessel *MSC Ilona* at the Pearl River estuary, the incident caused a breakage on the fuel tank of *MSC Ilona* and thus caused a leakage of 180 fuel in the amount of 1,280 tonnes (Incident 12.7). The incident occurred 11 nautical miles south of the land territory of Hong Kong. The leaked fuel oil, affected by the winds and sea currents, drifted south-westward and formed an elongated drifting strap with a length of 9 nautical miles. Guangdong Maritime Safety Administration (GMSA) received a report concerning this accident at 22:23 and immediately initiated the oil spill emergency response plan. GMSA deployed personnel, vessels and aircraft to participate in the treatment, forming three workgroups: the oil spill cleanup team, the maritime investigation team and the aerial surveillance and supervisory team. The Automatic Identification System (AIS)

---

* Li is with the Zhuhai Maritime Safety Administration (ZMSA). He is the director general of ZMSA and a senior engineer.

** Zhou is with the Department of Supervision, Zhuhai Maritime Safety Administration. He is the deputy department director and a senior engineer.

Michael G. Faure, Han Lixin & Shan Hongjun, *Maritime Pollution Liability and Policy: China, Europe and the US*, 413–422.

was employed to track the incident-related vessels, emergency treatment vessels and other vessels involved in the effort. The OSCAR model was employed to estimate and evaluate the dynamic state of the oil spill and remedy of emergency treatment. At 12:00, 8 December 2004, the Ministry of Transport of the People's Republic of China (PRC) established the 'Incident 12.7' Leading Group for Oil Spill Treatment, led by the deputy minister and consisting of leaders from the Water Transportation Department, the Institutional Reform and Legal System Construction Department, the Public Security Bureau, the Maritime Safety Administration and the Rescue and Salvage Bureau. The Ministry of Transport appointed GMSA as the in situ headquarter and its director general as the in situ commander in chief. Leaders from Guangdong Administration of Work Safety, South China Sea Bureau of Salvation, Guangzhou Salvage and Rescue Bureau, Shanghai Salvage and Rescue Bureau, and Shenzhen Maritime Safety Administration were appointed deputy commanders. According to the superior's order and the practical situation, the in situ headquarter set up a strategy of 'make best effort to avoid the oil stains from landing or entering sensitive water; employ physical approaches mainly and chemical ones secondarily'. The emergency treatment operation began the night of 7 December and finished in the afternoon of 16 December, lasting nine days. Altogether, twenty-two legal independent entities, including sixteen professional anti-oil-spill entities such as Daya Bay Hangpeng Environmental Service Company and six maritime system entities, joined the oil-spill cleanup operation. A total of 6,000 person-time of personnel, eighteen sorties of cruise helicopters and 500 vessel-time of different kinds of vessels were mobilized. Also, 2,500 metres of spill containment boom, 6,000 metres of oil absorbent hawser, three sets of oil skimmers, 120 tonnes of oil absorbent mat, 110 tonnes of oil spill dispersant, 2 tonnes of absorbing substance, and a great deal of other cleanup materials were employed. Finally, 400 tonnes of slop and 360 tonnes of oil refuse were retrieved.

2 CRITICAL ANALYSIS TO THE INCIDENT EMERGENCY TREATMENT

2.1 Potential Efficiency of Regional Emergency Resources Has Not Been Fully Deployed

The emergency treatment to 'Incident 12.7' was assessed as successful by an expert evaluator group because there was not an oil spill landing or wide area of pollution. However, GMSA determine this success had a lot to do with the winds and sea current at the time. If the wind direction had been southerly instead of northerly, the challenges would have been greater. In order to justify this analysis, in 2005, GMSA set up an oil spill emergency treatment rehearsal, which was named 'Blue Ocean Operation'. This rehearsal's time and location were kept confidential in advance. Participants were requested to keep their normal status and were kept from arriving at the rehearsal location before rehearsal. The instruction to start the mission was given during the night, and all operations were to be as real as in actual combat. 'Blue Ocean Operation' imitates the situation in 'Incident 12.7' with wind direction reversed and oil leaked amount as 300 tonnes. According to the simulation and evaluation of the rehearsal, it was found that, without the support from Hong Kong, the emergency treatment force from Guangdong region would have had difficulty in preventing the oil spill from entering Hong Kong water or stopping it from landing on Hong Kong land territory. Oil stain was assumed to enter Hong Kong water in thirteen hours, as shown in Figure 24.1.

Figure 24.1. Thirteen Hours Elapsed

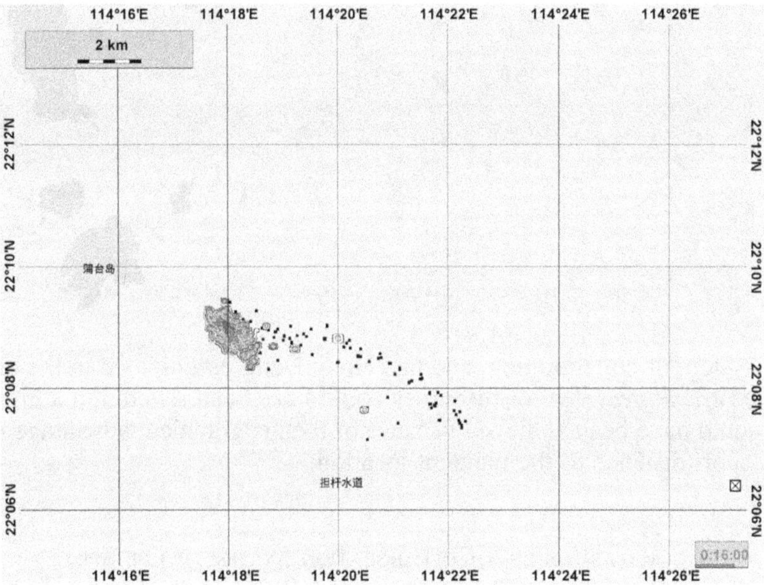

The oil stain was assumed to land on Hong Kong in sixteen hours, as shown in Figure 24.2.

Hong Kong is geographically close to the incident, or rehearsal, location (i.e., 11 nautical miles compared to 49 nautical miles for Huizhou, 106 nautical miles for Guangzhou and 90 nautical miles for Zhuhai's emergent resources); the arrival of an emergency treatment force from Hong Kong should take two hours from Huizhou, eight hours earlier than from Guangzhou, theoretically. Also, within these two to eight hours, the oil film expands very quickly. The area of oil film increases non-linearly with the elapse of time, and the pollution retrieval speed decreases non-linearly with the elapse of time. By simulating the relations between oil spill expanding time and the functionalities of oil absorbent mat, spill containment boom and oil skimmers, 70% of emergent resources could be saved if Hong Kong deploys its emergency treatment force to achieve the same effect as what was achieved by Guangdong arrived after two hours. Seemingly, should there be a

Figure 24.2. Sixteen Hours Elapsed

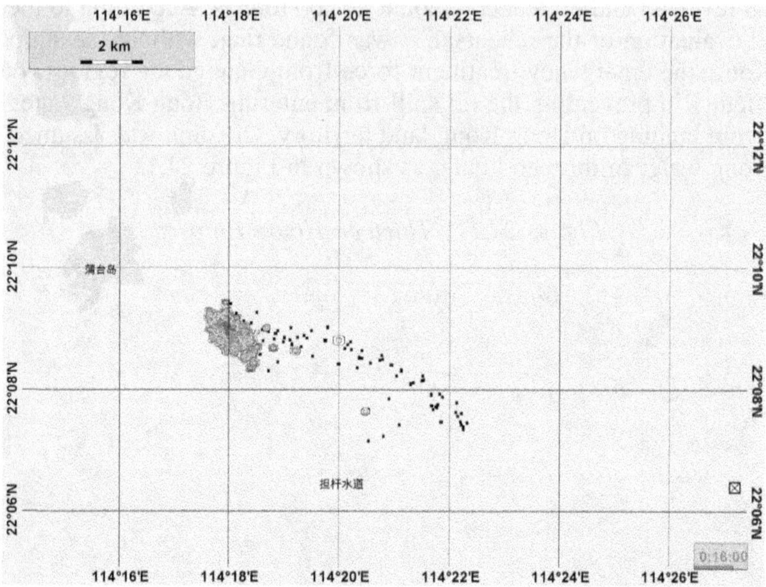

mature system of collaboration, and had Hong Kong participated in the operation Incident 12.7, a great deal of resources could have been saved and a higher efficiency could have been achieved because of the geographical advantage of Hong Kong's short distance to the incident location.

2.2 Establishing a Collaboration System of Oil Spill Emergency Treatment at the Pearl River Estuary Could Greatly Enhance the Oil Spill Emergency Treatment Efficiency

Realizing the uniqueness of Pearl River estuary's navigation situation, the high risk of oil spill and the special geographical relationship among Guangdong, Hong Kong and Macau, in order to unleash the advantage of oil spill emergency treatment collaboration, GMSA and the maritime departments in Hong Kong and Macau should make arrangements in advance to collaborate on developing an emergency treatment collaboration system and resisting the high risk for oil spill under relevant international treaties and the 'One Country, Two Systems' legal frames. In September, 2008, the 2008 Pearl River Estuary Maritime Safety Forum, on the theme 'Oil Spill's prevention, reaction, and collaboration' was held in Zhuhai. In this forum, maritime entities from Guangdong, Shenzhen, Hong Kong and Macau signed the 'Arrangement on the Collaborations on Emergency Treatment to Oil Spill for Maritime Vessels at the Pearl River Estuary Region'

(Arrangement). Arrangement requested that, in the case of a serious oil spill, the maritime entities mentioned earlier should collaborate with the guidance of the 'emergency plan', fully mobilizing emergency resources within the region and neighbouring engaged parties to prevent the spill from expanding, to reduce the damage and influence of pollution and to protect the ecological environment of the ocean near the Pearl River estuary. The signing of this arrangement demonstrates that the oil spill emergency treatment resources are to be fully exploited and the advantage of collaborating is to be fully realized.

3 LODGING OF CLAIMS EXPOSES THE DISTEMPERENESS OF COMPENSATION LAWS AND SYSTEMS

3.1 The Process of Claim Lodging

After the emergency treatment of Incident 12.7, in late December 2004, Guangdong Province established a claim lodging group headed by GMSA and consisting of Guangdong Provincial Oceanic and Fishery Administration (GPOFA) and Environment Protection Bureau, to negotiate compensation with the entity responsible for the incident and to handle the process of litigation. In March 2005, the vessel *Hyundai Advance* applied to maritime court that a fund for limitation of liability be established, which prompted the claim lodging into judicial proceedings. During May to June, GMSA, which represents the cleanup operators, GPOFA, which represents the aggrieved party, and a number of piscicultural enterprises from Hainan Province, summed to ninety entities and individuals, jointly registered claims and filed lawsuits concerning the incident at the Guangzhou Maritime Court. Along with the judicial proceedings, out-of-court settlement was not yet abandoned. The entity responsible for the incident insisted that the discussion on compensations should be comprehensive, including all entities from Hainan that registered claims. In the first half-year of 2006, via many periods of negotiations between both sides of claims, agreement of an out-of-court settlement was achieved. Compensation was ultimately settled via agreements.

3.2 Divergence on How Entity Responsible for Incident Should Assume Liability Is the Main Barrier to Compensation Negotiations

The two sides of Incident 12.7 held many talks and negotiations during the eighteen months between late 2004 and June 2006, among which the central topic has always been on how the two vessels responsible for the incident should assume liability for the pollution and damage. The claim lodgers, and the vessels *MSC Ilona* and *Hyundai Advance* held different viewpoints on how the liability should be assumed based on different legal grounds that benefit themselves respectively.

The claim lodger, based on the General Principles of the Civil Law, proposed that the two vessels bear joint responsibilities to the pollution incident. The spiller vessel *MSC Ilona*, based mainly on the China Maritime Code (CMC), proposed that the two vessels bear responsibility according to their proportion of degree of fault. The other vessel, *Hyundai Advance*, based mainly on 'Marine Environmental Protection Law' and a number of typical international cases, proposed that the principle of assuming liability be 'who leaks, who is responsible' and that the maximum amount of compensation from *Hyundai Advance* be subject to the limit of liability defined by CMC.

3.3 Analysis on How to Assume Liability

As to liability, assuming either of the two vessels should be jointly liable or bear responsibility according to their proportion of degree of fault or 'who leaks, who is responsible', depends first on which laws are applicable to the case, and second on the priority ranking of the applicable laws. The priority ranking of environment protection law system of our country is that applicable international laws have precedence over domestic laws, as well as higher-level laws have precedence over lower-level laws and special laws have precedence over comprehensive laws.

The author holds that the liability assumption of this case should follow rigorous logic and hierarchical relation. Because the case involves pollution damage the clauses especially state that pollution damage to environment of Marine Environmental Protection Law and General Principles of the Civil Law should take precedence over other clauses. The oil spilling vessel should bear pollution damage liabilities (except that case applicable to the exception clause of Marine Environmental Protection Law). Still, if the case is applicable under Article 130 of General Provisions of the Civil Law, the other party vessels should bear joint responsibility. The application of the two laws does not conflict. In other words, the spiller vessel bears pollution damage liabilities and does not exclude that the other party vessels should bear joint responsibilities. Furthermore, if damage is caused by collision, the vessels involved should bear responsibility according to their proportion of degree of fault to determine damage, that is, the party vessels determine damage according to their proportion of degree of fault. This cannot go against the interest of the claim lodgers. That is to say, the total amount compensated by the party vessels according to their proportion of degree of fault should not be less than the total amount compensated by the spiller vessel or by the party vessels bearing joint liability.

One thing to note is whether the divergence mentioned earlier could be settled along with the coming-into-force of the International Convention on Civil Liability for Bunker Oil Pollution Damage (Bunker Convention). The Bunker Convention came into force in China on 9 March 2008. Paragraph 2, Article 3 of the Bunker Convention, which differentiates it from the Civil Law Convention 1992

(CLC 1992) on how to ascertain liability when a ship spills oil by a crash, states that, 'If more than one individual are responsible for the incident, they share the joint responsibility. ... Some legal professionals argue that 'individual' here means different vessels that are concerned in the incident. If that is the case, the Bunker Convention holds the same point as the General Provisions of the Civil Law. However, it contradicts Article 5 of the Bunker Convention. Accordingly, 'more than one individual' here is to be understood as 'multiple owners', which is to mean that the vessel's owner is not single. Instead, it includes the chartered owner, the bareboat charterer, the vessel agent and the vessel operator. Hence, the author holds that the Bunker Convention is coherent to CLC 1992 on the criterion of liability to multiple-vessel pollution incident, which means joint responsibility should be assumed if the damage is 'rational un-dividable', whereas it is not recommended in the case in which it is 'rational dividable'.

3.4 Divergence on Whether Cleanup Cost Has the Priority of Gaining Compensation Leads to Dispute on the Distribution of Compensation

After the out-of-court settlement of Incident 12.7, parties in the claim lodger develop a dispute on the distribution of compensation. The dispute came mainly from two sides: GPOFA, which represents the aggrieved party, and GMSA, which represents the cleanup operators. Because the actual amount of compensation obtained via agreement is far less than the amount of claim (about one-third of it), no party could obtain the expected amount in its claim. The two sides expressed their viewpoint in order to gain a higher portion of the compensation. GPOFA asserted that, according to many previous oil spill cases, the portion of compensation for oceanic environmental and ecological loss has been always greater than that for cleanup cost. GMSA asserted that cleanup cost is an incurred cost, and that although an oil spill damage compensation system had not been established, there had been no other approaches to obtain compensation for cleanup cost. Should the cleanup cost be unreasonably compensated, the cleanup operators' benefit is actually damaged and its motivation reduced, which will result in a lowered effect of emergency treatment and thus a possibility to impair the protection of the oceanic environment. The Maritime Safety Administration is in charge of cleanup operations. It faces the risk of being the defendant in the case of an oil spill. In sum, incurred cost should have the priority to compensation.

4 COUNTERMEASURES AND SUGGESTIONS

According to the critical analysis to 'Incident 12.7', it could be found that the problem that the efficiency of oil spill emergency treatment is not enhanced fully because of the lack of mature regional oil spill emergency treatment collaboration has been addressed and improved in the 2008 Pearl River Estuary Maritime Safety Forum via signing the Arrangement on the Collaborations on Emergency Treatment to Oil Spill for Maritime Vessels at the Pearl River Estuary Region. For pollution incidents with multiple vessels involved, divergence on how to assume liability to pollution damage was once again raised. According to analysis, the 'Bunker Convention' is coherent to '1992 CLC' on the criterion of liability to multiple-vessel pollution incident. On another hand, problems such as that the emergency power is yet to be improved, that the operations are directed multiplex, that there is lack of an assessment system for cleanup, and that there has not been guarantee for cleanup cost compensation. In light of these problems, countermeasures and suggestions are given in the following section.

4.1 To Accelerate the Completeness of Relevant Legislations Such as 'CMC' in Order to Effectively Protect the Benefit of Aggrieved Party of Pollution and Participants of Cleanup Operations

The divergence on how entities responsible for oil spill resulting from collision should assume liabilities is not new. The barriers in negotiations caused by this divergence have been significant in previous judicial cases. Reviewing the three viewpoints, vessels in collision should hold joint responsibility, because this is the most beneficial approach to guarantee the legitimate interests of the aggrieved party in oil spills and thus protect the oceanic environment. Although many judgments and significant academic discussions work in vain to settle this dispute in order to exclude this divergence, it is necessary to revise legislation such as the CMC and Marine Environmental Protection Law to standardize this field. Article 9 in Guangdong Regulations on Collision Responsibilities of Fishery Vessels has made a step. It states that when an incident damages the benefits of a third party that is not responsible for the incident, the entities responsible for the incident should be jointly liable to compensating the third party.

Considering that the cleanup cost is an incurred cost (which is different to the oceanic ecological loss and fishery loss, part of which happens in the future, and thus recovery measures to which are employed), and that an incurred cost is unchangeable while the calculation of long-term loss and its justification are unclear, if the cleanup cost is not reasonably compensated, motivation of cleanup entities is reduced and hence the marine environment is subject to under-protection. Oil spill cleanup, in essence, is similar to a compulsory removal and rescue operation; whereas the latter saves persons and properties, the former saves environment. The compensation for cleanup cost could therefore refer to that for compulsory removal and rescue. In sum, incurred cost should have priority to gain compensation. According to the principles for being compensated, this view should be supported in the CMC.

4.2 To Implement Measures as Establishing Exclusive Funds for Oil Spill Damage to Increase the Emergency Power to Deal with Oil Spills

For years, it has been a knot to lodge claims for cleanup cost. Some cleanup entities reported under-payment after the emergency treatment. Therefore, a national compulsory oil spill insurance and fund system should be established according to the requirements in Oceanic Environmental Protection Law to guarantee the benefits of cleanup operators and their motivation.

The government should enforce guidance and protection to ensure completeness of the anti-pollution emergency system. It should help carry out the responsibility of pollution-responsible entities to remove pollution. It is expected to push oil-related vessels to sign agreements on oil spill cleanup and to make arrangements in advance to adopt cleanup operations. It should also establish an evaluation system to assess treatments. The government could refer to the measures employed in developed countries or Hong Kong SAR, consulting the maritime department before approving projects and withdrawing approval to those who offend the anti-pollution requirements. Referring to advanced managing modes, it could mobilize civilian resources in the case of emergency treatments, for example, by training volunteers. By rewarding and honouring volunteers, large-scale cleanup operations could be made possible.

4.3 To Enhance Efficiency of Emergency Treatment by Rationalizing the Managing System and Completing Emergency Evaluation System

The need to rationalize the current system and form a unified and powerful maritime administration force is needed. Complex direction should be avoided. Orders and requests for information from all directions are also dangerous to emergency treatment. Interference to in situ operations should be reduced in order to enhance efficiency. Repeated investments are also subject to reduction.

The evaluation system is to be perfected. A third-party evaluator should be employed to follow and assess cleanup operations. The effects and efficiency should be linked to compensation when compensating. The same entity should not be responsible for both cleanup and monitoring. A third-party evaluation is helpful to prevent sabotage or cheating, enhancing the overall effect and efficiency of the cleanup operation. It is also necessary to prevent filing of false reports of costs or exaggeration of the actual investment.

References

Wang, S.M. *Tutorial to Oil Spill Emergency Training*. Beijing: People's Transport Press, 2004.

Yan, Z.Y., D.Q. Xiong & P.M. Yin. 'Review on Maritime Oil Spill Models under Winds'. *Academic Journal of DMU* 4 (2001).

He, L.X., et al., 'Review on the 6th Chinese Maritime Law Symposium'. *Transaction of Chinese Maritime Law* 16 (January 2006).

Xu, G.Y., X.M. Lin & S.T. Wang. Legal and Practical Discussion on Maritime Vessels Oil Spill Emergency Planning. 2008 Pearl River Estuary Maritime Safety Forum, September 2008.

Yang, Y.F. & C.Z. Lin. The Range of Compensation of Oil Spill Damage on Chinese Water. *Chinese Law and Jurists* 1, no. 7 (December 2005).

Zheng, L.D. Review on Oil Spill of Vessel 'Hebei Spirit'. 2008 Pearl River Estuary Maritime Safety Forum. September 2008.

Chapter 25

A Case Study of Compensation for Marine Oil Pollution Damage: Centring in the Compensation Claim of the *Amorgos* Oil Pollution Accident

*Wang Yuh-Ling**

1 *MV AMORGOS* CLAIMS PROCESS

1.1 BACKGROUND

On 14 January 2001, the Greek-registered freighter *MV Amorgos* (hereinafter '*Amorgos*') ran aground and spilled oil off the coast of Lungkeng, in Kenting, Taiwan, causing great damage to Taiwan's marine environment. On behalf of the Taiwanese government, the Environmental Protection Administration of Taiwan sought compensation from the shipowner and the shipowner's mutual Protection and Indemnity (P&I) Club, Gard, for ecological and economic damages. The case was tried from 1 to 18 November 2004, in the district court of Arendal, Norway. The court's decision is broadly as follows:

(1) Gard and the owner of the *Amorgos* are liable for indemnity.
(2) Gard and the owner of the *Amorgos* should compensate the Taiwanese government the amount of United States Dollar (USD) 288,889

* Pacific Star Taipei Office, Taiwan.

Michael G. Faure, Han Lixin & Shan Hongjun, *Maritime Pollution Liability and Policy: China, Europe and the US*, 423–442.

(approximately Taiwan Dollar (TWD) 9,533,337) and pay interest at a rate of 5% of the annual interest rate from 17 March 2003, until the debt is paid in full.

(3) The Taiwanese government claim that Gard and the owner of the *Amorgos* should jointly pay litigation charges in the amount of Norwegian Krone (NOK) 3,340,892 (approximately TWD 16,704,460), should pay the court NOK 34,300 (approximately TWD 171,500), should defray the fees for court-appointed associate judges and should pay delayed interest on the aforementioned costs until they are paid in full.

Although the Norwegian court has ruled that Gard and the owner of the *Amorgos* are liable for damages to the Taiwanese government, the compensation was minimal, covering only the initial cost that 'had already been paid' by the Taiwanese government to assess the damage at the time of the incident. As for the costs the Taiwanese government has requested to restore coral reefs, revive the fishing industry, recoup lost tax and tourism revenue and so forth; however, the court deems there is either insufficient evidence to prove that such damage occurred or that it is impossible to prove a direct connection between the damage and the *Amorgos* oil spill incident and thus refused the request for compensation.

Regarding the cost of coral reef restoration, the court deems that it is unnecessary to take any action to stabilize or restore the coral reef in the effected area. The court's opinion is that nature is running its course to gradually restore this area; while the current growth rate is not ideal, this area will eventually recover. Considering the inclement weather in the effected area, the court deems that the restoration process of transplanting coral and releasing coral larvae may be unsuccessful.

Regarding the cost of reviving the fishing industry, the court deems there is insufficient evidence to prove a direct connection between the *Amorgos* oil spill incident and damage to marine resources; had the *Amorgos* indeed caused limited damage to marine resources, the damage could be rectified naturally without the installation of artificial reefs or the release of fish seedlings.

Regarding the loss of tourism revenue, the court deems that the decrease in local tourism could be attributed to factors unrelated to the *Amorgos* oil spill incident; because the area has seen a steady decline in local tourism over the past several years and the popularity of tourist spots can be capricious, there is insufficient evidence to attribute the loss in revenue at tourist spots inside Kenting National Park to the *Amorgos* oil spill incident.

Regarding the loss of tax revenue, the court deems that consideration should be given to income from the National Museum of Marine Biology and Aquarium (NMMBA) in Kenting, because it is an important source of revenue within Kenting National Park. If the income earned and taxes paid by NMMBA are taken into account, there was an increase in the amount of taxes paid in 2001 compared to 2000; therefore, there is insufficient evidence to prove that the *Amorgos* oil spill caused a decrease in governmental tax revenue.

It is the court's opinion that there was a need to assess the damage in the initial stages immediately following the running aground of the *Amorgos*; thus, as long as the plan to assess and predict possible damage from the incident was adequate and necessary, the shipowner is obligated to compensate Taiwan for all assessment costs. Taiwan has the jurisdiction to conduct overall assessments of damage in affected areas after incidents occur. The assessment costs spent in 2001 and 2002 were reasonable and necessary, and thus these costs should be fully compensated. As for future assessment costs, the court deems them unnecessary; assessments made in 2002 indicated that the area was recovering naturally, and therefore future assessment costs should not be compensated.

The Taiwan Environmental Authority regrets that the evidence submitted by Taiwanese governmental entities and expert witnesses regarding the damage caused by this pollution case has been met with disbelief by the Norwegian court. Because the compensatory items for this lawsuit, including the restoration of the coral reef, the revival of the fishing industry, the loss of tax revenue and the loss of tourism revenue, concern various departments, the Taiwan Environment Authority summoned attorneys and related government entities to analyse the court's ruling; they concluded that the ruling was detrimental to Taiwan, and that the court ruling could be attributed to the fact that international claims practices and submitted materials were insufficient and inadequate to prove the connection between the damage and the pollution; this time the appeal strategy is to focus on strengthening the evidence.

The Taiwan environmental authority emphasizes the reason for the appeal. After considering international claims practices, the substantial cost of a lawsuit and the possibility of strengthening evidence, a partial appeal with a likelihood of wining was proposed. The appeal was made following the advice of Taiwanese attorneys and the input of various government entities to focus on coral reef restoration, lost tourism revenue and so on. Taiwan's Norwegian attorneys accepted the court verdict for this case on 10 January, 2005, with any appeal to be filed by 9 February 2005; the appellate court and the district court are both located in Arendal, Norway. After the appeal was filed, the court required several months for evidence to be presented before it decided the court date, in session for two to three weeks. Instructed counsel presented the claims petition during the appeal, and the evidence was prepared.

1.2 Disputes

1.2.1 An International or Domestic Case

Because this incident took place in and the damage occurred in Taiwan, this nexus dictates that this should be considered a domestic case and be handled by the Taiwanese court under Taiwanese law. However, because the shipowner (Liberian), the ship operator (Greek) and the ship liability insurer (Norwegian) are all foreign nationals, they possess no assets in Taiwan; even if the Taiwanese court ruled that the compensation be a large amount, it is questionable whether the foreign courts of these foreign nationals would accept such a ruling and whether the ruling would be enforced. From a legal standpoint, this is a classic example of a 'foreign' case, for which there is no precedent to define the meaning of 'foreign claims'. Even if the domestic court were to win, it would be unable to enforce its decision.

1.2.2 Time-Consuming Nature of International Maritime Claims

Claims filed with the International Tribunal for the Law of the Sea (ITLOS) are time consuming because they involve international treaties, different laws from different countries, differing demands for evidence and language barriers. Even straightforward maritime cases, such as ship collisions and salvages at sea and so forth cannot be processed in a single day, because gathering evidence of damage and preparing for arbitration procedures demands time. Oil spill incidents, which are more complicated, naturally take much longer. The *Amorgos* oil spill incident exposed that Taiwan was ill-prepared to handle maritime oil spills, and consequently this lack of experience guarantees that preparing evidence for filing a claim will also be more difficult.

1.2.3 The High Cost of International Maritime Claims

Because the filing of international maritime claims is very specialized, procedures such as gathering evidence are time consuming and are normally handled by highly experienced but expensive British law firms. Consequently, the cost of hiring attorneys and filing for international maritime claims is quite high. For example, as of June 2004, the British maritime law firm hired by the *Amorgos* shipowner's liability insurer reportedly earned more than USD 1 million in fees. The Taiwan Environment Authority has reportedly spent over TWD 4 million in litigation fees on law firms experienced in handling international maritime cases. These law firms are able to offer wise consultation and knowledge about claims procedures. If this case were considered an international dispute, in which case it must be handled by foreign attorneys overseas, the Taiwan Environment Authority's budget for litigation fees would greatly exceed TWD 4 million. Under normal circumstances, maritime claims, though not bound by law, allow for the cost of hiring attorneys and specialists to be included in the claim, and therefore such costs can also be compensated by the shipowner or the ship litigation insurer.

1.2.4 The Amount Claimed for Damages

The Taiwanese government will file claims exceeding TWD 6 billion, including claims for damage to the national reputation. It is reported that while other departments have previously filed claims for close to TWD 100 million, the Taiwan Environment Authority is currently proposing that lost national tax revenue exceeds TWD 500 million, and damage to the coral and fish ecosystem and its restoration exceeds TWD 200 million, totalling over TWD 1 billion. However, after several months, the amount of the claims dropped from TWD 6 billion to TWD 1 billion. The law does not specify which kind of damages can be claimed, and therefore what is considered a lawful claim is disputable. Because this case involves foreign countries, there are international regulations and court cases regarding the assessment of oil spill damage and its affected area that must be considered. For example, in 1994, the Comité Maritime International (CMI) passed guidelines on oil pollution damage. Point 7 explicitly states that claims for oil pollution damage do not include the 'loss of taxes and similar revenues by public authorities'. There are international examples for the handling of these cases, such as the French *Tanio* case of 1980 and later the French *Haven* case. France filed claims for loss of tax revenue; however, the Convention for Civil Liability for Oil Pollution Damages (CLC) and the Fund Convention for Oil Pollution Damages both ruled they should not be compensated, the main reason being the lack of documentation because the estimation was too general. More important, the French government could not prove a direct connection between their loss of tax revenue and the oil spill incident. Regarding the domestic regulations of other countries, for example, in the United States in 1990, it was explicitly stated that the net loss of taxes can be included in damage claims; however, when there is a clear connection between net revenue and oil spills, victims are still responsible for the collection of evidence; thus, the outcome was the same as the two French cases mentioned earlier. Although the Taiwanese government filed the claims, the question is whether it bore the responsibility to gather adequate evidence to prove that 'there was a clear connection between the loss of tax revenue and the oil spill incident' and to provide 'concrete proof of net loss of tax revenue', not merely rough estimates from a finance department. If not, in the future there will be no opportunity to file any claim for compensation. Also, the Taiwanese people who do not understand the situation well will blame the government for not trying its best to claim compensation.

1.2.5 Regulations Limiting the Liability of the Country in Which the Shipowner and the Ship Operator Are Registered

Taiwan's maritime laws and ocean pollution prevention laws stipulate that oil pollution is not relevant to the limitation of liability. Because this is a 'foreign case', Taiwan must now consider international laws and the domestic laws of the countries involved. Currently, the international law regulating shipowner liability is the 1976 Convention on Limitation of Liability for Maritime Claims. Under the framework of this convention, the wreckage removal costs of this case, oil pollution damages and so forth, are relevant to the limitation of liability. According to the standards of this convention, the shipowner, ship operator and insurer for the shipowner's liability should be held liable for approximately no greater than USD 6.5 million (approximately TWD 220 million). The individuals involved in this case, from Liberia, Greece and Norway, are all from countries that signed and ratified the convention in 1986 and 1991. In other words, whether this case was filed in these countries or in Taiwan, the court verdict could still be enforced in these countries. These individuals must claim limitation of liability no greater than USD 6.5 million. These regulations involving overseas maritime cases must be taken seriously and cannot be ignored.

1.2.6 Wreckage Removal

The responsibility for wreckage removal belongs to the Taiwan Ministry of Transportation and Communications. Currently, the wreckage has been separated into several sections by typhoons, so the cost of removal is much higher than originally estimated. If the memorandum of agreement between the Ministry of Transportation and Communications and shipowners' liability insurer was still valid at the time of the incident, undoubtedly the current increase in the cost of removal would be covered by the (Taiwan) National Treasury, because the memorandum explicitly states:

> The Taiwanese government is responsible for the part exceeding the amount of Taiwan law limitation of liability.

If residual oil and iron sand have already been removed, whether it is necessary for the ship to be removed should be seriously considered. Therefore, the best recommendation is to have a comprehensive evaluation of the ship's potential hazards and dangers to maritime travel, the ecosystem, the fishing industry and tourism. If international standards are not met and the wreckage is deemed harmful, it should be removed.

1.3 Relevant International Regimes

Currently, major international regimes that are established to ensure that all parities affected by oil spill or discharge from ships can be appropriately compensated include the following: CLC 1969 and 1992 (which is the Protocol to the 1969 version); the CLC and the International Convention on the Establishment of an International Fund for Compensation for Oil Pollution Damage and its Protocols; or the Fund Convention 1971. The P&I Clubs were formed to provide shipowners with insurance for third-party liability incurred during transportation. However, CLC and the Fund Convention and its Protocol only regulate compensation liability of oil tankers. Most countries base their national compensation frameworks for oil pollution on these two conventions, but most oil pollution incidents are caused by non-oil tankers. In order to fill the gaps in the international oil pollution compensation regimes, the International Maritime Organization (IMO) adopted the International Convention on Civil Liability for Bunker Oil Pollution Damage (Bunker Convention) in 2001, (CLB) which went into effect on 21 November, 2008. The Bunker Convention will cover oil pollution incidents caused by non-oil tankers. The oil pollution by the *Amorgos* is a typical case caused by a non-oil tanker, which used to be excluded from international conventions. The implementation of the Bunker Convention provides affected parties a way forward.

Unfortunately, because Taiwan did not adopt CLC, it is doubtful it will adopt CLC to protect environmental rights.

Below is a discussion on the content of Bunker Convention, 2001.

Modelled on CLC 1992, and taking into consideration the Hazardous and Noxious Substances (HNS) Convention 1996, the Bunker Convention 2001 is designed to clarify liability of oil pollution incidents and establish a reasonable compensation mechanism. Highlights of its content are discussed in the following section.

1.3.1 Scope of Application

1.3.1.1 Ship

CLC 1992 defines 'ship' as any seagoing vessels and seaborne crafts that are built to carry or able to carry oil in bulk as cargo. International Convention on Civil Liability for Bunker Oil Pollution Damage (CLB) has a broader definition of ships, including not only ships as defined in CLC 1992, but also any seagoing vessels or seaborne crafts of any type whatsoever. However, the provisions of the Convention shall not apply to warships or other ships owned or operated by a State and used, for the time being, only on the governmental non-commercial service, nor to wrecked ships that cause oil pollution as a result of fuel oil leakage.

1.3.1.2 Bunker Oil

The Bunker Convention applies to fuel oil, that is, any kind of hydrocarbon mineral oil that is used or intended to be used for the operation or propulsion of the ship, including lubricating oil and any residuals of such oil. Therefore, it covers not only heavy fuel oil but also diesel for ship generators, lubricating oil for main engines, waste oil generated after the replacement of lubricating oil, oil stain from fuel tanks and so forth.

1.3.1.3 Geographical Coverage

When pollution damage takes place in the territory, including the territorial sea, of a State Party, exclusive economic zones (EEZs) established in accordance with international laws, or an area beyond and adjacent to the territorial sea of a State and extending not more than 200 nautical miles from the baselines from which the breadth of its territorial sea is measured, the Convention shall apply. This is a continuation from the Protocol to CLC 1992.

1.3.2 Party or Parties Liable for Compensation

According to the Bunker Convention, the 'shipowner' is the party that shall be subject to compensation liability. The definition of 'shipowner' includes not only the registered owner as stipulated by CLC, but also bareboat charter, manager or operator of the ship and so forth. Compensation claims cannot be made against parties other than those described here. Those affected by pollution now have more parties to go after, and difficulties to file claims against single shipping company can be minimized. When more than one person is liable for an oil pollution incident, liability shall be joint and several. When an accident involving two or more ships occurs and pollution damage results therefrom, the shipowners of all of the ships concerned, unless exonerated under Article 3, shall be jointly and severally liable for all such damage that is not reasonably separable.

When oil pollution is caused, intentionally or unintentionally, by people other than the shipowner, such as employees and agents of the shipowner, crews and so forth, the Bunker Convention does not spell out as clearly as CLC 1992 whether those affected can file claims against them, thus creating a grey area. Under strong request by the International Salvage Union (ISU), the Legal Committee of the IMO has had a resolution passed by the Convention, in which States are urged, when implementing the Convention, to consider the need to introduce legal provision for protection for persons taking measures to prevent or minimize the effects of bunker oil pollution. It recommends that persons taking reasonable measures to prevent or minimize the effects of oil pollution be exempt from liability unless the liability in question resulted from their personal act or omission, committed with the intent to cause damage or recklessly and with knowledge that such damage would probably result. It also recommends that States consider the relevant provisions of HNS 1996 as a model for legislation.

Article 7.5 of HNS 1996 echoes Article 4.2 of CLC 1992, stipulating that compensation claims can only be made against the shipowner, not against the shipowner's employees, agents, or crews; navigators or service providers who are not crews; charterers; ship managers or operators; rescuers and so forth, unless the damage is done by intention or resulted from an act or omission when they foresee such damage. Unfortunately, the resolution is not a part of the official body of the Convention and thus is not legally binding as international law. Whether parties other than the shipowner can be exempt from liability depends on whether such resolution is reflected in national laws of various countries.

1.3.3 Basis of Liability and Criteria of Exclusions

The strict liability basis is continued from CLC to the Bunker Convention; that is, the shipowner should be liable for pollution incidents caused by intention or negligence, unless exclusions apply. Also, when damage is partially or wholly caused by an act or omission or negligence of the person who suffered the damage or from the negligence of that person, the shipowner may be partially or wholly exonerated from liability to such person.

1.3.4 Scope of Compensation

The scope of compensation stipulated in the Bunker Convention covers pollution damage caused by fuel oil spill or leakage and costs of preventive measures to prevent or mitigate such pollution damage. 'Pollution damage' means (i) loss or damage caused outside the ship by contamination resulting from the escape or discharge of bunker oil from the ship, wherever such escape or discharge may occur, provided that compensation for impairment of the environment other than loss of profit from such impairment shall be limited to costs of reasonable measures of reinstatement actually undertaken or to be undertaken; (ii) the costs of preventive measures and further loss or damage caused by preventive measures. 'Preventive measures' mean any reasonable measures taken by any person after an incident has occurred to prevent or minimize pollution damage. No claims for compensation for pollution damage shall be made against the shipowner other than in accordance with the Convention.

This Convention shall not apply to pollution damage as defined in CLC 1992, whether or not compensation is payable in respect of it under that Convention. In other words, CLC 1992 regulates obligations of oil tankers (with or without oil cargo on board) whose oil cargo or fuel oil spills, while the Bunker Convention covers fuel oil spills by non-oil takers and any other kinds of ships. As long as an incident comes under the definition of pollution damage stipulated in CLC 1992, the Bunker Convention shall not apply, regardless of whether the party or parties that suffer can file claims in accordance with CLC 1992.

What is called into question is that once both HNS 1996, which has also been passed by IMO, and the Bunker Convention come into effect, when an accident happens to a ship carrying dangerous and toxic cargo, causing leakage of both oil and a toxic substance, how the two conventions interact, what legal framework the party or parties that suffer should appeal for and the scope of compensation the party or parties that suffer can claim are the issues to be further clarified.

1.3.5 Limitation of Liability

Nothing in the Bunker Convention shall affect the right of the shipowner and the person or persons providing insurance or other financial security to limit liability under any applicable national or international regime, such as Limitation of Liability for Maritime Claims (LLMC) and its Protocol 1976, as amended. This is different from CLC in terms of limitation regimes. CLC adopts an independent standard to decide on the limits of liability of the shipowner, whereas the Bunker Convention allows direct applications of various national or international frameworks on limitation of liability as appropriate. With the adoption of the resolution, the Bunker Convention urges all States that have not yet done so, to ratify or accede to the Protocol of 1996 to amend LLMC 1976, so as to unify the legal regimes of liability limitation. Nevertheless, the resolution is not mandatory. There are also doubts on whether limitation of liability can be invoked based on LLMC 1976 for pollution damage defined in Bunker Convention. In addition, except for property damage and consequential loss, LLMC 1976 is not fully aligned with CLB in terms of the definition of costs of cleanup and removal, costs of preventive measures, pure economic loss, costs of recovery of environment and so forth. If the Bunker Convention allows the direct application of LLMC 1976, the shipowner can invoke limitation of liability for certain compensatory items. For those that the shipowner cannot invoke limitation, the Bunker Convention does not specify whether the shipowner shall bear unlimited liability. It will be even more complicated and difficult to determine whether limitation of liability can be invoked for compensation claims when non- States Parties of LLMC 1976 join other conventions with limitation of liability or adopt individual legal frameworks. This shows that the Bunker Convention needs to be further strengthened.

1.3.6 Compulsory Insurance or Financial Security

The Bunker Convention requires ships over 1,000 gross tonnage to maintain insurance or other financial security to cover the liability of the registered owner for pollution damage in an amount equal to the limits of liability under the applicable national or international limitation regime, but in all cases, not exceeding an amount calculated in accordance with the Convention on Limitation of Liability for Maritime Claims 1976, as amended. This means that the registered owner, rather than the shipowner, is the party that has to purchase insurance or provide other financial security. Yet, the definition of the shipowner in Article 3.1 of the Bunker Convention covers the registered owner. Both of them are the parties liable for compensation liability, but only the registered owner is required to purchase insurance or provide other financial security. The reasons could be that, in practice, claimants usually pursue the registered owner or the insurer rather than other parties for compensation under normal circumstances or certificates issued or certified by the appropriate authority of a State Party have so stipulated. Furthermore, only ships with a gross tonnage greater than 1,000 shall be required to maintain insurance or other financial security in an amount equal to the limits of liability under the applicable national or international regime, but in all cases, not exceeding an amount calculated in accordance with the LLMC 1976, as amended. Also, a State may declare that the article stipulating compulsory insurance and financial security does not apply to ships registered in the State and operating exclusively within its territory. The definition of exclusive operation within the State's territory is that the ship stays within the State's territory, including the territorial sea but excluding the EEZ, throughout the whole voyage from departure to arrival.

1.3.7 Direct Litigation Regime

According to the Bunker Convention, any claim for compensation for pollution damage may be brought directly against the insurer or other person providing financial security for the registered owner's liability to pollution damage. This is the direct litigation regime carried over from CLC. In such a case, the defendant may invoke the defences the shipowner would have been entitled to invoke, including limitation pursuant to Article 6. Furthermore, even if the shipowner is not entitled to limitation of liability according to Article 6, the defendant may limit liability to an amount equal to the limitation under the applicable national or international limitation regime, but in all cases, not exceeding an amount calculated in accordance with the Convention on Limitation of Liability for Maritime Claims, 1976, as amended.

1.3.8 Time Limits for Claiming Compensation

Rights to compensation shall be extinguished unless an action is brought thereunder within three years from the date when the damage occurred. However, in no case shall an action be brought more than six years from the date of the incident that caused the damage. Where the incident consists of a series of occurrences, the six-year period shall run from the date of the first such occurrence. This is in line with CLC.

1.3.9 Jurisdiction

The principle of territorial jurisdiction for oil pollution compensation claims is carried over from CLC to the Bunker Convention. The jurisdiction will be decided based on where an incident causing pollution damage takes place and where preventive measures are taken. Because the article on jurisdiction applies only to actions for compensation against the shipowner, insurer or other person providing security for the shipowner's liability, litigations against other parties are excluded.

1.4 Analysis and Assessment of the Case

A few aspects are worthy of discussion in terms of the serious pollution damage the *Amorgos* caused to the waters and coastal areas of Taiwan.

1.4.1 Applicable Laws and Court with Jurisdiction

Given that the shipmaster and the chief engineer are both Greek, the insurer is a Norwegian company, the ship is registered in Liberia and the incident occurs in Pingtung, Taiwan, does the court of Taiwan have jurisdiction? According to Article 9.1 of the Law of Application of Law on Foreign Civil Relation, 'liability resulted from tort should be subject to laws of the place it takes place'. The incident resulted from tort, so the laws of Taiwan should apply. However, the ship that causes pollution is a foreign ship, so it is a case involving foreign civil relations. Compensation regimes for marine pollution are different from country to country; thus it is still a question mark whether verdicts of the Taiwan court will be upheld and enforced by foreign countries.

1.4.2 Who Should Be the Claimant

According to Article 4 of Taiwan's Marine Pollution Control Act,

> the competent authority referred to in this Act means the Environmental Protection Administration, Executive Yuan, at the central government level, the

municipal government in special municipalities and the county or city government in counties or cities. The scope of jurisdiction for a special municipality, county or city competent authority shall be its administrative jurisdiction within the territorial seas.

The area polluted by the *Amorgos* is in Hengchuen, which is under the jurisdiction of Pingtung County Government, but the site of the incident is within Kenting National Park, a public entity in Taiwan: Who should file the claim? The Environmental Authority filed the case, but can it do so on behalf of Taiwan? Who would be the party or parties impacted?

1.4.3 Claim for Compensation and Scope of Liability

Taiwan agrees that the shipowner can invoke limitation of liability in accordance with Article 21 of Maritime Law of Taiwan on three grounds: (i) no facts that can prove the occurrence of such incident resulted from the intention or negligence of the shipowner are found; (ii) if the shipowner or the insurer does not go by the Maritime Law of Taiwan to conduct salvage, Taiwan can do so and later claim for compensation against the shipowner in accordance with the Marine Pollution Control Act and Law on Commercial Ports. In this case, the shipowner is willing to pay not only for the full costs of pollution removal, but also a maximum salvage cost of USD 2.46 million, the excess of which shall be borne by Taiwan; and (iii) the shipowner and the insurer are willing to remove the shipwreck that they still own and are obliged to at their expenses, indicating their good faith to get this resolved. Therefore, it is agreed that the shipowner invokes limitation of liability to minimize social costs.

Some Taiwanese scholars point out that, according to LLMC 1976, limitation of liability can be invoked in principle for the pollution damage resulted from the *Amorgos*, costs of cleanup, costs of salvage, fines and so forth. However, the erroneous amendments to the Maritime Law have excluded the pollution damage and costs of cleaning from limits of shipowner liability, and the legal authority to instruct the shipowner to salvage is not clear. Therefore, the owner of the *Amorgos* can invoke limitation of liability to the removal of the wreckage in accordance with Maritime Law. Furthermore, although Article 33.2 of the Marine Pollution Control Act mandates the appropriate authority to take necessary responding measures, the measures should be aimed to 'prevent, eliminate, or mitigate pollution'. In other words, unless it is proved that the ship or the cargo on board is the cause of the pollution defined by the Act, it will be a daunting challenge, technically or legally, to order the shipowner to remove the shipwreck from the waters of Taiwan. From the international precedents in dealing with pollution incidents, no controversies are foreseen in terms of the cleanup of oil cargo and fuel oil, but there is no precedent to follow for the liability to remove iron ore fines and the ship itself, which may or may not be the cause of pollution.

1.4.4 Party or Parties against Which Pollution Damage Is Claimed

Taiwan has claimed for compensation against the shipmaster, the chief engineer, the shipowner and the insurer, but the basis for compensation seems inappropriate. First, according to Article 31 of the Law on the Taiwan's Exclusive Economic Zone and the Continental Shelf and Article 32.1 of the Marine Pollution Control Act, the shipmaster and chief engineer should bear joint liability to oil pollution damage that results from negligence. However, according to Article 57 of the UN Convention on the Law of the Sea 1982, the EEZ shall not extend beyond 200 nautical miles from the baselines from which the breadth of the territorial sea is measured. Article 76 of the same Convention stipulates that the 'continental shelf' comprises the seabed and subsoil of the submarine areas that extend beyond its territorial sea throughout the natural prolongation of its land territory to the outer edge of the continental margin, or to a distance of 200 nautical miles from the baselines from which the breadth of the territorial sea is measured where the outer edge of the continental margin does not extend up to that distance, and the outer edge of the continental margin wherever the margin extends beyond 200 nautical miles from the baselines from which the breadth of the territorial sea is measured shall be established by stipulated rules. Article 2 of the Law on the Taiwan's Exclusive Economic Zone and the Continental Shelf is aligned with the UN Convention on the Law of the Sea. As the baselines of the territorial seas of Taiwan are about 1 kilometre from the Lungkeng Natural Conservation Area, and they are straight lines in this area, it is considered the inner waters of Taiwan. Therefore, it is doubtful whether the application of the Law on the Taiwan's Exclusive Economic Zone and the Continental Shelf is appropriate.

Moreover, Article 32.1 of the Marine Pollution Control Act stipulates 'the shipmaster and shipowner shall promptly adopt measures to prevent, eliminate or mitigate pollution'. This is a statutory obligation, different from liability to damage compensation. Also, the Act does not require the shipmaster or other crew to be responsible for damage compensation, so extensive interpretation of the law, that is, making the shipmaster, the chief engineer or other crew members jointly and severally liable in accordance with the earlier mentioned two laws, does not seem appropriate.

1.4.5 Appropriateness to Restrict the Crew from Leaving Taiwan

It seems inappropriate for the Environmental Authority to restrict the crew members from leaving Taiwan based on Article 35 of the Marine Pollution Control Act. Because the crew are employees of the operator, such restriction has undoubtedly infringed on the crew's dignity and personal freedom. Seizure of the ship on the ground of preservation of evidence is reasonable, but personal freedom should not be trespassed.

2 DISCUSSION ON APPLICABLE TAIWAN LEGAL REGIMES OF COMPENSATION FOR MARINE OIL POLLUTION

According to Article 33.1 of the Marine Pollution Control Act, 'the shipowner shall bear liability for compensation for marine pollution damage caused by the ship'. Therefore, as long as the ship causes pollution damage to waters, the shipowner should be liable for compensation, regardless of whether the damage is done by intention or as a result of negligence by the shipowner. The principle of strict liability is upheld. Furthermore, in accordance with Taiwan's Civil Law, as the causal relationship between damage and danger has to be clear for damage compensation claims; this is examined in the following discussion on tort from the perspective of the Civil Law of Taiwan.

2.1 Liability to Damage Compensation for Tort in the Civil Law of Taiwan

When pollution damage is the result of an act or omission done with intent or from the negligence of the person causing oil pollution, thus impacting the party/parties or its/their legal rights, it shall be defined as a result of tort. In the Civil Law of Taiwan, tort is categorized as general tort, stipulated in Article 184, and special tort, as in Articles 28, 185 and 191.3. The Maritime Law and the Marine Pollution Control Act are both civil special laws. Article 33.1 of the Marine Pollution Control Act provides:

> The shipowner shall bear liability for compensation for the damage caused by the pollution of the sea by the ship.

It is generally interpreted that when an action is defined by law to be that of a legal person, as in Article 28 of the Civil Law 'legal person's liability in tort', it falls into the category of special tort. In pollution incidents caused by the ship, the shipowner usually is not the actor; the shipowner hires crews to act on its behalf, and the employer's liability is specified in Article 188 of the Civil Law. Therefore, this is also special tort. If a pollution incident is caused by a party or parties responsible for tort resulting from intent or negligence, the party or parties shall be liable for damage compensation in accordance with Article 184.1 of the Civil Law. If damage is caused by employees, such as the shipmaster or the crew, the employer shall be liable for damage compensation as well in accordance with Article 188 of the Civil Law. If the employer has done a reasonably good job in selecting and overseeing the employees, the employer can be exonerated. Moreover, as the Maritime Law and the Marine Pollution Control Act are civil special laws, they shall be applied first.

Intent or negligence is still the criteria to determine liability for general civil tort and the principle of fault liability is upheld. However, for special tort or special incident, such as marine pollution, regardless of the applicable law being the

Maritime Law, Marine Pollution Control Act, or CLC 1969, strict liability is upheld in principle, regardless of whether it is by intent or negligence.

2.2 Application of International Conventions on the Oil Pollution Cases in Taiwan

Taiwan is not a State Party to international conventions. According to the Maritime Law of Taiwan, unlimited liability could be invoked upon the shipowner. However, if the shipowner cannot fulfil its compensation liability and there is no support from the fund set up by the Convention or from insurance, the party or parties affected by oil pollution will not be compensated. To prevent such a dilemma, can the court of Taiwan base its verdicts on international conventions?

Different views on whether the court of Taiwan can invoke CLC 1969 or the Fund Convention 1971 to support its verdict are as follows:

2.2.1 Restricted Affirmation

According to Article 5 of the Maritime Law of Taiwan:

> Maritime cases are ruled by the Law. Other laws shall apply where the Law does not stipulate.

Other laws should include international laws and the Civil Law. Should the two laws contradict, the principle that countries should not invoke national laws as an excuse not to abide by international laws should be upheld. Because oil pollution by the ship is categorized as a maritime case, and Taiwan is not a State Party to the Convention, rules that are in line with international practices could be applied.

2.2.2 Affirmation

The court of Taiwan could file for the indirect application of the Convention on the basis that the country of the ship's registry is a State Party to the Oil Pollution Convention or Fund Convention so as to avoid difficulties in enforcing verdicts of Taiwan court.

2.2.3 Negation

Only domestic laws shall apply for oil pollution incidents because the Convention is binding only in States Parties, not in non-States Parties.

It seems that the affirmation view is more credible among the three. We can use the country of the ship's registry as the connecting factor to link with the International Convention of which it is a State Party as the basis for our court verdicts. This will be more objective and easier for enforcement by foreign courts.

Article 21 of Taiwan's Maritime Law limits liability of the shipowner. However, in this case, the shipowner has unlimited liability because Article 22.4

of the Law excludes damage compensation by the ship carrying toxic chemicals or causing oil pollution from limited liability. Because Taiwan is not a State Party to the Tankers Owners Voluntary Agreement Concerning Liability for Oil Pollution 1969; the International Convention on Civil Liability for Oil Pollution Damage 1971; the Interim Supplementary Agreement on Oil Pollution Damage Liability 1971; CLC 1969; the International Convention on the Establishment of an International Fund for Compensation for Oil Pollution 1971 and so on, if we claim that the Convention is not applicable and invoke local laws for tort, we may indeed invoke unlimited liability in the case of the *Amorgos*, but in fact, if the company the ship belongs to is not financially optimal and only owns this ship, there may be unexpected development in terms of civil compensation. Even if we go by Taiwan laws and invoke unlimited liability against the *Amorgos*, we do not know whether we can ask for mandatory enforcement of Taiwan verdicts by the court of the registry country. Oil pollution damage is excluded from limitation of liability because once it occurs, compensation amounts are mostly astronomical, so dedicated international conventions or domestic laws should be established and calculation of compensation amounts should be formulated as well. Otherwise, if limitation of liability is invoked, it is no different from zero liability. This is the underlying rational for exclusions. It is not that unlimited liability always attaches to pollution damage. Article 3.3 of the Convention on Limitation of Liability for Maritime Claims 1976 states that the provisions of this Convention shall not apply to claims for oil pollution damage within the meaning of CLC 1969 or of any amendment or Protocol thereto that is in force. This clearly indicates that pollution damage is not limited liability and clams of such should be made in accordance with the Protocol to the CLC 1969. Given that Taiwan is not admitted to the International Convention and lacks relevant laws, Article 22 of the Maritime Law, which directly translates exclusion of pollution damage from limitation of liability to unlimited liability, is a complete departure from the LLMC 1976. Under such circumstances, how could we expect verdicts by the Taiwan court be recognized internationally? We should base on the Law of Country of Ship's Registry and further cite CLC 1969 and FI 1971 to substantiate compensation and win international recognition.

Most shipowners purchase insurance from the insurer members of the Shipowners Protection and Indemnity Association, or P&I Club. CLC 1969 requires all ships to carry oil pollution liability insurance or provide other financial security, the amount of which is to be decided by the limits stipulated by the Convention. According to Articles 33.2 and 33.3 of the Marine Pollution Control Act, compulsory insurance or financial security is required for all ships of Taiwan, in case no compensation can be sought after pollution occurs against those companies that own only one ship. In addition, Article 34 of the Marine Pollution Control Act provides:

> Claimants for compensation for pollution damage may seek compensation directly from the liability insurer or seek to secure the guarantee as compensation.

In this case, claims should be made directly against the insurers of P&I Clubs.

When dealing with marine oil pollution that involves foreign parties, Taiwan should use the country of the ship's registry as the basis and take into consideration international regimes to derive appropriate governing laws for courts to decide on civil compensation claims. International conventions should be used as the blueprint so that Taiwan's domestic regimes are aligned with international conventions, and when other countries apply Taiwan laws among other private international laws, they are in line with international conventions, minimizing difficulties Taiwan may face because of its lack of access to international conventions.

2.3 Discussion and Review of Taiwan's Legislation on Marine Oil Pollution

Taiwan's legislation on marine oil pollution should be established by taking into account international legal practices and relevant international conventions. Because most of such cases are trans-boundary in nature, Taiwan's legislation should better align with international laws and regimes to avoid obstacles and inability to enforce damage compensation. Taiwan's Maritime Law and Marine Pollution Control Act only regulate pollution damage caused by the ship in a general sense; they do not further elaborate on pollution caused by different sources, that is crude oil or fuel oil. Although the Convention on Civil Liability for Bunker Oil Pollution Damage 2001 classifies oil pollution into different categories, Taiwan, in the absence of clear regulations in national laws, can establish national regulations based on customary laws that can be applied as the internationally accepted source of authority to align itself with international practices.

Article 21 of the Maritime Law provides:

> The liability of the shipowner in the following aspects is limited to the total value of the ship, freight, and other accessorial charge.

The shipowner shall bear limited liability. However, Article 22.4 differs from the Conventions and international practices. It is concluded from the two articles mentioned earlier that once an oil pollution incident occurs, the shipowner has to bear full compensation liability, even though the owner can invoke limitation of liability. The principle of strict liability is upheld.

Another legal framework on marine oil pollution is the Marine Pollution Control Act which provides high-level regulations on sources of marine pollution. According to Article 33.1 of the Marine Pollution Control Act:

> The shipowner shall bear liability for compensation for the damage caused by the pollution of the sea by the ship.

Article 14 of the Act further provides

> Those that cause the production of pollution due to one of the circumstances in one of the following subparagraphs shall not be punished.

The text limits the exclusions to 'emergency evacuation', 'force majeure', 'acts performed in accordance with regulations or orders', 'acts performed with the permission of the competent central authorities' and so forth. As a result, it seems that under the Marine Pollution Control Act, liability cannot be exempted for pollution damage compensation that falls under civil liability. In other words, as long as the ship pollutes waters and causes damage, damage compensation liability is imposed, regardless of whether it resulted from intent or negligence. The principle of strict liability is upheld again.

From Article 3.11 of the Marine Pollution Control Act:

> Polluter means a natural person or the statutory responsible person, manager or representative of a public or private premise that causes a polluting act; when on a ship or aircraft, this means the owner, lessee, agent or operator;

Article 33.1:

> The shipowner shall bear liability for compensation for the damage caused by the pollution of the sea by a ship.

And Article 33.4:

> The shipowner designated in the foregoing article and Paragraph 1 of this article includes the owner, lessee, agent and operator of a ship.

It can be concluded that the liability for oil pollution by the ship is on the shipowner, which includes the owner, lessee, agent and operator of the ship. This is a broader definition compared with the Maritime Law, and it is different from the International Convention on Civil Liability for Oil Pollution Damage, in which only the shipowner is considered liable. Because the Marine Pollution Control Act has no regulations on limitation of liability and Article 22.4 of the Maritime Law excludes oil pollution from limitation of liability, unlimited liability may be imposed for oil pollution damage in Taiwan.

Article 22.4 of Taiwan's Maritime Law is established, taking into account LLMC 1976. The underlying meaning of Article 3.2 of the Convention is

> Limitation of liability as stipulated by the Convention (LLMC) is not applicable to damage of pollution by toxic substance or oil that is regulated by national laws or international conventions as limited liability. Limitation of liability by LLMC only APPLIES to damage of pollution by toxic substance or oil where limitation of liability is ABSENT in national laws or international conventions.

Therefore, the Maritime Law is applied because international convention or national laws with limitation of liability are absent.

Because Taiwan cannot be acceded into international conventions and there are no laws stipulating limitation of liability for oil pollution by the ship, the Maritime Law is applied, disallowing the shipowner to invoke limitation of liability in oil pollution incidents. This has caused a severe burden to the shipowner. The international society has imposed strict liability to oil pollution in respective legal frameworks, but they all come with exclusion criteria and limitation of liability. The Maritime Law of Taiwan imposes unlimited liability on the shipowner, which is completely against the international practices set to encourage participation in marine trade. With the advent of powerboats that replace sailboats, the value of cargo on board grows substantially. Because the risks at sea are high and uncontrollable, once any incident takes place, the shipowner could suffer enormous loss. If unlimited liability is imposed and the shipowner has to give every thing as owns, onshore and offshore, for compensation, bankruptcy is inevitable and in turn marine trade will be severely hit as people fear going into the industry. Therefore, laws should be established to limit liability of the shipowner to the extent that only the assets at sea are subject to subsequent obligation payoffs.

The establishment of the Marine Pollution Control Act merely mandates the Environmental Protection Administration in the protection of the marine environment and the Coast Guard Administration in the enforcement, evidence collection and case transfer to the prosecution. However, the Marine Pollution Control Act does not classify marine pollutions. Article 33 of the Act provides

> The shipowner shall bear liability for compensation for the damage caused by the pollution of the sea by the ship.

The principle of strict liability is upheld, and there are no exclusions or limitation of liability for the shipowner. Once a marine oil pollution incident occurs, the shipowner has to bear unlimited liability, which is a complete departure from the spirit of international regimes. Moreover, because Taiwan did not actually sign to the international conventions or treaties, there is still a question as to whether Taiwan can seek compensation for damage against the shipowner under the framework of international conventions. Also, under applicable laws of Taiwan, the Maritime Law or Marine Pollution Control Act, unlimited liability is imposed on the shipowner. When the claimant receives favourable civil verdicts, if the shipowner is not willing to pay compensation by the verdict and does not have any onshore assets, how is the verdict legally binding? Will courts of foreign countries accept and execute the verdict? Although Taiwan's Maritime Law intends to help the affected party or parties obtain full compensation by imposing unlimited liability on the shipowner, civil rulings may not be able to be effectively enforced. As a result, there is still room for discussion and improvement of such regulations.

3 CONCLUSION

Based on analysis on the *Amorgos* case, it seems that Taiwan's authorities are less experienced in dealing with marine pollution incidents, and many aspects in terms of its legal regimes should be strengthened. Given that Taiwan has no access to international conventions and its sovereignty is not recognized, the enforcement of domestic court verdicts against the shipowner of foreign nationality is difficult in practice, not to mention that liability of the shipowner will eventually be transferred to the insurer. However, it is still an encouraging development that the International Convention on Civil Liability for Bunker Oil Pollution Damage 2001 took effect on 21 November 2008, providing a legal basis for compensation against oil pollution by non-oil tankers and giving the cargo owner one more layer of protection against risks at sea.

Part VII
Comparative Conclusions

Chapter 26
Comparative Conclusions

Michael Faure, Han Lixin** & Shan Hongjun****

1 INTRODUCTION

In the introduction to this book we mentioned that oil pollution damage poses increasing challenges to the legal community, not only in Europe and the United States, but increasingly in China as well. Even though China has not (yet) been the victim of a major pollution incident, other (smaller) pollution incidents occur on a regular basis and, moreover, the coastal areas of China are heavily polluted. A comparative analysis of the prevention and compensation mechanisms available in different jurisdictions is highly interesting because some legal systems (such as in the United States) choose to develop their own (elaborated) compensation system, whereas other countries (more particularly the European Member States and China) largely rely on the international regime. The position of China is an interesting one because it did join the Convention for Civil Liability for Oil Pollution Damages (CLC), but (with the exception of Hong Kong) not the Fund Convention for Oil Pollution Damages (Fund Convention). This of course poses particular interesting questions concerning the adequacy of the current legal regime in China to prevent oil pollution incidents and provide compensation to victims. An issue addressed by many contributors to this volume was what the optimal prevention

* Maastricht University and Erasmus University Rotterdam, the Netherlands.
** Dalian Maritime University, China.
*** Dalian Maritime University, China.

Michael G. Faure, Han Lixin & Shan Hongjun, *Maritime Pollution Liability and Policy: China, Europe and the US*, 445–456.

and compensation mechanism would be, whereby often explicitly the question was addressed whether it would be in China's interest to follow more strongly the international regime or rather pursue the US example. As the US example showed, not following the international regime does not necessarily mean that compensation to victims would be lower; quite the contrary may be true.

Given that research into these important questions was one of the goals of this book, more than twenty contributions of, in some cases, a rather varied nature, provided insights into these issues. Reflections were provided by both academics and practitioners. In some cases, experts related to the (Chinese) Maritime Safety Administration provided a contribution on the effectiveness of the current legal regime.

Damage caused by oil pollution will unfortunately still be an important topic for the years to come. Even though, as many contributors show, the legal regime has been quite successful in reducing the number of oil spills (at least in the United States and Europe), oil spills still may occur, which raises questions concerning the effectiveness of the prevention regime and the adequacy of compensation. These questions especially play a role in the context of China, where the legal regime is not (yet) as refined as in the United States and Europe. Moreover, increasingly both the international community and legal maritime and environmental scholars are aware that damage cannot only be caused by oil (although the quantity of damage caused by oil is usually the largest) but is also caused by other noxious substances and by bunker fuel, not transported by oil tankers. These insights have increasingly led to adaptations of the legal regime that are also reflected in some of the contributions to this volume.

The contributions dealt specifically with prevention, compensation (Part I), the Bunker Convention (Part II) and the use of criminal law (Part III). Some contributions focused on differences between the international regime and the US regime and between the US regime and that of China (Part IV), whereas others focused more specifically on damage assessment, claims handling and compulsory insurance for damage in China (Part V). Many contributions have paid attention to empirical evidence resulting from case studies (Part VI). A few general conclusions from these contributions and suggestions for the way forward are equally formulated (Part VII).

2 PREVENTION

Although perhaps most scholars would focus the attention on compensation for oil pollution damage (probably because many legal questions arise after an oil spill accident and thus the question arises as to how adequate compensation can be provided to victims), a simple logic remains that 'prevention is better than cure'. Hence, both the international legal regime as well as national legal systems have to an important extent paid attention to the design of optimal mechanisms to prevent oil pollution damage. Even though first steps in that respect were taken in the International Convention for the Prevention of Pollution of the Sea by Oil in

1954 (the so-called OILPOL 1954), Wang Hui shows in Chapter 2 that most important steps in fact followed with the implementation of the International Convention for the Prevention of Pollution from Ships (known as MARPOL 73/78). MARPOL not only imposed specific discharge standards but also imposed important rules concerning the design of tankers. An obvious way to avoid contamination by oil carried as cargo or fuel was the introduction of segregated ballast tanks, ensuring that ballast water could not be contaminated. Of interest, the United States unilaterally enacted in the Oil Pollution Act 1990 (OPA 1990) a compulsory requirement on double hulls for tankers visiting its ports. This led, as Wang Hui shows in Chapter 2, to an interesting race to the top. Indeed, an effect of banning single hulled tankers from US waters was that these older and more dangerous tankers were hence used in European waters. Not surprisingly, the well-known incidents with the *Erika* in 1999 and the *Prestige* in 2002 both occurred with single hull tankers. As a result the European Commission also took action to further accelerate the phasing-out of single hull tankers.

Important issues can arise when prevention has failed and an incident takes place. Even before arriving at the compensation issue, questions arise concerning the competence to take emergency measures. The chapters dealing with case studies in Part VI show that in practice there is a lot of uncertainty concerning the specific powers and competences to take these emergency measures. Moreover, Eddy Somers addresses a point of particular interest that can arise after an incident, being the right of entry in maritime ports for ships in distress. It may be clear that refusing that right of entry could increase the damage (caused by oil pollution), but on the other hand allowing access for ships in distress may also increase the likelihood of damage to coastal states. Not surprisingly, the latter are not always enthusiastic to allow ships in distress to enter. Somers sketches that this, for example, played a role in the well-known *Erika* and *Prestige* cases. He equally shows that there is in fact no multilateral treaty that regulates a right of entry for ships in distress, nor is there a rule of customary international law that would force coastal states to accept a ship in distress. Somers therefore encourages parties within the International Maritime Organization (IMO) to come to a balanced solution, regulating the conditions under which coastal states could refuse access to ships in distress, because the current (unregulated) situation could potentially lead to an even increased scope of the damage.

3 COMPENSATION ISSUES

Many contributions of course deal with the question of whether the existing legal rules and more particularly the international compensation regime are adequate in the sense that they are able to provide compensation to victims in most serious pollution cases. Most contributors agree that, specifically taking into account the most recent changes to the international conventions as a result of which available amounts have substantially increased, the regime will in most cases be able to cover the damage suffered. However, many contributors also show that this, to an important extent, depends on how one defines the notion of damages. Robert Force makes clear in Chapter 15, dealing with the situation in the United States that the approach under the conventions is that compensation is payable only when the damage can be quantified in monetary terms. The question for the conventions is therefore how much it would cost to restore the damaged resource to baseline. This, so Robert Force shows, is different in the US approach under OPA 1990, more particularly because OPA realizes that assessing damages to the environment is in some cases very difficult. Therefore, some scientists have advocated techniques such as 'contingent valuation' to capture non-economic values and translate them into a monetary figure. Even though this may be a debated technique, James Boyd shows in Chapter 4 the possibilities that exist to make an accurate economic assessment of damage to natural resources. In Chapter 4, James Boyd proposes to use an ecosystem services assessment by examining what services a particular (damaged) resource provided and analysing how this can be quantified. The measurement of these 'ecosystem goods and services', so Boyd argues, will become increasingly important, also from a legal perspective, even though many practical problems may arise in the precise economic assessment.

Within the US context, both Robert Force (in Chapter 15) and James Boyd (in Chapter 4) make clear that under the US Oil Pollution Act, it is the National Oceanic and Atmospheric Administration (NOAA) that is the federal trustee for claims arising from marine injuries. Hence, within the specific US context the problem does not arise that a suit could not be brought on behalf of the damaged environment if no one would have property rights on the damaged natural resource. However, Proshanto Mukherjee shows in Chapter 5 that whereas this issue may not pose a problem under the OPA 1990, the question of who can sue (known in legal terms as the issue of *locus standi*) has not been regulated in the international conventions and is therefore basically left to domestic law. Mukherjee shows that the prevailing rule in most jurisdiction is that the suing entity must have a proprietary interest in the environment in order to establish *locus standi.* A possible solution to avoid this restriction is to use the so-called public trust doctrine in case damage is caused to non-proprietary interests. Mukherjee suggests incorporating this notion into the international conventions in order to fill the lacuna that presently exists and bring more uniformity.

An issue equally not clearly regulated in the international conventions is the extent to which economic losses can be compensated. This issue is addressed in detail by Baris Soyer in Chapter 6. Soyer shows that under the Funds' admissibility criteria, pure economic loss can under strict conditions be compensated, more particularly if there is a sufficiently close link of causation between the contamination and the loss of damage. However, he equally shows that the criteria applied by British courts to pure economic loss are similar, but not necessarily the same. In fact, he shows that in many cases the Fund is even more generous than British courts. But at a more general level, Soyer shows in his chapter that these differences are illustrative for the fact that claims for pure economic loss resulting from oil spills are not dealt with in a similar manner in all jurisdictions that are party to the conventions.

Many of the chapters dealing with the situation in China and included in Part V also point to the fact that compensation today in China may be inadequate as a result of the fact that a domestic fund for oil pollution damage is lacking and that China is (with the exception of Hong Kong) not a party to the Fund Convention. Many Chinese contributors, for example, also James Hu, discussing the *Hebei Spirit* oil spill incident in Chapter 23, therefore argue that China should set up a domestic oil pollution compensation fund, because otherwise inadequate compensation will be available for victims of major oil spills in China.

Damage can of course not only arise as a result of oil pollution. Wang Hui therefore discusses in Chapter 2 pollution caused by oil and compares it with damage caused by bunker oil and by hazardous and noxious substances (HNS). Today, damage caused by both bunker and oil are subject to international conventions; as far as HNS is concerned, a draft convention exists, which has, however, not yet entered into force. Wang Hui discovers that the three liability regimes all set up a strict liability of the shipowner, all have a compulsory insurance requirement and all have a limitation of the liability of the shipowner. However, channelling of liability can be found only in the The Convention on Civil Liability for Oil Pollution Damage (CLC) and HNS, but not in the Bunker Convention.

Jason Scott Johnston makes clear in Chapter 7 that, at least on paper, environmental liability aims not only at compensation but also at providing incentives to potential polluters for prevention. It should, in other words, lead to deterrence. Jason Johnston argues that this deterrence approach has been quite successful as far as the US OPA 1990 is concerned. After the introduction of OPA the volume of oil spilled from tankers into US waters has fallen from an average of 70,000 barrels per year to an average of 4,000 barrels per year, which is a decrease of 95%. This, according to Jason Johnston, shows that the passage of OPA did indeed have a deterrent effect on polluters. These numbers once more underline the necessity of exposing potential polluters (tanker owners) fully to the damage their activities can cause, because this may provide adequate incentives for prevention, as the US example shows.

4 THE BUNKER CONVENTION

Shan Hongjun shows in Chapter 8 that the existence of the Bunker Convention next to the CLC can, in combination with domestic law, lead to a very complex legal situation and to complex questions concerning the precise delineation between the various legal regimes. The Bunker Convention is particularly important because it has been implemented in China (differently than the International Oil Polluters Compensation (IOPC) Fund and the Supplementary Fund Convention). Shan Hongjun therefore points at the paradoxical fact that today in China victims from bunker enjoy a more adequate protection (as a result of the Bunker Convention) than cases in which the CLC 1992 would apply.

Issues of delineation between the CLC and the Bunker Convention are also discussed in Chapter 9 by Han Lixin and Wang Dapeng. A particular problem that arises and to which they pay attention is that China is not a contracting party to the Limitation of Liability for Maritime Claims (LLMC) 1976. As a result of that the compulsory insurance or financial security requirement imposed in the Bunker Convention can only be determined (as far as the amount for which security has to be provided) by Chinese domestic law. The result is, again, as Han Lixin and Wang Dapeng show, that the question of how to determine the amounts of the limitation (and the related duty to insure) has become rather complex under Chinese law.

5 CRIMINAL LAW

The chapters in Part III deal with a very topical issue – the application of criminal law to oil pollution incidents. Michael Faure addresses in Chapter 10 why, from an economic perspective, criminal law should be used for oil pollution damage. He argues that a low probability of detection and potentially high gains from pollution make non-monetary sanctions (which have to be imposed through criminal law) necessary. However, the case of the *Erika* incident and the subsequent decision of the Tribunal de Grande Instance de Paris showed that in fact the civil damages imposed by the criminal court may be at least as important as the criminal sanctions. Michael Faure also discusses the strong belief in the criminal law that exists today especially at the European Union (EU) level that forces Member States to introduce criminal sanctions for oil pollution damage.

This tendency in the EU toward forcing Member States to impose criminal sanctions is also (critically) discussed by Liu Nengye and Frank Maes in Chapter 11. They show that the priority of the EU's decision-making is clearly in favour of the coastal states. However, they argue that this practice is potentially a violation of international law and more particularly of the MARPOL Convention. The European Court of Justice held in a recent decision that because the European Community is not a party to MARPOL, it is not bound by the convention. The EU now forces Member States to criminalize cases of 'serious negligence', which may go further than the requirements under MARPOL. Marc Huybrechts is therefore very clear in Chapter 12 that this EU Directive violates international law. He addresses more generally the necessity of using criminal law for oil pollution incidents and fears that seafarers may be in danger of being prosecuted for oil pollution incidents even when it can be doubted that the incident is due to their *mens rea.* A similar concern is formulated in Chapter 13, in which Yuechuan Jiang criticizes the decision in the *Hebei Spirit* case whereby the captain and a chief officer were sentenced to eighteen months and eight months, respectively, in jail by a court in Korea. That decision was also critically discussed in Chapter 23 by James Hu. Hu argues that the master and chief officer were prosecuted merely for criminal negligence, but actually sentenced to jail. Accepting criminal liability for oil pollution incidents merely on the basis of negligence is severely criticized, so James Hu shows, in the international community, which reacted very strongly against the decision of the Korean authorities.

These cases and examples show that notwithstanding the fact that in principle criminal liability could be used for oil pollution incidents (as shown by Michael Faure in Chapter 10) there is a great danger in practice that authorities impose criminal liability also in cases in which it is doubtful that seafarers could actually have prevented the oil spill from occurring by taking preventive measures. Moreover, it seems as if in practice (not only in these cases but also in the EU Directive) the threshold for criminal liability is seriously lowered. The result can be that criminal liability is imposed on seafarers even when it is doubtful that the behaviour for which they are held criminally liable was actually under their control. This is a problem also addressed by Lian Jun Li and Mu Qin in Chapter 14 with respect to the situation in Hong Kong. They show, however, that if the seafarer showed that he took all reasonable steps to ensure that the regulations were complied with in principle, no criminal liability should be imposed.

Some chapters also ask the question of whether under current law criminal liability for oil pollution is possible in China and, if not, whether it would be desirable to introduce this. In Chapter 11, Liu Nengye and Frank Maes show that in theory Chinese law could be applied to oil pollution incidents, but that the Penal Code does not contain any specific offence for pollution of the sea. In theory it may be possible to apply some other articles from the Penal Code or from the Marine Environmental Protection Law (MEPL), but the latter law provides criminal only liability in the case of 'significant pollution' and 'major damage'. The authors rightly argue that this standard is very subjective and that the threshold for criminal liability may hence be very high. The same issue is also mentioned in Chapter 13 by Jiang Yuechuan. He argues strongly against the introduction of criminal liability if marine pollution were merely the result of negligence. This would, so he argues, lead to a strict liability in criminal law, which violates essential principles of criminal law in China.

These contributions show that the recent *Hebei Spirit* case has given rise to a lot of worry in the international maritime community that seafarers may be unreasonably exposed to criminal liability, even when it is doubtful that they could have taken reasonable measures to prevent the pollution incident. Moreover, most contributors hold that traditional criminal law principles (e.g., the requirement of *mens rea*) should still be applied, in order to avoid that criminal liability is imposed merely when a seafarer acted negligently. Even though the consequences of this negligence can be far reaching (causing serious environmental pollution) it may be too far reaching to impose imprisonment merely for negligent behaviour.

6 THE UNITED STATES AS EXAMPLE?

It is well-known that the United States often takes a remarkable position in the international legal community. Chapter 15 by Robert Force makes clear that the United States purposefully did not join the international conventions, but rather choose deliberately after the *Exxon Valdez* incident to introduce its own legal regime with OPA 1990. In Chapter 11, Liu Nengye and Frank Maes make clear that the United States followed a similar approach as far as the United Nations Convention on the Law of the Sea (UNCLOS) is concerned – the United States often supports the coming into being of these international conventions, but does not join the convention out of fear to abandon its sovereignty. The fact that the United States often does not join international conventions, does, however, not at all mean that it would provide protection at a level lower than that provided in the conventions or would totally disregard the conventions. For example, as far as the UNCLOS is concerned, Liu Nengye and Frank Maes show in Chapter 11 that the United States has in their view an obligation 'to refrain from acts which would defeat the objective and purpose of the treaty'.

As far as OPA 1990 is concerned, Robert Force makes clear in Chapter 15 that in many respects the protection provided under it goes further than the international regime. For example, he points to the fact that compensation for non-economic losses can be recovered under OPA for injury to natural resources. OPA also provides that replacement of the damaged resource may be an adequate remedy as well and under OPA compensation is also recoverable for interim losses of a resource. That means that compensation can be provided for the loss of the use of the resource while it is damaged and is being restored. These types of compensations are, so Robert Force argues, not available under the conventions. Also, a few specific procedural aspects of US law are mentioned by many contributors as being advantageous for adequate restoration of the environment. For example, James Boyd points in Chapter 4 to the important role of NOAA acting as federal trustee for claims arising from marine injuries. This role as trustee avoids complicated questions on *locus standi* if damage would be caused to natural resources that are not privately owned. For that reason, Mukherjee also argues in Chapter 5 that a public trust doctrine (as it has been developed in the United States in OPA 1990) may be worthwhile to introduce in the international conventions as well to fill a lacuna that presently exists.

It is therefore not surprising that some Chinese scholars equally point to the US regime as a model worth exploring for China. Both Li Tianshang and Han Lixin, in Chapter 16, as well as Guo Ping, in Chapter 17, point to the fact that the situation of China today is comparable to the situation of the United States. Also, China has become not only a country with a large amount of oil transportation, but also a major oil importer. The likelihood that China would therefore be victimized by a major oil pollution incident is very realistic. They also show that coastal industries have become increasingly important in China and thus an adequate compensation may today be more important in China than merely the protection of the shipping industry. For these reasons, Li Tianshang and Han Lixin, as well as Guo Ping, argue in Chapters 16 and 17 that the US compensation system can provide valuable insights for the situation in China as well.

It should also be noted that Wang Hui showed in Chapter 2 that the United States was the first, with OPA 1990, to ban single hull tankers from US waters, which was only followed years later by Europe, when they decided to accelerate the phasing-out of single hull tankers. Also as far as prevention is concerned, OPA 1990 does provide an interesting example. That is the case not only because of the regulation of preventive measures but also because of the liability regime. In that respect we should once more refer to Chapter 7, in which Jason Scott Johnston shows that the introduction of OPA had a considerable deterrent effect on oil pollution and de facto led to a reduction of oil pollution incidents of 95%.

7 DAMAGE ASSESSMENT, CLAIMS HANDLING AND INSURANCE IN CHINA

The many chapters in Part V deal with various aspects of the compensation regime in China. The starting point is the interesting chapter by Song Ying (Chapter 18), in which she examines to what extent China is participating in the international regime aiming at the protection of marine environmental pollution. Song Ying shows that various marine environmental problems pose great challenges to China, including oil spills, habitat destruction and overfishing. China is participating in most of the principal environmental treaties and has incorporated those in its domestic laws. However, Song Ying argues that notwithstanding this formal implementation, two major problems arise. One is related to the fact that public participation is not well developed in China's marine environmental protection law. She argues that increasing public participation could also improve public awareness and thus lead to a reduction of pollution. A second issue is that much of the legislation may be focused on marine pollution, whereas Song Ying argues that there is an urgent need in China to effectively deal with land-based pollution as well.

An issue that has been debated in China for a long time is the specific application of CLC convention to purely domestic oil pollution cases. Traditionally, most authors argue that CLC is not applicable to domestic oil pollution cases. That is the opinion that is, for example, represented by Liying Zhang in Chapter 20. She argues that an international convention such as CLC cannot be applied to domestic pollution cases. Therefore, to a domestic pollution case domestic law should simply be applied. However, Chapter 19 by Chen Qi shows that this is an issue on which not all authors agree, because Chen Qi strongly argues in favour of applying the CLC also to purely domestic pollution cases. Both sides provide interesting arguments in favour and against applying CLC to domestic pollution cases and, as Liying Zhang shows in Chapter 20, case law is also divided on this issue. There is, however, one issue on which there is no debate, as made clear in Chapter 21 by Chu Beiping and Zhang Jinlei, which is that in all cases compulsory insurance will have to be provided. Indeed, both the CLC and the Chinese ordinance of transportation in coastal and inland waters impose a duty to provide compulsory insurance. Notwithstanding the major advantages of compulsory insurance, the authors argue that in reality not all vessels have obtained insurance because the insurance market may in some cases be insufficiently developed and not offer a wide variety of products.

The final chapter in Part V, from Sun Guang, deals with the judicial authentication of oil pollution. She discusses the way in which judicial authentication (expertise) takes place to assess the damage and shows, *inter alia*, by discussing the Tasman Sea oil pollution case, how damages caused by oil pollution are de facto assessed in China.

8 CASE STUDIES

Many contributions in this book provide empirical insights on the way in which oil pollution cases are dealt with in practice. For example Wang Hui shows in Chapter 2 that according to the Fund a majority of claims are fully covered. It would only be a few catastrophic incidents that are not covered. This would confirm that the working of the fund is relatively effective in providing compensation. The same would be true, so she argues, as far as hazardous and noxious substances are concerned. Apparently the current limits of liability of the shipowner under the HNS Convention are considered sufficient to cover most incidents involving HNS.

Michael Faure discussed empirical data in Chapter 10 showing that whenever a new piece of legislation concerning oil spills was enacted, the number of oil spills effectively decreased.

Empirical data are also discussed in the three specific case studies presented in Part VI. Even though the cases that were discussed (*Hebei Spirit* in Chapter 23, *Ilona* in Chapter 24 and *Amorgos* in Chapter 25) were quite different, the conclusions of the authors are remarkably similar. They all point to the necessity for China to set up a domestic oil pollution compensation fund; they equally argue that there is a danger in imposing a too harsh criminal liability on seafarers and argue that the conditions for criminal liability of seafarers should absolutely be fine-tuned in order to avoid situations such as that in the *Hebei Spirit* case, in which seafarers end up in jail when their guilt in the oil pollution incident is doubtful. A third recommendation from the cases is that the emergency response system should also be improved. That was a conclusion, not only introduced by James Hu in Chapter 23, resulting from the *Hebei Spirit* case, but also from Zhonghua Li and Zhujun Zhou from their discussion of the *Ilona* oil spill incident in Chapter 24.

9 THE WAY FORWARD

This project and the resulting book are the follow-up of an earlier collaboration between Maastricht and Dalian Maritime University (leading to the book edited by Michael Faure and James Hu on *Prevention and Compensation of Marine Pollution Damage: Recent Developments in Europe, China and the US*, 2006). Compared to the earlier conference the discussions were brought further, for example, by focusing on new and different topics (e.g., criminal liability and the Bunker Convention), but also by engaging in a more critical and integrated comparative assessment, analysing, for example, whether the US model is preferable, compared to the international conventions and whether it thus could constitute an interesting example for China as well.

Even though we hope that with this book we have contributed to a better understanding of the legal complexities of marine pollution damage and we hopefully also may inspire the (international and national) policy-makers to improve legislation, leading to better prevention and compensation of marine pollution damage, we also realize that, as is often the case with academic research, many issues still need to be resolved. Indeed, the contributors to this book identified many issues that still need more attention that have maybe not been sufficiently addressed.

One issue, touched upon in this book but not further developed, is the competence of states to inspect, control, enforce and sanction in the case of violations of regulations and oil pollution incidents. A tendency can be noticed away from the traditional powers of the flag state towards awarding more powers to port states and in some cases even to coastal states. The question arises as to which powers are most effective for adequate protection of the marine environment while at the same time equally safeguarding maritime interests, such as free passage in coastal waters.

The relationship between the international conventions and national domestic laws may need further clarification and research. For example, the role played by the EU, with its possibility to enforce directives upon Member States, shows that in some cases the EU is better able to enforce the implementation of international obligations upon its Member States, thus curing the traditional problem of non-compliance with international conventions. Also the particular positions of China and the United States towards the international conventions raises interesting questions. Many contributions in this volume showed that even though the United States is not party to international conventions, its regime concerning oil pollution damage is in many respects leading in the world and even inspires the international conventions. This raises the important question for China whether it should take further steps towards the implementation of international conventions (e.g., the Fund Convention or the Convention on Supplementary Funding) or whether it should rather follow the US example by going its own way (e.g., with a domestic compensation fund).

Many contributors also showed that the conventions do not regulate all possible problems. For example, the issue of who has standing to sue and whether economic losses can be compensated are to an important extent left to the legislation of national laws. Here the question arises as to what extent this is desirable or whether further steps should be taken in the international conventions. Given the important differences with US law (which provides a wider scope of protection than the international conventions) the question also arises as to whether the conventions should be reformed along the lines of US law.

Finally, further research is absolutely needed into the different legal instruments used to prevent and compensate oil pollution damage. Non-compliance with regulation aiming at prevention is now increasingly sanctioned by using the criminal law. However, the extent to which the criminal law is used in practice in specific cases (e.g., in the *Hebei Spirit* case) raises questions with respect to the effectiveness of the criminal law. If seafarers have to live under constant fear of possible imprisonment in the case of a pollution incident, even if they were not to blame for the accident, the costs of sea transport could be substantially increased because working on an oil tanker may become a risky job, requiring higher remuneration. These potentially perverse consequences of increasing criminal liability should be carefully addressed as well, taking into account alternative remedies such as administrative sanctions or other remedial measures.

Notwithstanding the fact that many issues may need further research, in particular, the legal issues related to offshore facility pollution, we hope that the contributions in this book may inspire practitioners, academics and policy-makers towards further improvement in the prevention and compensation of marine pollution damage, and of course we are planning to continue this collaboration to examine these and other remaining questions in the future.

ENERGY AND ENVIRONMENTAL LAW & POLICY SERIES

1. Stephen J. Turner, *A Substantive Environmental Right: An Examination of the Legal Obligations of Decision-makers towards the Environment*, 2009 (ISBN 978-90-411-2815-7).
2. Helle Tegner Anker, Birgitte Egelund Olsen & Anita Rønne (eds), *Legal Systems and Wind Energy: A Comparative Perspective*, 2009 (ISBN 978-90-411-2831-7).
3. David Langlet, *Prior Informed Consent and Hazardous Trade: Regulating Trade in Hazardous Goods at the Intersection of Sovereignty, Free Trade and Environmental Protection*, 2009 (ISBN 978-90-411-2821-8).
4. Louis J. Kotzé and Alexander R. Paterson (eds), *The Role of the Judiciary in Environmental Governance: Comparative Perspectives*, 2009 (ISBN 978-90-411-2708-2).
5. Tuula Honkonen, *The Common but Differentiated Responsibility Principle in Multilateral Environmental Agreement's: Regulatory and Policy Aspects*, 2009 (ISBN 978-90-411-3153-9).
6. Barbara Pozzo (ed.), *The Implementation of the Seveso Directives in an Enlarged Europe: A Look into the Past and a challenge for the Future*, 2009 (ISBN 978-90-411-2854-6).
7. Henrik M. Inadomi, *Independent Power Projects in Developing Countries: Legal Investment Protection and Consequences for Development*, 2010 (ISBN 978-90-411-3178-2).
8. Nahid Islam, *The Law of Non-Navigational Uses of International Watercourses: Options for Regional Regime-Building in Asia*, 2010 (ISBN 978-90-411-3196-6).
9. Yasuhiro Shigeta, *International Judicial Control of Environmental Protection: Standard Setting, Compliance Control and the Development of International Environmental Law by the International Judiciary*, 2010 (ISBN 978-90-411-3151-5).
10. Katleen Janssen, *The Availability of Spatial and Environmental Data in the European Union: At the Crossroads between Public and Economic Interests*, 2010 (ISBN 978-90-411-3287-1).
11. Henrik Bjørnebye, *Investing in EU Energy Security: Exploring the Regulatory Approach to Tomorrow's Electricity Production*, 2010 (ISBN 978-90-411-3118-8).
12. Véronique Bruggeman, *Compensating Catastrophe Victims: A Comparative Law and Economics Approach*, 2010 (ISBN 978-90-411-3263-5).
13. Michael G. Faure, Han Lixin & Shan Hongjun, *Maritime Pollution Liability and Policy: China, Europe and the US*, 2010 (ISBN 978-90-411-2869-0).

www.ingramcontent.com/pod-product-compliance
Lightning Source LLC
LaVergne TN
LVHW020646300926
849161LV00004B/844
* 9 7 8 9 0 4 1 1 2 8 6 9 0 *